Chemical Evolution

Chemical Evolution
Origin of the Elements, Molecules, and Living Systems

Stephen F. Mason
Emeritus Professor of Chemistry
King's College, University of London

CLARENDON PRESS · OXFORD
1991

Oxford University Press, Walton Street, Oxford OX2 6DP
Oxford New York Toronto
Delhi Bombay Calcutta Madras Karachi
Petaling Jaya Singapore Hong Kong Tokyo
Nairobi Dar es Salaam Cape Town
Melbourne Auckland
and associated companies in
Berlin Ibadan

Oxford is a trade mark of Oxford University Press

Published in the United States
by Oxford University Press, New York

British Library Cataloguing in Publication Data
Mason, Stephen F.
Chemical evolution : origins of the elements, molecules
and living systems.
1. Organisms. Evolution. Biochemical aspects
I. Title
575
ISBN 0–19–855272–6

Library of Congress Cataloging in Publication Data
Mason, Stephen Finney, 1923–
Chemical evolution : origins of the elements, molecules, and
living systems / Stephen F. Mason.
p.   cm.
Includes bibliographical references and index.
1. Chemical evolution.   I. Title.
QH325.M32   1990   577—dc20   90–7321
ISBN 0–19–855272–6

For information about our audio products, write us at:
Newbridge Book Clubs, 3000 Cindel Drive, Delran, NJ 08370

Printed in Great Britain by Courier International Ltd,
Tiptree, Colchester, Essex

The progress of science is threatened today not only by loss of financial support and of good people—which is bad enough; not only by diversion of too much of its energy to applied or engineering work that may not yet be bolstered by enough basic knowledge; and not only by the confusion and disenchantment of the wider public—and that, too, demands our concern, because some of it is surely due to a lack of proper attention on the part of scientists. No, what seems to me to be the most sensitive, the most fragile part of the total intellectual ecology of science is the understanding, on the part of scientists themselves, of the nature of the scientific enterprise, and in particular the hardly begun study of the nature of scientific discovery. In this pursuit, our own day-to-day experience as scientists will help us if we set it into the historic framework provided by those that went before us.

Gerald Holton (1975). Mainsprings of scientific discovery. In *The nature of scientific discovery*, (ed. Owen Gingerich). Smithsonian Institution Press, Washington, DC.

Preface

'There is but one chemistry' affirmed Adolphe Wurtz (1817–84) in 1862, in a comment on the apparent merging of mineral with organic chemistry at the time. This perceived unity of chemical science in the 1860s, however, did not outlast the nineteenth century, despite the development of a general and testable theory of molecular structure, and other conceptual consolidations. The core subject, split into the fossilizing divisions of inorganic, organic, and physical chemistry, then became an ever-growing collection of empirical observations and recipes, linked to a greater or lesser degree by general principles taken increasingly from advances in atomic and molecular physics.

Although the chemical sciences during the second half of the twentieth century have vastly increased their expertise in the discovery of new substances and the more efficient production of known materials, they have lost something of the *general* intellectual interest and esteem acquired and enjoyed over the previous two hundred years. Journals covering the whole of natural science, such as *Nature* or *Science*, now carry a smaller proportion of chemical articles than they did fifty or a hundred years ago, and those articles tend to come from fields divergent from the main chemical domain—from cosmochemistry, geochemistry, biochemistry, and molecular biology.

These now-peripheral fields were part of the general chemical science of the day throughout the nineteenth century. In addition to his numerous other innovations, Robert Bunsen (1811–99) made pioneering contributions to cosmochemistry and geochemistry. Bunsen introduced not only the spectral analysis of the elementary composition of the Sun and the stars, with Gustav Kirchhoff (1824–87), but also the primary geochemical division between the granites and the basalts in crustal rocks (from their silica/metal oxide ratio). The biological chemistry of Emil Fischer (1852–1919), through his researches on the sugars, purines, amino acids, polypeptides, and enzyme–substrate specificity, was ancestral to both biochemistry and molecular biology.

Many textbooks published during the nineteenth century, such as the *System of chemistry* by Thomas Thomson (4th edition, 1818) or the *Treatise on chemistry* by H. E. Roscoe and C. Schorlemmer (1877), review historically what was then known of the chemical composition of the Sun and recovered meteorites, as well as terrestrial minerals and organic natural products. These texts include discussions of biochemical processes, such as fermentations, along with accounts of industrial and laboratory chemical processes. Subsequently the historico-encyclopaedic tradition of textbook writing became progressively supplanted by the more specialist and ephemeral 'state-of-the-art' mode of text production, unrooted in the context of either the development of the subject or its 'external' applications or relations with other fields.

It is generally agreed that the production of 'state-of-the-art' texts with a life expectancy of a decade or so is inevitable, in view of the continuous growth of the vast volume of scientific knowledge. Finer subdivision of each science into ever-narrowing fields appears to be a natural development, as in the specialist texts considered necessary for efficient science training, unencumbered by inessentials. But the distinction between the essential and the inessential is a strong function of the specific context, dependent upon the particular time, place, and purpose. The judgement must always be local and short term.

Equally it is generally agreed that the probability of significant scientific advance tends to be significantly higher at the borderlines between specialist fields than within an individual field. This is illustrated in the chemically based articles published in journals such as *Nature* and *Science*. The perception of a potentially fruitful borderline area between two fields generally requires much more than a knowledge and understanding of the 'state of the art' in the two fields individually. The judgement of scientific promise in a boundary or frontier region may well require a discriminatory historical and contextual perspective; a general appreciation of the development and the salient features of the scientific enterprise in one or more of its major domains. Thus Louis Pasteur's vision of a natural world pervaded by chiral forces stimulated the studies on the differential electroweak stabilization of the D-sugar and L-amino acid series that dominate the biochemistry of living organisms (Chapter 14).

The most delicate and vulnerable section of the scientific enterprise lies in the educational process, the training of future generations of scientists in the schools and universities. The sheer volume of the science, and the external pressures towards short-term economic utility, have virtually eliminated the historico-encyclopaedist tradition from chemical education. In its later stages, chemical education presents the growing points of the science as a disconnected series of specializations, without much coherence or application to other scientific fields. Yet chemistry remains a central science, and its thought-style and procedures dominate the analysis of the structure and dynamics of all materials in the universe at the molecular level, whether cosmic or terrestrial, of a physical or a biological origin, natural or synthetic.

Surveys of the principal discoveries in the fields divergent from nineteenth-century chemical science, in cosmochemistry, geochemistry, biochemistry, and molecular biology, restore some coherence and provide a wider chemical view of the world, particularly when set in an evolutionary context. In recent years such surveys have tended to emphasize either the inorganic or the organic aspects of chemical evolution; the nucleosynthesis of the chemical elements or the molecular development of the solar system on the one hand, and the origin and development of terrestrial living organisms on the other. The studies presented here as a 'state-of-the-art' survey are grounded on the once-common view that 'there is but one chemistry' and informed by the venerable historico-encyclopaedist tradition. This book brings together the inorganic and the organic rationale of chemical evolution as the knowledge and the ideas themselves evolved, until the later 1980s.

I thank the Leverhulme Trustees for the award of an Emeritus Fellowship and Wolfson College Cambridge for an Extraordinary Fellowship, both held over the period 1988–90, during which this book was written.

Department of History and Philosophy of Science, University of Cambridge.

Cambridge
November 1989

S. F. M.

Contents

Contents

1
Introduction

Around the middle of the twentieth century, the studies of chemical evolution in the late-nineteenth-century tradition diverged into specialist branches and changed emphasis from the inorganic to the organic. The progressive synthesis of the chemical elements from hydrogen by nuclear transformations in the stars became relatively well understood, in broad outline, giving detailed substance to the general conjectures of the nineteenth-century chemists and physicists on the generic relation of hydrogen to the heavier elements. New instrumentation gave a fresh impetus to studies, dating back to the early 1800s, of the chemical composition of recovered meteorites, and the organic meteoritic material attracted more attention than it had during the century before. As had been surmised, the inorganic compositions of meteorites provided a measure of the relative cosmic abundances of the chemical elements, overlapping and complementing the corresponding abundances determined by stellar spectroscopy. More astonishing was the characterization in carbonaceous meteorites of a profusion of organic molecules, including the amino acids and other biomolecules.

Prebiotic chemistry as a distinct discipline devoted to the study of the chemical origins of living organisms dates back only to the 1950s, with more tenuous roots in the evolutionary conjectures of the previous century than the nucleosynthetic theory of the transmutation of the elements. The rise of a prebiotic branch of largely organic chemistry was principally stimulated by the pioneering studies of the 1950s on the molecular structures of the biopolymers (the proteins and the nucleic acids), and on their role in metabolism, biosynthesis, and biological replication. A number of the basic prebiotic reactions turned out to be so facile, as required for their presumed biogeochemical role, that they had been discovered by chemists long ago. Indeed early workers, such as Döbereiner (1780–1849), Butlerov (1828–86), and Strecker (1822–71), had been practising prebiotic chemistry unwittingly, like Molière's Monsieur Jourdain speaking prose.

The studies presented here are historical in a double sense. They are concerned both with the development of evolutionary concepts in modern chemistry, and with current theories of the chemical history of the natural world. The book deals with the discoveries and theories, from around 1800, concerning the origins of the chemical elements, of terrestrial and extraterrestrial molecules, and of living organisms. Discussions of the early hypotheses mainly serve as an introduction to our current conclusions on the historical evolution of the universe, from around the time the main nucleosynthesis of the elements of the periodic table began some 100 million years after the primary creation of hydrogen and helium in the Big Bang.

Cooling of the elmentary atoms produced the molecules of the immense dark clouds in the interstellar medium. These clouds consist largely of molecular hydrogen, enriched with compounds of the heavier elements ejected during the demise of exhausted stars. Around 5 billion (5×10^9) years ago, the gravitational collapse of a dark molecular cloud, triggered possibly by a shock perturbation from a nearby supernova, formed the proto-Sun and the solar nebula. Here the molecules of the dark cloud were reworked into conglomerates of iron−nickel alloy, silicate and aluminate minerals, and organic assemblies of hydrocarbons and their oxygen and nitrogen derivatives, with much carbonaceous polymer.

Recovered meteorites provide the evidence. Near-original samples of these conglomerates and assemblies, dating back to around 4.5 billion years ago, have bombarded the Earth throughout geohistory. The early and more intense bombardments contributed to the organic and salt content of the oceans, which go back to about 4 billion years before the present; the oldest sedimentary rocks are not much younger. About 2 billion years ago or earlier, the atmosphere was transformed from anoxic to oxygen containing by photosynthetic organisms which used water in the photoreduction of carbon dioxide, liberating molecular oxygen. Microfossils and traces of the organic remains of their photosynthetic ancestors, which used hydrogen sulphide or organic substances for the photoreduction under a reducing atmosphere, go back some 3.5 billion years.

The chemical mechanisms by which the earliest forms of life arose from prebiotic organic materials remain a puzzle. Near the end of the twentieth century, these mechanisms are still as enigmatic to us as the generative mechanisms for the chemical elements were to chemical physicists a century earlier. At that time, from the approximation of so many atomic weights to whole numbers, it was clear that the various atomic species had some generic or constitutional connection, first expressed in Prout's hypothesis of 1815 that the heavier elements are made up from integral numbers of hydrogenic units. The relationship between the different chemical elements remained obscure until the discovery of radioactivity provided access to the study of atomic structure, particularly of the atomic nucleus in the hands of Rutherford.

As the question of the genesis of the chemical elements became settled, the expectations of the chemical evolutionists moved on from the inorganic to the organic. Now a broad consensus holds that physico-chemical models of simple organisms will eventually be constructed in the laboratory, with the form of self-replicating systems subject to evolution by natural selection. Indeed, enzymatic model systems of this form, but remote from probable prebiotic conditions, have been investigated since the mid-1960s.

The construction of more probable models may depend on the normal piecemeal development of the chemical sciences; that is, on the progressive unravelling of biochemical, geochemical, and cosmochemical processes in finer detail, evermore elegant and ingenious prebiotic syntheses and kinetic analysis of model systems, supported by increasingly sophisticated instrumentation and analytical techniques.

From the historical analogy of the role of radioactivity in solving the problem of the genesis of the chemical elements, it is tempting to suppose, however, that a fuller understanding of prebiotic chemistry awaits the discovery of systems in the natural world operating on novel principles that, when characterized, turn out to be relevant to the origin of life. Such systems may well lie somewhere along the inorganic–organic interface between geochemistry and biochemistry, perhaps among the ubiquitous clay minerals. These versatile catalysts of organic (and inorganic) reactions were among the first minerals to crystallize, and reproduce from seed crystallites, in the primordial seas.

2

The chemical elements in nineteenth-century science

2.1 From Lavoisier to Mendeleev

The Chemical Revolution traditionally dates to the publication in 1789 of the first reconizably modern textbook, *The elements of chemistry*, by Antoine Lavoisier (1743–94). The chemical elements were defined empirically as the 'actual terms whereat chemical analysis had arrived', and Lavoisier listed some 23 authentic elements known to him. The new science defined a domain of compound materials characterized by constancy of elementary composition, as opposed to physical mixtures of indefinite composition. The objectives of its practitioners were to be the preparation of new substances and materials with novel properties. The elements taken two at a time might yield hundreds of new compounds, or combined three at a time could give thousands, while the possible combination of four different elements ran to millions. The followers of Lavoisier assumed that each chemical element constitutes a distinct and permanent species, unconnected constitutionally with other elementary species, in contrast to the transformist assumptions of the earlier proto-chemists. By systematic weighings before and after the experiment, Lavoisier refuted the contention of Jan Baptista van Helmont (1579–1644) that he had converted elementary water into earth, by boiling water for several weeks in a soda-glass flask (which produced a siliceous deposit from the partial dissolution of the glass).

The assumption of fixed species was basic to the atomic theory of John Dalton (1766–1844), published in his *New system of chemical philosophy* (1808). The theory postulated that each chemical element is a homogeneous assembly of atoms which differ in relative weight, number per unit volume, and in combining number (valency) or set of combining numbers, from the atoms of other elements. Problems appeared immediately. The atomic theory derived its initial appeal from the discovery of exact integer quantization in chemical combination, embodied in Dalton's law of multiple proportions (1804). The two compounds formed by a pair of elements, A and B, might be formulated as A_2B and AB or alternatively as AB and AB_2, if the weights of A which combine with a fixed weight of B lay in a 2:1 ratio. An independent determination either of the relative atomic weights or of the combining numbers distinguishes between the two possibilities; but methods for such determinations, to satisfy Dalton and most other chemists, were not available until the 1860s.

Whole numbers, less exact than those of the law of multiple proportions, appeared in the law of gaseous combination by integer volumes (1808) of Joseph Gay-Lussac (1778–1850), whose results led Amedo Avogadro (1776–1856) to propose in 1811 that the elementary gases, such as nitrogen or oxygen, are composed of diatomic molecules, with an equal number of molecules per unit volume, under the same conditions of temperature and pressure. More generally, as A. M. Ampère (1775–1836) pointed out in 1814, the results obtained by Gay-Lussac, such as the formation of two volumes of hydrogen chloride from one of hydrogen and one of chlorine, implied only that a molecule of an elementary gas contains an *even* number of atoms. Ampère himself proposed that even the simplest gas molecules should be tetra-atomic, with a tetrahedral form: since crystals are three-dimensional, their molecular building blocks must be so too. The Avogadro–Ampère hypothesis implied that relative atomic or molecular weights could be inferred, without reference to the corresponding equivalent weights, from measurements of vapour (gas) density relative to that of hydrogen. However, most chemists resisted this implication, with a few notable exceptions, until the 1860s.

Dalton and most of his contemporaries, and even the succeeding generation of chemists, could not accept the implication of homoatomic molecules, such as O_2 or N_2. It was held that like atoms repel one another, either on account of the atmosphere of caloric (heat) surrounding each atom (required by the thermal expansion of gases), or by their like electrical charges, postulated from the results of the electrolytic experiments on salts and hydroxides carried out by Humphrey Davy (1778–1829) and Jöns Jakob Berzelius (1779–1848). The formation of heteroatomic molecules, such as NO or HCl, presented no problems from either viewpoint. Dalton's law of partial pressures (1801) suggested that each gas is a vacuum to every other gas, without the mutual repulsion of the like-atom case; and the electrochemical *dualistic theory* of Berzelius (1811) ascribed attractive electrical forces between atoms with different degrees of electronegativity. From the law of gaseous combination by integral volumes, Berzelius concluded that a given volume of different *elementary* gases at the same temperature and pressure contain the same number of *atoms*. The conclusion, although rejected by Dalton, was not ruled out by the supposed repulsion between like atoms, and it enabled Berzelius to formulate water as H_2O and ammonia as NH_3, in contrast to Dalton's formulations, HO and NH.

For Dalton and Berzelius, the dire consequences of the Avogadro–Ampère view appeared to be evident in the protyle hypothesis (1815) of William Prout (1785–1850). Prout found that the densities of a range of different gases, measured relative to hydrogen, approximated to whole numbers: 14 for nitrogen and 16 for oxygen. He concluded that the atoms of all other chemical elements must have been formed by multiple condensations of hydrogen atoms. A few years earlier in 1812, Davy had supposed that hydrogen might be the primordial matter from which all other chemical elements derived, for he could obtain hydrogen from most metals and such unlikely materials as the black ammonium amalgam produced by the electrolysis of ammonium salts with a mercury cathode. Proposals of this kind threatened to restore the doctrine of the transmutation of the elements, a doctrine that some, like Michael Faraday (1791–1867), never relinquished.

While mineral chemistry could make do with the empirical equivalent weights of the elements for half a century, organic chemistry could not. Confined to a handful of elements, and with an ever-increasing array of isomers, organic chemistry required its own ordering principles, distinct from the electrochemical notions which Berzelius and his followers endeavoured to transfer from inorganic to organic chemistry. During the 1840s, the French chemist Jean Baptiste Dumas (1800–84) and his associates, August Laurent (1808–53) and Charles Gerhardt (1816–56), developed a theory of organic molecular *types*. The members of a given type, the derivatives of a given parent substance, have molecular properties independent of the electrochemical properties of the constituent atoms. The substitution of electropositive hydrogen by electronegative chlorine in acetic acid gave a series of chloroacetic acids which were all recognizably vinegars.

The rival merits of the *type theory* and the electrochemical *dualistic theory* gave rise to much Gallic–Teutonic contention during the 1840s, going beyond the organic–inorganic division. Advocates of the type theory tended to be more favourably disposed towards the hypotheses of Avogadro and Prout, and less inclined to regard the elements as unconnected fixed species, or an atom as indivisible. Both Laurent and Gerhardt in the later stages of their work adopted vapour densities as the 'sole guide' for the determination of molecular weights and, by the early 1850s, they had a set of atomic weights implicitly based upon Avogadro's hypothesis. Since equal volumes of gases (at the same temperature and pressure) contain equal numbers of molecules, according to the hypothesis, the density of a gas (or vapour) is proportional to the molecular mass. For compounds composed of common elements, the set of molecular weights gave the atomic weights of the constituent elements.

A decade earlier, Dumas, assisted by Jean Stas (1813–91), had redetermined the relative combining weights (the equivalents) of the basic organic elements, hydrogen, carbon, and oxygen, finding the ratios to be whole numbers to three or four significant figures, as Prout's hypothesis required. The equivalent weights of some other elements were undoubtedly fractional, Dumas noted, so that the basic protyle unit might well be a half or a quarter of a hydrogen atom. The subatomic units could be arranged in different ways to give isomers (different elements with the same atomic weight), as appeared to be the case for the nickel group or for the platinum metals. Inverting the Berzelian seniority order, Dumas applied organic concepts to mineral chemistry, supposing the alkali metals to form a homologous series analogous to the paraffins, with the alkaline earths, the halogens, and the chalcogens forming similar series. It was these views, aired by Dumas at the Ipswich meeting of the British Association for the Advancement of Science in 1851, that gave Faraday a renewed hope of interconverting the chemical elements.

The later versions of the type theory developed into the molecular structure theory of the 1860s when a reliable basis for the relative atomic weights of the chemical elements, and the combining numbers of the atoms (their valencies), became generally agreed. The international conference of chemists at Karlsruhe in 1860 marked a turning point. The meeting was arranged by August Kekulé (1829–96) to discuss the problem of atomic weights. No major decisions were reached at the conference, but many of the 140 participants went away with a copy of a text by Stanislao Cannizzaro (1826–1910), who showed that the systematic application of Avogadro's hypothesis to related series of compounds gave, with the analytical data, a set of atomic weights and valencies for the elements consistent within a series and between different series. Cannizzaro's text proved to be convincing, particularly to the younger generation of chemists, both organic and inorganic.

The first outcome of the new chemical consensus was a testable theory of organic molecular structure, namely Kekulé's (1865) hexagonal ring formula for the benzene molecule. The hexagonal structure rationalized the known features of aromatic chemistry, and served as a guide for further explorations of the field.

Next came a *tour de force* in inorganic chemistry—the periodic table of the chemical elements which, unlike earlier classifications, had an open and incomplete form that led to the discovery of new elements. In 1869 Dmitry Mendeleev (1834–1907) and J. Lothar Meyer (1830–95) produced a definitive group and period classification of the 67 chemical elements then known. They showed that the valencies, atomic volumes, and other properties of the elements follow a common periodic function of increasing atomic weight, with gaps corresponding to elements as yet unknown, and a few anomalies ascribed to misordering. Mendeleev cleared up some of the anomalies and worked out detailed expectations for the properties of the elements remaining to be discovered, and even the probable mode of discovery, spectroscopic or otherwise (Fig. 2.1).

2.2 Spectrochemistry and astrophysics

After Lavoisier, 31 new elements were discovered in the period 1790–1830, but few others were found before the development of spectrochemical

THE PERIODS OF THE CHEMICAL ELEMENTS, TAKING THE ATOMIC WEIGHT OF O = 16.

Higher saline oxides.	Groups.	Elements of even series.				
R_2O	I	K = 39·1	Rb = 85·4	Cs = 132·9	—	—
RO	II	Ca = 40·1	Sr = 87·6	Ba = 137·4	—	Rd = 224
R_2O_3	III	Sc = 44·1	Y = 89·0	La = 139	Yb = 173	—
RO_2	IV	Ti = 48·1	Zr = 90·6	Ce = 140	—	Th = 232
R_2O_5	V	V = 51·4	Nb = 94·0	—	Ta = 183	—
RO_3	VI	Cr = 52·1	Mo = 96·0	—	W = 184	U = 239
R_2O_7	VII	Mn = 55·0	? = 99	—	—	
	VIII	Fe = 55·9	Ru = 101·7	—	Os = 191	
		Co = 59	Rh = 103·0	—	Ir = 193	
Typical elements.		Ni = 59	Pd = 106·5	—	Pt = 194·9	

Gaseous hydrogen compounds. | Higher saline oxides. | Groups.

Gaseous hydrogen compounds.	Higher saline oxides.	Groups.						
	R_2O	I	H = 1·008	Li = 7·03	Na = 23·05	Cu = 63·6	Ag = 107·9	— Au = 197·2
	RO	II		Be = 9·1	Mg = 24·3	Zn = 65·4	Cd = 112·4	— Hg = 200·0
	R_2O_3	III		B = 11·0	Al = 27·0	Ga = 70·0	In = 114·0	— Tl = 204·1
RH_4	RO_2	IV		C = 12·0	Si = 28·4	Ge = 72·3	Sn = 119·0	— Pb = 206·9
RH_3	R_2O_5	V		N = 14·04	P = 31·0	As = 75·0	Sb = 120·0	— Bi = 208
RH_2	RO_3	VI		O = 16·00	S = 32·06	Se = 79	Te = 127	— —
RH	R_2O_7	VII		F = 19·0	Cl = 35·45	Br = 79·95	I = 127	— —
0	0	0		He = 4·0	Ne = 19·9	Ar = 38	Kr = 81·8	Xe = 128 — —

Elements of uneven series.

D. MENDELÉEFF.
1869–1902.

Fig. 2.1 The periodic table of the chemical elements of Mendeleev from the preface to the third English edition of his *Principles of chemistry* (1905). Notable additions to his first table (1869) in his last revision (1902) were the elements scandium (eka-boron), gallium (eka-aluminium), and germanium (eka-silicon), predicted from his early periodic classification; the new Group 0 of inert gases; and the first chemically characterized new radioactive element radium (group II, since barium-like), represented by Rd rather than the current Ra symbol.

In footnotes to the 1902 table, Mendeleev indicated that the values of the atomic weights then current for argon (39.9) and potassium (39.1), cobalt (59) and nickel (58.7), and tellurium (127.7) and iodine (127) placed these pairs of elements in the 'wrong' relative order, judging by their physical and chemical properties, and he assigned them the atomic weights listed. Further footnotes predicted the discovery of a metal resembling manganese (eka-manganese) with an atomic weight of 99, corresponding to technetium (atomic weight 98) which lacks stable isotopes, and a whole new long period between cerium (atomic weight 140) and tantalum (atomic weight 183), represented in part by the rare earth elements then accepted as well characterized (Pr, Nd, Gd, Er, and Yb). Many of the rare earth elements claimed (more than 30) turned out to be mixtures: the number was fixed at 14 by Moseley's X-ray determination of the atomic numbers of heavier and lighter elements (1913).

analysis in 1860 by the chemist Robert Bunsen (1811–99) and his physicist colleague at the University of Heidelberg, Gustav Kirchhoff (1824–87). The improved spectroscope showed that the characteristic colour imparted to the flame of Bunsen's new gas burner by a metal salt consisted of bright sharp lines at wavelengths characteristic of the elements in the substance. In this way, Bunsen and Kirchhoff discovered the new alkali metals, caesium and rubidium, in 1860. During the following year William Crookes (1832–1919), a freelance chemical consultant in London, found another element spectroscopically, thallium, and Ferdinand Reich (1799–1882), at the Freiburg School of Mines, discovered indium similarly in 1863.

The name of each of these elements derived from the colour of its most prominent visible spectral line; blue for caesium, red for rubidium, green for thallium, and indigo for indium, but the next element, helium (1869), discovered from the yellow D_3 line in the prominences of the solar chromosphere spectrum by Pierre Janssen (1824–1907) in Paris and Norman Lockyer (1836–1920) in London, was named after its first source (Greek, *helios*, the Sun). The first two D lines had been characterized as dark sharp lines in the yellow region of the solar spectrum by Josef Fraunhofer (1787–1826), a partner in an optical instrument institute at Munich, who required monochromatic reference sources to determine the light dispersion of optical glasses. Over the period 1814–23, Fraunhofer distinguished 574 dark lines in the solar spectrum and measured their wavelengths with a grating instrument, classifying the more prominent lines alphabetically from the red to the violet region of the spectrum. Fraunhofer pioneered stellar spectroscopy too, finding similar dark lines in the spectra of the brighter stars.

By 1860 it was generally known that Fraunhofer's two D lines in the solar spectrum corresponded in wavelength to the two yellow lines emitted by sodium salts in a flame. On propagating continuous radiation from the oxyhydrogen limelight through a sodium flame, Bunsen and Kirchhoff were astonished to see the dark sodium D lines against the bright continuous spectral background, and using sunlight in place of the limelight, the normal D lines became even darker. Kirchhoff concluded that radiation from a hot source is selectively absorbed at the characteristic wavelengths of the elements in a cooler gaseous medium, the Fraunhofer dark lines arising in this way from selective absorption by atoms in the solar atmosphere.

The perspective opened up by the discoveries of Bunsen and Kirchhoff stimulated further collaborations between astrophysicists and spectrochemists. With Edward Frankland (1825–99) of the Royal College of Chemistry in London, Lockyer looked for the solar D_3 yellow line in the laboratory spectra of hydrogen and other gases, but without success. In 1895 William Ramsey (1852–1916) at University College London isolated terrestrial helium from the lead uranate mineral, cleveite, and identified the elusive yellow line in the spectrum of the gas. Lockyer, who began his career as an amateur astronomer, attained professional status in 1879 as the first director of the Solar Physics Laboratory, established in South Kensington, London, having founded in 1869 the scientific journal, *Nature*, which he edited for half a century. He had much in common with William Crookes, who founded the *Chemical News* in 1859 and edited the journal actively until 1906 and in an honorary capacity until his death in 1919. Both Lockyer and Crookes were enthusiastic exponents of the evolution of the chemical elements.

From 1863 another amateur astronomer, William Huggins (1824–1910), who built his own observatory at Tulse Hill, south London, worked with his neighbour, Walter Allen Miller (1817–70), professor of chemistry at King's College London, on the line spectra of the stars. In a footnote to a reprint of their major report of 1864 (Hearnshaw 1986), Huggins added the reflection:

One important object of this original spectroscopic investigation of the light of the stars and other celestial bodies, namely to discover whether the same chemical elements as those of our Earth are present throughout the Universe, was most satisfactorily settled in the affirmative; a common chemistry, it was shown, exists throughout the Universe.

The work of the nineteenth-century spectrochemists and stellar spectro-scopists effectively answered the contention of the French philosopher Auguste Comte (1798–1857), in his *Cours de philosophie positive* (1835), that there are questions which must remain forever inaccessible to scientific enquiry, a salient example being the chemical composition of the Sun and the stars. Astronomy must remain confined to macroscopic physics, to studies of the shapes, sizes, distances, and motions of the heavenly bodies, Comte supposed, 'whereas we would never know how to study by any means their chemical composition, or their mineralogical structure' (Hearnshaw 1986). At the same time, Comte felt confident enough in his own intuition and the general beliefs of the period to affirm a plurality of inhabited worlds; the inhabitants of other planets in our solar system 'being in some sense our fellow citizens ... whereas the inhabitants of other solar systems will be complete strangers' (Crowe 1986).

During the nineteenth century the doctrine of the plurality of inhabited worlds, particularly the belief in extraterrestrial creatures equivalent to the whole range of lifeforms on the Earth, declined in popularity in scientific circles. The range of the putative extraterrestrial organisms became limited to the cosmic microbes, the seeds of life, and their supposed habitat was widened from the various planetary systems of the universe to the vast realms of interstellar space. Here, according to the panspermists (the residuary legatees of the world pluralists), the vital germs of protein (or

DNA after 1953) floated freely, or rode upon the stellar winds (as the physical chemist Svante Arrhenius supposed), or dwelt on meteorites, comets, and other cosmic debris (Kamminga 1982).

2.3 Spectroscopic classification and inorganic evolution

Within a few years of Comte's death, many of the terrestrial elements were detected spectroscopically in stellar sources, and other celestial elements were proposed, in addition to the authentic case of helium, such as nebulium, postulated by Huggins from a green line of unknown origin he observed in the spectrum of the Orion nebula. Lockyer and other spectroscopists found that the laboratory spectrum of a given gaseous element changes with temperature from the relatively cool flame excitation, through the intermediate arc case, to the hot spark excitation. The spectral changes were ascribed to a 'dissociation' of the atom, analogous to the thermal dissociation of molecules, such as ammonium chloride, which had been detected from vapour density determinations. Atomic lines of all three types and molecular band spectra were observed in the spectrum of the Sun and other stars to varying degrees (strong, weak, or absent), thus providing a basis for the classification of star types in addition to the spectrochemical detection of the elements.

From a survey of some 4000 star spectra, Angelo Secchi (1818–78) at the Roman College Observatory identified four types in 1868: (a) white stars like Sirius and Vega, with four prominent hydrogen lines; (b) yellow stars, such as Arcturus or Capella, with a solar-type spectrum; (c) orange–red stars, exemplified by Betelgeuse and Antares, with a series of bands of dark and bright lines; and (d) faint red stars. Subsequent stellar classifications, carried out notably by Edward C. Pickering (1846–1919) and his cohort of lady astronomers ('Pickering's harem') at the Harvard College Observatory, subdivided Secchi's types alphabetically, with A to D for type (a), E to L for type (b), M for type (c), and N for type (d), with the addition of O, P, Q for stars with bright spectral lines not observed in other classes. The Harvard alphabetical classification was reorganized from 1901 by Annie Jump Cannon (1863–1941) to the present sequence of decreasing temperature—O, B, A, F, G, K, M—with numerical subdivisions for continuity, based by 1915 upon the spectra of some 225 300 stars. (There are rival claims from Princeton and Harvard for the mnemonic for the main stellar sequence: Oh Be A Fine Girl/Guy Kiss Me.)

The spectroscopic sequence of the stellar classification was widely regarded as an evolutionary order for some fifty years from the 1870s when Lockyer proposed a theory of the evolution of the stars and the chemical elements. The theory, based upon the Dumas version of Prout's hypothesis, was presented in his books *Chemistry of the Sun* (1887) and *Inorganic evolution* (1900). In order to account for the absence of laboratory atomic lines from some stellar spectra, Lockyer suggested in 1873 that the atoms of the chemical elements are dissociated into their subatomic units at the elevated temperatures of the blue—white stars. When these stars cooled, the chemical elements evolved along a homologous series of increasing complexity, as Dumas had supposed, from light to heavy atoms. Developing proto-elements, or elements not yet discovered on the Earth, were assumed to give the stellar spectra which had no known laboratory counterparts. The stars themselves evolved along an 'arch of temperature' from cool red giants, taken to be loose conglomerates of gas, dust, and meteorites which became more compact and heated up through gravitational contraction. During the initial heating phase, the red giants evolved through orange and yellow forms to the hottest blue—white stars; and these in due course, on cooling and further contraction, passed again through yellow and then orange stages to become the small, faint, dense red stars. In the later versions of the scheme Lockyer named each stage, in an evolutionary geological style, after a prominent representative star (Fig. 2.2).

William Crookes arrived at similar ideas of chemical evolution from his spectroscopic studies of gases at low pressure subjected to high-voltage electric discharges, and his chemical investigations of the closely related group of rare earth elements, the lanthanides. At the Sheffield meeting of the British Association for the Advancement of Science in 1879, Crookes argued that the plasma in the low-pressure discharge tube constituted a 'fourth state of matter', analogous to conditions in the stars where the atoms appear to be dissociated into the primary matter, Prout's protyle, or the protyle subdivisions of Dumas. At a subsequent meeting of the British Association (1886), Crookes proposed that the chemical elements had evolved in the stars from a plasma of subatomic particles under the influence of oscillatory electrical forces, like those of the discharge tube, as the stars cooled. The theory was developed by Crookes in a Friday Evening Discourse at the Royal Institution (1887) and his presidential address to the Chemical Society (1888), with additions (1898) following the discovery of the group 0 of inert gas elements.

Crookes supposed that the principal electrical force had an amplitude corresponding to a period in Mendeleev's classification of the elements, while a subsidiary oscillatory 'chemical' force separated the elements into 'electropositive' and 'electronegative' types and divided each long period of Mendeleev's table into an 'even' and an 'odd' set, corresponding to the a and b subgroups. As the primary matter cooled, according to Crookes, the chemical elements evolved in the order of increasing atomic weight and, up to the turning points of the elements of group IV, in the sequence of increasing positive ionic valency. At a tetravalent element the primary electrical force changed phase to swing back through negative ionic valencies, from trivalent, to divalent, and then monovalent, continuing onwards through zero valence (unrepresented in the 1880s) to increasing positive

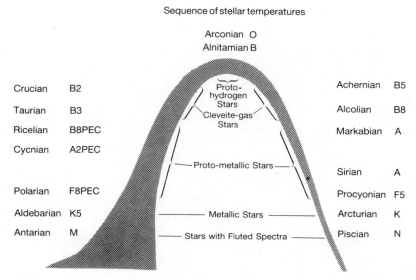

Sequence of stellar temperatures

Arconian O
Alnitamian B

Crucian	B2	Proto-hydrogen Stars	Achernian B5
Taurian	B3	Cleveite-gas Stars	Alcolian B8
Ricelian	B8PEC		Markabian A
Cycnian	A2PEC		
		Proto-metallic Stars	Sirian A
Polarian	F8PEC		Procyonian F5
Aldebarian	K5	Metallic Stars	Arcturian K
Antarian	M	Stars with Fluted Spectra	Piscian N

Fig. 2.2 The evolution of the stars and the chemical elements through an 'Arch of temperature', according to Lockyer (1914, *Nature*, **94**, 618). Stars begin (lower left) as red giant aggregates of meteoritic materials and evolve under gravitational collapse, through the successively hotter orange, yellow, and white stages, to the blue–white O and B stars at the peak of the temperature arch. Here all atoms are dissociated into protyle material from which the chemical elements evolve during the cooling down of the star to the red dwarf stage (lower right); first through proto-element stages, then to atoms of progressively larger atomic weight, and ultimately to molecules. The hottest 'Protohydrogen Stars' have line spectra containing the Pickering series, then ascribed to a precursor of hydrogen because of its close similarity to the Balmer series, but shown by Niels Bohr to arise from ionized helium (He^+). The 'Cleveite-gas Stars' give the line spectrum of neutral helium (isolated from cleveite by Ramsey and by Cleve). The 'Proto-metallic Stars' exhibit line spectra subsequently found to arise from singly and multiply ionized metal atoms, while the spectra of the 'Metallic Stars' are those of neutral metal atoms. The 'Stars with Fluted Spectra' are relatively cool and show band spectra characteristic of a range of diatomic molecules.

ionic valencies once more, up to the next limiting turning point of tetravalence.

Recapitulations of the electrical oscillation gave the 'giant pendulum swings' which, with the progressive increase in atomic weight, generated the elements of the periodic table, with vacant spaces for some heavier elements as yet unknown. The amplitude of the electrical pendulum swings decreased as the primary matter cooled, giving the heavier elements less diverse properties than the lighter elementary bodies, e.g. in electronegativity and metallic character. Crookes supposed that the 'even' phases of the

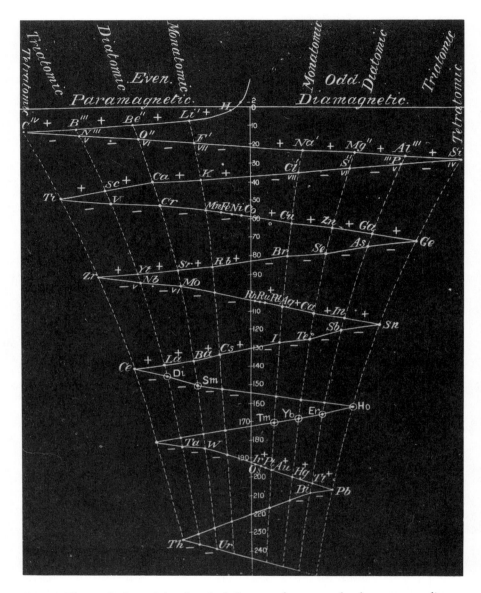

Fig. 2.3 The evolution of the chemical elements from protyle plasma, according to Crookes (1886–7). The vertical scale represents both a decreasing plasma temperature and an increasing atomic weight, with which the scale is calibrated. The generation of elements differing progressively in electronegativity and valency was ascribed to cosmic electrical oscillations, producing swings of diminishing amplitude extending from one group IV element to the next. The decline in amplitude followed 'a mathematical law' (unspecified), provoking Mendeleev (1889) to a

electrical oscillations generated paramagnetic elements, attracted towards a magnetic field, whereas the 'odd' phases produced diamagnetic elements, repelled by a magnetic field (Fig. 2.3). The paramagnetism of molecular oxygen, then recently discovered, was cited as evidence for this inference.

Crookes illustrated his theory of the evolution of the chemical elements with a three-dimensional model, first presented in his address to the Chemical Society in 1888. The model had the form of a doubled helical structure, described as a 'lemniscate spiral', in which the vertical scale, with hydrogen at the top and uranium at the bottom, represented both the increase in atomic weight and the decrease in plasma temperature during the evolution (Fig. 2.4). Two mutually perpendicular oscillatory forces in the horizontal plane produced a repeating figure-of-eight pattern, with a primary amplitude spanning the separation between two successive group IV elements (e.g. carbon and silicon), and a secondary amplitude separating an element with a positive ionic divalency from a neighbour with the corresponding negative ionic divalency (e.g. magnesium and sulphur).

In the evolutionary scheme of Crookes the atomic weights of the elements were determined by the rate of cooling of the primary matter. Each element or set of elements represented a 'platform of stability', and the several platforms were linked on the cooling curve by ladders with unstable rungs, representing labile elementary bodies. It was probable, he supposed, that the cooling rate was not always uniform and varied with location. Rapid cooling would produce sets of closely related elements, like the group VIII triads (Fe, Co, Ni; Ru, Rh, Pd; and Os, Ir, Pt) or, more particularly, the remarkably similar rare earth elements, which are representatives of the 'cosmical lumber-room where the elements in a state of arrested development—the unconnected missing links of inorganic Darwinism—are finally aggregated'.

In general, Crookes maintained, 'the original protyle contained within itself the potentiality of all possible atomic weights', and 'our atomic weights merely represent a mean value around which the actual atomic weights of the atoms vary within certain narrow limits'. On a given platform of stability, 'the stable element appropriate to that stage would absorb, as it were, the unstable rungs of the ladder which led up to it'. Thus calcium consists largely of atoms of weight 40, but a few have weights of 39 and 41 or even of 38 and 42, averaging to the observed bulk atomic weight of 40.08. In 1915 when Soddy coined the term *isotope* to describe

---

criticism of 'geometers' of chemical evolution. Crookes supposed that the elements of the 'even' and the 'odd' subgroups in Mendeleev's periodic classification are 'paramagnetic' and 'diamagnetic', respectively, citing the paramagnetism of molecular oxygen as evidence. (Reproduced from 1970, *The Royal Institution Library of Science: Physical sciences*, (ed. W. L. Bragg, and G. Porter), Vol. 3, p. 419, Elsevier, Amsterdam.)

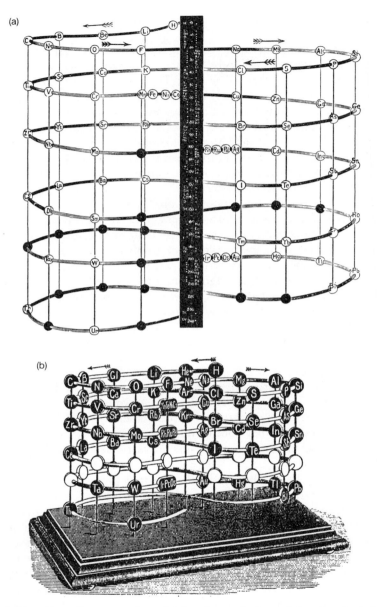

Fig. 2.4 The three-dimensional model for the evolution of the chemical elements presented by Crookes in his Presidential Address to the Chemical Society (1888) and subsequently modified (1898) to accommodate the newly discovered group 0 of inert gas elements at the crossover points of the figure-of-eight evolutionary sequence: (a) from W. Crookes (1888). *J. Chem. Soc.*, **53**, 503; (b) from W. Crookes (1898). *Proc. R. Soc.*, **63**, 409: here the symbol Gl (glucinium, formerly an alternative name for beryllium) is employed instead of Be for the fourth element. The 1898 model is exhibited at the Science Museum, South Kensington, London.

atoms of a given chemical element with different atomic mass and distinctive radioactive properties, Crookes pointed out that he had anticipated the concept some 30 years earlier.

Radioactivity, discovered by Henri Becquerel (1852–1908) in 1896, appeared to exemplify the 'unstable rungs' of labile elements in the ladder of inorganic evolution, and Crookes took up chemical studies of uranium salts with enthusiasm. Marie Curie (1867–1934) observed in 1899: 'One could, perhaps, refer radioactivity to Crookes's theory of the evolution of the elements, by attributing radioactivity to elements of high atomic weight recently formed, whose evolution may not yet be complete'.

Crookes was delighted too by Ramsey's discovery during the 1890s of an entire new group in the periodic table—that of the inert gases with an apparent zero valence. Indeed it was Crookes who identified helium spectroscopically in the gas Ramsey isolated by heating the uranium mineral, cleveite. The inert gas elements fitted neatly into the evolutionary scheme of Crookes. These elements represented successive dead-centres of the 'giant pendulum swings' at zero valence (Fig. 2.3), and the sequence of central crossover points in the figure-of-eight repeat pattern of his doubled-helix periodic table (Fig. 2.4a). Crookes added helium, neon, argon, and krypton to his doubled-helix model in 1898 and predicted the discovery of further inert gas elements, required to occupy the remaining vacant crossover positions in his model (Fig. 2.4b). Like Mendeleev, Crookes predicted that a whole, new, long period of elements remained to be discovered between cerium and tantalum, represented by a complete figure-of-eight of vacant positions in his model.

Mendeleev, in his Faraday lecture to the Chemical Society in 1889, pointed out that the match between the amplitudes of the 'giant pendulum swings' in the evolution of the elements postulated by Crookes and the empirically based periods and their subdivisions in the classification of the elements was far from perfect, particularly for the heavier elements. 'The periods of the elements have a character very different from those which are so simply represented by geometers,' Mendeleev observed, 'they correspond to points, to numbers, to sudden changes of the masses, and not to a continuous evolution.' Mendeleev himself doubted the scientific utility of all theories of the evolution of the chemical elements, compared with his own procedure of predicting the properties of elements as yet unknown from the established properties of the elements in his periodic classification.

Like Crookes, but for wholly different reasons, Mendeleev was delighted by the discovery of radioactivity and a whole new group of inert gases, placed in the new group 0 of his periodic table (Fig. 2.1). After visiting the Paris laboratories of Becquerel and the Curies, Mendeleev added their names to the distinguished list of 'Reinforcers of the Periodic Law', namely those who had discovered the missing chemical elements he had predicted. Mendeleev (1902) supposed that the lightest of his predicted elements, the cosmic luminiferous ether with an atomic weight of 10^{-6} relative to unity for hydrogen, was continuously taken up and emitted by the heaviest of the elements in the phenomenon of radioactivity.

Early theories of the origin of atomic spectra were influenced by the protyle hypothesis, and it was hoped that an analysis of the series of lines appearing in the spectrum of a given element might provide some information on the subatomic units of the atom. The evolutionary views of Lockyer and Crookes suggested that each subatomic unit in an atom might give a distinctive spectral line, but the immense number of lines exhibited by elements such as iron made this approach unattractive. George Johnstone Stoney (1826–1911) proposed in 1871 that the units of an atom vibrated harmonically, emitting a fundamental and a series of harmonics in a flame or arc spectrum, and absorbing light at the corresponding wavelengths in the Fraunhofer spectrum. But the harmonic relations imposed on even the simplest line series, that of the hydrogen atom over the visible wavelength region, were unconvincing and also unfruitful, for the fits claimed were poor and many expected harmonics were not observed.

Without a physical model, a mathematician at Basle, J. J. Balmer (1825–98), found the basic relationship between the wavelengths of the visible hydrogen lines by a graphical fitting procedure in 1884. Balmer's relationship

$$\lambda = \lambda_0[n^2/(n^2 - 4)] \tag{2.1}$$

with n taking integral values of 3, 4, 5, ..., predicted further lines of the hydrogen series in the ultraviolet region. These additional lines, unknown to Balmer, had recently been found in both laboratory and stellar spectra, as Balmer's colleague Hagenbach pointed out. Attempts to fit the spectral line series of other elements in both stellar and laboratory spectra soon followed.

In 1886 Pickering at the Harvard College Observatory observed a series of lines in the spectrum of the star zeta-Puppis to be almost identical with Balmer's hydrogen series, and Pickering together with Lockyer and other stellar evolutionists ascribed the new series to proto-hydrogen, the supposed subatomic precursor of the hydrogen atom. (In 1913 the Pickering series was shown by Niels Bohr to be due to singly ionized helium, isoelectronic with the neutral hydrogen atom.)

At Cambridge in 1879, G. D. Liveing (1827–1924) and J. Dewar (1842–1923) analysed the laboratory emission spectrum of sodium into three series, the lines being sharp in one and diffuse in another, while the third, the principal series, was so named because it also appeared in the absorption spectrum of sodium vapour. Each component of the series consisted of a close pair of lines, exemplified by the D_1 and D_2 sodium lines of the Fraunhofer spectrum, and Liveing and Dewar found that the line spectra of other alkali metals consisted of similar series of doublets. The separation between the two lines in each doublet of a given series was shown by W. N. Hartley (1846–1913), at Dublin in 1883, to be constant if measured in terms of the frequency (v) rather than the wavelength (λ), related by $v = c/\lambda$, where c is the velocity of light. A similar constant frequency separation was found by Hartley in the three-line groups, the triplets, observed in the spectrum of magnesium and the alkaline earth metals.

These spectroscopic analyses suggested that the spectroscopic line frequency, or the more convenient wavenumber ($\tilde{v} = 1/\lambda = v/c$, i.e. the number of wavelengths λ per centimetre, in units of cm^{-1}), might be a more fundamental property than the line wavelength. In 1890, J. R. Rydberg (1854–1919) at Lund inverted Balmer's equation (2.1) for the visible hydrogen spectral series from wavelength to wavenumber form and generalized the relationship to cover the spectral series known for the other elements:

$$\tilde{v} = \tilde{v}_\infty - R/(n + \mu)^2 \qquad (2.2)$$

where $\tilde{v}_\infty$ is the series limit as the sequence of integers n becomes infinite, R is the universal Rydberg constant ($109\,737$ cm^{-1}), and μ is a fractional 'defect' which is zero for the case of hydrogen and constant for each of the series of other elements.

Rydberg had long been searching for a unifying principle underlying the ordering of the chemical elements in the periodic table. Indeed it was Rydberg who first suggested that the ordinal position of an element in the table, the atomic number Z, is more significant than the atomic weight W, since some elements lie patently in the 'wrong' order in terms of weight, as in the case of the long-known inversion of tellurium ($Z = 52$, $W = 127.6$) and iodine ($Z = 53$, $W = 126.9$). The periodic variation of the atomic volume and other physical and chemical properties of the elements in the table suggested to Rydberg that the oscillations of the light waves emitted in the line spectra of the elements held the key, and that spectrum analysis would provide the solution. In 1890 Rydberg maintained that his general equation (2.2), showing that a modified form of the Balmer relation (2.1) for the visible hydrogen spectrum governs the spectral line series of other elements, confirms 'without doubt the hypothesis that the elements are compound, and that hydrogen forms the matter of which they are built in the first place' (McGucken 1969).

3
Atomic dissociation and transmutation

3.1 The discharge tube particles and waves

The new field of radiation physics was opened up by the studies of the optical spectra of gases as a function of their pressure by J. Plücker (1801–68) and J. H. Hittorf (1821–1914) from 1858. Ultimately their pioneering work led to the characterization of three types of 'radiation' from the electric discharge tube: X-rays, and cathode and anode rays. Plücker and Hittorf at Bonn employed the mercury vapour pump, recently invented by their colleague, H. Geissler (1815–79), to achieve pressures as low as 10^{-6} atmospheres, then unprecedented. A gas in the discharge tube at 10^{-3} atmospheres produces a luminous column with a colour and spectrum individual to the particular gas, but the column breaks up and disappears as the pressure is reduced. At the same time, a glow confined at first to the cathode advances down the discharge tube towards the anode, ultimately filling the tube completely with a blue–green luminescence from the glass envelope. In 1869 Hittorf reported that all gases give low-pressure cathode rays with the same properties. The cathode rays travel in straight lines, objects in their path casting shadows in the overall envelope luminescence, and the rays are deflected by a magnetic field, as shown by displacements of the shadows with a magnet.

William Crookes suggested in 1879 that the low-pressure and high-excitation conditions of the discharge tube must dissociate the atoms, so that the cathode rays are streams of particles, protyle, or its subatomic units. Indeed they appeared to have a momentum sufficient to turn a small paddle wheel in the discharge tube. The motion of the paddle wheel was later shown to be due to a heating effect, since a restrained wheel rotated when released after the electric discharge was cut off. The particle theory of cathode rays was sustained in Britain, but elsewhere, particularly in Germany, a wave theory prevailed.

Heinrich Hertz (1857–94), who discovered the radiowaves and microwaves predicted by James Clerk Maxwell's theory (1861–4) of electromagnetic radiation in 1888, supported the wave theory. Hertz searched without success for the magnetic field cathode rays should produce (if they are charged particles in motion) and also for the expected deflection of cathode rays by an electrostatic field. His assistant at Bonn, Philipp Lenard (1862–1947), showed in 1892 that cathode rays pass through an aluminium foil window in the discharge tube and penetrate through 8 cm of air to excite a luminescent screen, and later (1902) that cathode rays are emitted by metals under ultraviolet irradiation (the photoelectric effect). With equipment borrowed from Lenard, to Lenard's subsequent chagrin, W. K. Roentgen (1845–1923) discovered in 1895 that cathode rays striking a metal target in the discharge tube generate secondary radiation, X-rays, which similarly excite a luminescent screen, but at longer range. Another particle-or-wave problem arose from attempts to characterize X-radiation.

The anomaly that cathode rays were deflected by magnetic fields but appeared to be unaffected by electric fields was solved by J. J. Thomson (1856–1940) at Cambridge. Improved low-pressure control, by limiting the scatter due to particle collisions, brought to light the electric deflection expected for negatively charged particles. Thomson showed in 1898 that the value of the electric field required to cancel the deflection produced by a known magnetic field gave the ratio of the charge to the mass, e/m, of the 'corpuscles', as he termed them. Stoney had already coined the term *electron* for the unit charge, having argued in 1874 that, if matter is atomic, then electricity must be so too from Faraday's laws of electroequivalency (1834), the unit of charge being represented by the ratio F/N of the Faraday to Avogadro's number. The unit of charge and Thomson's ratio, e/m, implied that the electron has a smaller mass or else a larger charge than the hydrogen ion.

A similar value of the ratio e/m was found by Pieter Zeeman (1865–1943) from the broadening and splitting of atomic spectral lines in a magnetic field (1897), and by Thomson and his students with electrons from thermionic and photoelectric sources. With the latter source and a cloud chamber, Thomson obtained an approximate value for the electronic charge, much improved by the charged oil-drop measurements (1913–17) of R. A. Millikan (1868–1953) at Chicago, showing the mass of the hydrogen atom or ion (proton) to be some 1836 times larger than the mass of the electron. Additionally, in 1916 Millikan determined Planck's constant, h, relating the quantized energy of a photon to its classical electromagnetic frequency, $E = h\nu$, from Einsten's (1905) theory of the photoelectric effect.

With the cathode ray particles now characterized as fundamental constituents of all substances, Thomson moved on with F. W. Aston (1877–1945) to the anode or positive rays, individual to the particular residual gas of the discharge tube. By 1913 they found that neon, with an atomic weight of 20.18, is composed of two main species (isotopes), with atomic masses of 20 and 22 in the approximate abundance ratio of 10:1; Aston partially separated the two components by diffusion, before leaving for war service in 1914–18. On his return to Cambridge in 1919, Aston transformed the positive ray apparatus into a precision mass spectrograph with an accuracy of one part in a thousand, and in 1920 he announced the vindication of Prout's hypothesis of integral atomic masses.

While the measured atomic mass (M) of the isotopes of nearly all the elements examined were integral relative to ^{16}O, there was a significant exception in the case of hydrogen, with $M = 1.008$. Aston accounted for the anomaly by taking up the old speculation of Marignac (1860), that Prout's hypothesis is not exact because mass is lost as energy in the condensation of protyle to other elements, given a new credibility by Einstein's (1905) mass–energy relation $E = mc^2$. By 1927 Aston had built a second mass spectrograph, with a precision of one part in 10^4, and measurements with the instrument showed that the atomic mass (M) of virtually all isotopes deviates to a small extent from the nearest integer, the mass number (A). Aston defined the weighted difference, $(M - A)/A$, as the 'packing fraction' of the isotope (subsequently termed the 'mass defect'), representing the fractional loss of mass as energy in the combination of A hydrogen atoms to form the isotope of atomic mass M. The fractional loss was found to be the largest for the iron group elements which, accordingly, were taken to consist of the most stable of all the atoms.

3.2 The radioactive elements and atomic structure

Henri Becquerel, with an expertise in luminescence effects, was prompted by Roentgen's discovery of long-range fluorescence excited by X-rays to investigate the possible production of X-rays by luminescent materials. Uranyl(VI) salts, long known as efficient fluorescent substances, produced an analogous effect, initially the blackening of covered photographic plates, even without prior photoexcitation (1896). Marie Curie monitored the effect quantitatively from the conductivity induced in the air surrounding a radioactive substance by means of a sensitive piezoelectric electrometer, which had been invented by her husband Pierre Curie (1859–1906) and his brother, Jacques, following their discovery of the piezoelectric effect in quartz and other polar crystals (1881). Marie Curie found two groups of radioactive substances, the compounds of uranium and of thorium. The radioactivity of each pure uranium compound was roughly proportional to the uranium content, but the crude ore, pitchblende, proved to be much more active.

With Pierre, Marie Curie applied the standard inorganic group separation procedure to pitchblende, isolating one new element, polonium, in the bismuth fraction and another, radium, in the barium group, both many orders of magnitude more active than uranium (1898). The radioactivity of these elements initially appeared to be constant in time, as well as independent of physical conditions and state of chemical combination, implying for the Curies, Becquerel, and others, such as Mendeleev and Kelvin, that radioactive substances collect, modify, and transmit a cosmic radiation continuously incident upon the Earth, just as the discharge tube converts electrical energy into X-rays and anode and cathode rays.

The analogy appeared to be close when the β-rays from radioactive elements were shown to be high-energy electrons, with the same e/m ratio as the particles of the cathode rays but with much higher velocities; while the γ-rays were found to be unaffected by electric and magnetic fields, resembling hard X-rays but with an even greater penetration. At first, the third type of radiation, the α-rays, with little penetration, were a puzzle, and so too were the 'emanations' of radon and its isotope, thoron, which spread radioactivity to nearby substances through the gas phase, depending on the air currents.

Over the period 1901–3, Ernest Rutherford (1871–1937) and Frederick Soddy (1877–1956), then at McGill University, Montreal, developed the radical theory that, in radioactivity, atoms of the heaviest elements are undergoing spontaneous decay into atoms of the lighter elements. The new decay theory was at variance both with the early Becquerel–Curie view that the radioelements themselves remain unchanged during their activity, being fixed species like the other elements, and with the Crookes–Lockyer theory of chemical evolution from lighter to heavier elements. Rutherford and Soddy found that the addition of aqueous ammonia to a thorium(IV) nitrate solution produced a precipitate of thorium(IV) hydroxide which retained 25 per cent of the original α-activity but none of the β-activity. The filtrate of 'thorium-X' (an isotope of radium) retained all of the β-activity and all of the thoron emanation production, with 75 per cent of the original α-activity, so that there was an apparent conservation of radioactivity overall.

Over the course of time the $Th(OH)_4$ precipitate gradually gained in radioactivity, following a first-order kinetic law with a half-life of some 4 days, while the thorium-X filtrate exponentially lost activity with the identical first-order rate constant, which likewise governed the thoron production, decreasing from the filtrate and increasing in the precipitate. The overall conservation of radioactivity was now seen as a steady-state sum of the different activities in a decay series, from thorium to thorium-X and thence to thoron. Thoron and radon were characterized as isotopes of a noble gas element, with the same boiling point ($-150°C$), inert to all chemical reagents then available, and differing only in radioactive half-life.

In 1903 Soddy moved to University College London to join William Ramsay, the discoverer of argon (1895) and the other stable inert gases, for further study of the radioactive element of the noble gas group. Ramsay and Soddy released occluded radon from aged radium(II) chloride into a small discharge tube and observed the transmutation of radon into the lightest element of the group, helium, by the appearance and growth of the yellow D_3 line in the spectrum over successive half-lives (3.8 days) of radon.

Subsequently at Glasgow (1911–13), Soddy, with contributions from A. S. Russell and K. Fajans, formulated the radioactive displacement laws, specifying that α-activity transforms an element into another lying two places to the left in the long form of the periodic table, while β-activity converts an element into another lying one place to the right in the table. Soddy organized the 35 radioactive elements then known into three of the four radioactive series, each represented by successive values of the integer *n* in a general mass number formula: $4n$, for the series starting from ^{232}Th and ending with ^{208}Pb; $4n + 2$, for the series from ^{238}U to ^{206}Pb; and $4n + 3$, for the ^{227}Ac series. These 35 radioactive elements occupied only 12 spaces in the periodic table of the elements. The three known radioactive series overlapped chemically, and so too did some of the elements within each of the series. The overlapping elements could not be separated by chemical procedures, although they were distinct atomic species with their own specific radioactive half-life and particular mass number. Soddy coined the term *isotope* to describe such atoms with different weights but placed in the 'same box' of the periodic table. In 1913 Soddy predicted that lead from thorium ores would have a larger atomic weight than the average, and he confirmed this expectation a year later in an analysis of thorite from Ceylon.

Meanwhile, in 1902, Rutherford at McGill had detected the deflection of α-rays by electric and magnetic fields, an effect which had eluded Becquerel and the Curies. The electric and magnetic deflections showed that the α-particle, from its e/m ratio of $+(1/2)$, was either H_2^+ or He^{2+}. On moving to Manchester in 1907, Rutherford developed with Hans Geiger (1882–1945) a counter to detect a single charged particle from its generation of an ionization cascade in a low-pressure gas. They used the counter to verify that a screen of zinc sulphide, the spinthariscope invented by Crookes in 1903, did indeed record the arrival of a single α-particle by a scintillation. With this instrumentation, Rutherford and Geiger in 1908 measured the charge, and thus the mass, of individual α-particles, confirming the He^{2+} assignment, and then determined Avogadro's number by counting the number of α-particles producing a given volume of helium.

Rutherford then investigated the scattering of α-particles by thin metal foils and sheets of mica. Geiger, with Ernest Marsden (1889–1970) from 1909, found that the most probable angle of scatter was small, some 2–4° in the case of gold foil, but a small fraction of the incident α-particles, one in 20 000 from gold foil 0.4 μm thick, appeared to rebound from the foil at large scatter angles. The fraction of rebounding α-particles was approximately proportional to the atomic weight of the metal constituting the foil for a given equivalent thickness. After lengthy checking and analysis of the data, Rutherford announced his nuclear model for the atom in 1911. The small fraction of incident α-particles scattered at large angles implied that the atom, typically of 10^{-10} m in dimension, must be largely empty space with a small massive nucleus, some 10^{-15} m in size, bearing a charge proportional to the atomic weight. There was little immediate response to the new theory. In the short term, it was not so much opposed as ignored.

At Manchester, the α-scattering experiments brought a new atomic property into prominence—the atomic number Z, as foreshadowed by Rydberg and now representing the number of unit charges on the nucleus. In 1912 Max von Laue (1879–1960) with W. Friedrich and P. Knipping at Munich established the electromagnetic wave character of X-rays by photographing the X-ray diffraction pattern of crystals. The discovery was immediately applied by Rutherford's student, Henry Moseley (1887–1915), who used a potassium ferrocyanide crystal as a diffraction grating to determine the wavelengths of the characteristic X-ray spectral lines from each of the metallic elements placed, as an anticathode target, in a cathode ray discharge tube.

In 1913 Moseley reported that the wavenumber or frequency of each series of characteristic X-ray lines is proportional to $(Z - \sigma)^2$, where the integer Z represents the serial number of the element in the periodic table, increasing by unity from one column to the next along a period, and σ is a constant equal to 1.0 for the penetrating K series of X-ray lines and 7.4 for the less penetrating L series. Moseley's relationship cleared up the anomalous pairs with inverted atomic weights, such as cobalt/nickel and tellurium/iodine. Gaps in the series indicated that six elements remained to be discovered between aluminium ($Z = 13$) and uranium ($Z = 92$). Contemporaries of Moseley working with Rutherford at Manchester were Niels Bohr (1885–1962) and George de Hevesy (1885–1966). In 1922 Hevesy discovered one of the missing elements, hafnium ($Z = 72$), during the first of his two periods with Bohr at Copenhagen as a refugee from his native country, Hungary.

Bohr at Manchester in 1913 provided a solution for the problem of the classical instability of the Rutherford nuclear atom. An electron moving in a closed orbit around a positively charged nucleus, being an accelerated charge in an electrostatic field, is expected by Maxwell's theory to emit electromagnetic energy and to spiral down continuously into the nucleus. Bohr postulated that orbits in which the electronic angular momentum has the quantized values $n(h/2\pi)$, where n is an integer, are stationary and free from the classical instability. Electromagnetic radiation is absorbed or emitted by an atom only if the electron undergoes a sudden transition from one stationary state to another, with a difference in energy measured by the photon energy, $h\nu = E(n_1) - E(n_2)$. With his quantum postulate, Bohr was able to derive theoretically the wavenumbers of the spectral lines of the Balmer series in the visible region and the corresponding ultraviolet and infrared series of the hydrogen atom, expressing the Rydberg constant in terms of the electronic mass and charge (m and e, respectively) and other univeral constants (c, the velocity of light, and h, Planck's constant of action) as $2\pi^2 me^4/ch^3$.

While Bohr's theory held with precision only for the spectra of one-electron atoms, his concept of stationary electronic states and his quantum postulate provided guiding principles for the interpretation of multi-electron atomic spectra and for the development of quantum mechanics in the mid-1920s. In 1915 Arnold Sommerfeld (1868–1951) at Munich postulated a set of elliptical electron orbits, requiring another quantum number, in place of each circular Bohr orbit to account for the alkali metal spectral series (sharp, principle, diffuse, and fundamental), with a third space–orbit quantum number to explain the Zeeman splitting of a spectral line in a magnetic field. The Zeeman effect required in addition two values (±½) for electron-spin quantization, clockwise or counter-clockwise. Sommerfeld's student, Wolfgang Pauli (1900–58), proposed in 1924 an exclusion principle prohibiting identical values of all four quantum numbers, n, l, m, and s, for any two electrons in a given atom. Pauli's principle gave the periodic table of the elements a new rationale in terms of the progressive filling of electronic shells and subshells around the atomic nucleus, governed by the relations between the quantum numbers, up to the filled shell of a group 0 inert gas element, marking the completion of a period. The term 'inert gas' turned out to be a misnomer, and was replaced by 'noble gas' when xenon, and then other group 0 elements, were found to combine with fluorine and other highly electronegative elements.

In his Paris thesis of 1924, Louis de Broglie (1892–1987) suggested that a particle with a momentum p is guided by a pilot wave with a related wavelength, $\lambda = h/p$. Stimulated by the proposal, Erwin Schrödinger (1887–1961) at Zurich in 1925 developed his non-relativistic wave mechanics, in which the three space–orbit quantum numbers (n, l, m) rationally emerge as the fundamental and the overtone mode integers of spherical wave harmonics. The equivalent matrix mechanics theory of Werner Heisenberg (1901–76) at Göttingen was developed at the same time and, soon afterwards, the more complete relativistic quantum mechanics of Paul Dirac (1902–84) at Cambridge, specifying all four (n, l, m, s) of the quantum numbers of an electronic orbital in a given atom.

3.3 The structure of the atomic nucleus

On succeeding Thomson at Cambridge in 1919, Rutherford resumed his earlier studies, interrupted in 1914, of the products of collisions between α-particles and gas molecules. The hydrogen nuclei (protons), which had appeared in such collisions, were ascribed initially to knock-on events from hydrogen-containing impurities. However, Rutherford himself had found in 1917 that nitrogen alone of the atmospheric gases gave this effect and, two years later, he attributed the proton production to nuclear disintegration. With James Chadwick (1891–1974), Rutherford found by 1924 that protons are ejected, under high-energy α-particle bombardment, by all elements lighter than calcium except for helium, carbon, and oxygen.

The cloud chamber, developed from 1895 by C. T. R. Wilson (1869–

1959) at Cambridge, produced photographs of the tracks of the charged particles involved in the collision, and the products formed were characterized by a track analysis based on the conservation of energy and momentum. The probability of nitrogen disintegration under high-energy α-particle bombardment is small ($\sim 2 \times 10^{-5}$), and P. M. S. Blackett (1897–1974) obtained in 1925 photographs of eight collision events, indicating that the nitrogen nucleus absorbed the α-particle to form an oxygen nucleus with the ejection of a proton alone. There was no evidence for the expulsion of an electron, expected from the contemporary view that a nucleus of mass number A and charge Z consists of A protons and $A - Z$ electrons.

Although the view then current was consistent with the observation of electron ejection from the nucleus in β-radioactivity, Rutherford felt that the confinement of A protons and $A - Z$ electrons in so small a volume as the atomic nucleus was implausible on electrostatic grounds, and he looked for unusual events indicating the participation of a neutral massive nuclear particle. In 1930 Walther Bothe (1891–1957) and his student, H. Becker, at Berlin reported that an exceptionally penetrating radiation, assumed to consist of high-energy γ-photons, resulted from the bombardment of beryllium with energetic α-particles. Irène Curie (1897–1956) confirmed and extended the discovery with her husband, Frédéric Joliot (1900–58), showing that the radiation produced protons from paraffin wax. Rutherford suspected that the radiation was probably the sought-for neutron, and Chadwick established that this was indeed the case in 1932. The conservation of energy and momentum in collisions between the penetrating radiation from the α-ray/beryllium source and a range of target atoms, from hydrogen to argon, showed that the 'radiation' consisted of particles with zero charge and unit mass.

Although Irène and Frédéric Joliot-Curie missed the neutron in confirming the Bothe–Becker experiment, they discovered artificial radioactivity in 1934 by using an aluminium foil window to separate the α-source from the beryllium target. The foil itself became radioctive, with a half-life of 3 minutes, and the classical chemical separation techniques showed that the ^{27}Al had been converted to ^{30}P which subsequently decayed to ^{30}Si and a positron. The positron, e^+, as the antiparticle counterpart of the electron, e^-, had been predicted by Dirac in 1930, as a hole in an invisible ocean of negative energy electron states. The production of e^+ and e^- pairs in γ-ray and cosmic-ray showers was detected in 1932 by C. D. Anderson of the California Institute of Technology at Pasadena, and by Blackett independently.

The α-particle from natural radioactive sources served as the standard probe for studies of nuclear structure and reactivity over three decades, but such α-rays have a limited energy range. Rutherford and others sought to overcome the limitation in the late 1920s by returning to the principles of the electric discharge tube. At Cambridge, John Cockcroft (1897–1967) with E. T. S. Walton constructed the first linear accelerator, producing beams of protons with energies up to 3×10^6 electron volts (3 MeV). With a lithium target, they found in 1932 that α-particle scintillations on a zinc sulphide screen began to appear at proton beam energies of 10^5 eV, the number increasing rapidly at higher voltages. It was shown that absorption of a proton by ^7Li produces nuclear fission to give two α-particles moving apart in opposite directions, each with an energy of 8.6 MeV. The result was confirmed and extended by E. O. Lawrence (1901–58) at Berkeley, who had developed a circular particle accelerator, the cyclotron, with his student, M. S. Livingstone, over the same period. With a deuteron beam in the cyclotron and a beryllium target, Lawrence found that boron and 9 MeV neutrons were generated.

Following the report of the production of artificial radioactivity by the Joliot-Curies in 1934, Enrico Fermi (1901–54) in Rome began the study of nuclear transformations by neutron bombardment, following the principle that the neutron, lacking a charge and so free from electrostatic repulsion, forms an ideal probe. Fermi's group discovered that neutrons were particularly effective when slowed down to thermal velocities with paraffin wax or some other moderator. By 1937 they had produced some 47 new radioactive isotopes by the irradiation of 68 elements with thermal neutrons. Uranium was among the first of the elements subjected to neutron activation, and the β-radioactivity of the product suggested that a transuranic element with an atomic number of 93 had been formed.

The radiochemist Ida Noddack (1896–1978) had already questioned this conclusion in 1934, suggesting the possibility of nuclear fission, since Fermi's group had demonstrated chemically only the absence of elements between lead and uranium in the products. The chemical characterization of the pairs of lighter elements from the fission of the uranium nucleus under neutron activation was achieved in 1938 by Otto Hahn (1879–1968) and his student, Fritz Strassmann (1902–80), at the Kaiser Wilhelm Institute, Berlin, in collaboration with Lise Meitner (1878–1968), until she was obliged to take refuge in Sweden. The Joliot-Curies in Paris showed in 1939 that neutrons are produced by the disintegration of the uranium nucleus, suggesting the possibility of a chain reaction in uranium nuclear fission. The operation of a nuclear fission chain reaction on a major scale was achieved in 1942 by Fermi's group with the first artificial atomic pile under the Stagg Field stadium at Chicago University. Thirty years later it was discovered that, some 2 billion years ago in West Africa, a rich deposit of uranium ore had operated as a natural nuclear fission reactor for many years, leaving a characteristic range of isotopic fission products (Section 10.3).

4

The age and abundances of the elements

4.1 Stellar abundances

In 1913 Henry Norris Russell (1877–1957) at Princeton and the Danish astrophysicist Ejnar Hertzsprung (1873–1967) at the Potsdam observatory independently discovered a correlation between the absolute brightness of a star and the stellar surface temperature. For stars at a known distance from the solar system, or at the same but unknown distance, a decrease in absolute magnitude was related to a fall in the surface temperature along the Harvard classification sequence (O, B, A, F, G, K, M), based on the colour change of the black-body radiation from blue–white to red (Fig. 4.1). Russell, as a postdoctoral fellow in England at the Cambridge observatory in 1900–3, determined the trigonometric parallax of a large number of stars photographically, obtaining stellar distances which enabled him to convert the apparent brightness of the stars into the corresponding absolute magnitude. Hertzsprung studied the spectra of stars within a star cluster, all approximately equidistant from the Earth, so that the apparent brightness of the stars provided a relative measure of absolute magnitude.

Following a suggestion by Hertzsprung, Walter Adams (1876–1956) at the Mount Wilson observatory from 1914 developed a 'spectroscopic' parallax technique whereby the luminosity ratio of two stars of the same spectral type is gauged by the ratio of the intensities of 'sensitive' spectral lines. The method gave the absolute magnitude and distance of many more stars. The number of known star distances, 305 in 1910, increased to some 7000 by 1935, and the range was extended from a few hundred light-years, the limit for trigonometric parallax, to many thousands of light-years. Results from the magnitude–colour surveys indicated that the majority of stars, some 90 per cent, lie on the *main sequence* of Russell and Hertzsprung. A minority of stars proved to be exceptional, notably the red giants with a larger luminosity than analogous stars with the same colour on the main sequence (O, B, A, F, G, K, M), and the white dwarfs, for which the luminosity is smaller than their colour analogues.

Initially Russell took the main sequence of stars to represent an evolutionary order, following the theory of Lockyer, who added enthusiastic editorial comments to Russell's papers to *Nature* in 1914. But Russell and most astrophysicists abandoned the evolutionary view after the publication in 1926 of the *Internal constitution of the stars* by Arthur Eddington (1882–1944). With the assumption that the atoms are completely ionized in stellar interiors, where the ideal gas laws should be applicable, Eddington

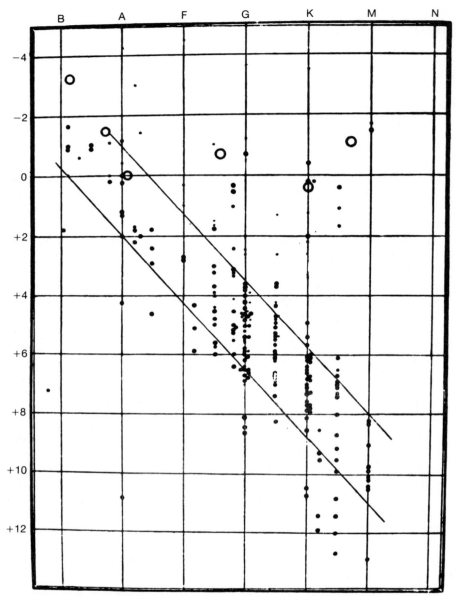

Fig. 4.1 The first stellar 'main sequence' diagram, correlating the fall in the surface temperature of a star from blue–white to red along the Harvard classification order (B, A, F, G, K, M) with the decrease in absolute magnitude for stars of known distance. The open circles represent the mean absolute magnitudes (averages of some 120) of distant bright stars ('giants') with small parallax. The Sun, a G-star wich an absolute magnitude of +4.7, lies at the middle of the main sequence area enclosed by the diagonal lines (the absolute magnitude scale runs from −5 to +14, or from 7500 to 1/5000 times the solar luminosity). The 'dwarf' stars lie below the main sequence, clustering on the lower right-hand side. (From H. N. Russell (1914). *Nature*, **93**, 252.)

demonstrated theoretically that absolute luminosity is proportional to total mass for stars of a given class, first the red giants and then the main sequence. A star cannot evolve along the main sequence, since its mass is approximately constant, although it may evolve on to the sequence by gravitational contraction from a nebulous gas cloud, or evolve off into another group as a giant of larger radius but with about the same mass. The main sequence registers essentially a mass order and thereby a lifetime order, since the more massive stars burn out more rapidly.

Eddington assumed that the stars are composed largely of the lighter elements from which, in the stellar interior, all the orbital electrons (Z in number) are ionized. The $Z + 1$ particles, composed of the atomic nucleus (containing Z protons and N neutrons with $N \sim Z$), and the Z electrons would have a mean particle mass of about 2, except for hydrogen, where ionization to a proton and an electron gives a mean mass of 0.5. At high temperatures, the average density of a star becomes independent of its particular chemical constitution if composed largely of elements other than hydrogen, owing to the complete ionization of the atoms, and luminosity becomes a general function of total mass. Eddington calculated the luminosity of the Sun, assuming a mean particle mass of about 2, and derived a theoretical value which proved to be too large by a factor of about 100. The predicted luminosity depends upon the fourth power of the assumed mean particle mass of the solar material, and the overestimate suggested that hydrogen might be the predominant element in the Sun. Subsequently Russell showed spectroscopically that hydrogen indeed constitutes the principal element of the solar atmosphere.

Stellar spectroscopy obtained a new perspective of the origin of atomic line spectra from the theory of Bohr (1913). Each series of spectral lines of an element converges towards a high-frequency limit, at which the excited electron ionizes and the escape of the electron leaves a positively charged ion with its own spectroscopic characteristics. These characteristics are often similar to those of a lighter element with a smaller charge, or neutral. Thus Bohr showed that the hydrogen-like line series in the spectrum of the star zeta-Puppis, discovered by E. C. Pickering in 1886, arises from the helium ion, He^+, isoelectronic with neutral hydrogen, each with a single orbital electron. Following Bohr's lead, the earlier theory that the stellar spectra differ on acount of the varying chemical composition and degree of dissociation of the elements in the stars became replaced by the view that the primary variation in the spectroscopically accessible stellar atmospheres lies in the degree of ionization of elementary atoms common to most stars, depending on the temperature and pressure.

The new approach was introduced by Meghnad Saha (1894–1956) at Calcutta, who worked out in 1920 the equilibria between the neutral atoms and their ions in the solar chromosphere, using the ionization energies of the gaseous elements and the Maxwell–Boltzmann energy distribution for the degree of ionization as a function of temperature. Saha ascribed the absence of the neutral rubidium and caesium lines from the normal solar spectrum to the low ionization energies of these elements, 4.18 and 3.89 eV respectively, and predicted that the missing lines might be observed in solar spectra restricted to the cooler sunspot regions where, additionally, the neutral potassium spectral lines should appear more strongly. The expected rubidium lines and stronger potassium lines in the sunspot spectra were soon confirmed by Russell, from measurements at the Mount Wilson observatory.

The use of Planck's black-body radiation law (1900), particularly by Hertzsprung, in conjunction with the spectral line series and ionization energies (IE) of the elements, allowed the characterization of the surface temperature and the chemical composition of the various stellar types during the 1920s. The surface temperature of the O-type stars, above 3×10^4 K, suffices to ionize helium (IE 24.58 eV), so that the Pickering series of He^+ appears in the spectra of these stars. The B-type stars $(T \sim 2 \times 10^4$ K) give the spectra of excited neutral helium and hydrogen, notably the Balmer series, which involves absorption from the second level $(n = 2)$ of the hydrogen atom, 10.20 eV above the ground state. The Balmer absorption series is strongest in A-type stars $(T \sim 9000$ K), and it is accompanied by the spectral lines of ionized heavier atoms, such as the Fraunhofer K line of Ca^+. In the spectra of the type F stars $(T \sim 7000$ K) the Balmer series has weakened, and the Ca^+ K line is stronger, while weak neutral metal lines appear and become more prominent in the G-type stars, like the Sun $(T \sim 5800$ K). The spectra of the cooler K-type $(T \sim 4400$ K) and M-type $(T \sim 3500$ K) stars contain bands due to diatomic species, CN, OH, CH, TiO, in addition to strong lines due to neutral metal atoms (Fig. 4.2).

The pressure of the stellar atmosphere, as well as the surface temperature of the star, was shown to influence the equilibrium between the neutral atom and the ion of an element. Antonia Maury (1866–1954) at the Harvard College observatory reported in 1897 a subdivision of the stellar spectral types into line 'qualities' of diffuse or sharp, and Hertzsprung found in 1906 that the stars with sharp line spectra are bright and distant, forming a small group of giant stars separated from the main sequence by the 'Hertzsprung gap'. Developing the Saha theory in 1921, Russell pointed out that the ionization of most or all of the elements in a stellar atmosphere generates a common total electron pressure, and the particular temperature at which one-half of the atoms of a given element are ionized is governed by the total pressure, not the partial pressure due to the element alone. In the extended atmosphere of the red giant stars, the spectral absorption lines are sharper because the effective total pressure is lower, and the temperature required to produce a given fractional ionization of an element is lower too, relative to main sequence stars of the same spectral colour type.

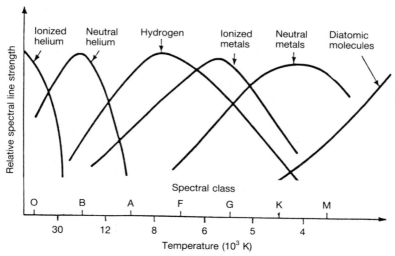

Fig. 4.2 The relative intensities of spectral line absorption in stellar atmospheres by the ions and neutral atoms of the elements, and their diatomic molecules, related schematically to the surface temperature of the star and its position in the Harvard classification sequence (O, B, A, F, G, K, M).

With astrophysical applications in mind, Russell at Princeton extended the Bohr theory of atomic spectra from one-electron to multi-electron systems with the spectroscopist, Frederick Saunders of Harvard, in 1923–5. The Russell–Saunders *L–S* coupling scheme, in which the electronic orbital angular momentum in space and the spin angular momentum around the electronic axis are each summed separately over the electrons of an atom, prior to the ultimate coupling of the total space (*L*) and the total spin (*S*) angular momentum, was first applied to the alkaline earth metals. By this means, Russell and Saunders discovered two-electron transitions from the appearance of 'anomalous' triplet terms in the ultraviolet spectrum of calcium at energies higher than the ionization potential of the element (6.09 eV).

Russell went on to investigate spectral line intensities, with the object of evaluating the solar abundance of the chemical elements, in collaboration with Walter Adams at the Mount Wilson observatory and Charlotte Moore at the Princeton observatory. In 1929 Russell published a classical report on the chemical composition of the Sun's atmosphere, based mainly on his development of the Saha ionization theory with the available intensity data. The relative abundances of 56 of the elements and six diatomic molecules were estimated, and Russell arrived at the new and striking conclusion, presaged by Eddington's theoretical overestimate of

the solar luminosity, that the solar atmosphere must be composed largely
of hydrogen.

Cecilia Payne (1900–79) in the first PhD thesis on astronomy from
Harvard, entitled *Stellar atmospheres* (1925), had tentatively concluded
earlier that hydrogen and helium are particularly abundant in stellar
atmospheres. Russell showed in 1929 that only a very small fraction of the
solar hydrogen at thermal equilibrium could be excited to the second and
third electronic stationary state (requiring approximately 10 and 12 eV,
respectively), from which the visible Balmer series and infrared Paschen
series originate. Yet these two series are strong in the solar spectrum,
implying that there is an enormous solar reservoir of hydrogen in its
ground electronic state. No other element was found to give a strong solar
spectral line from an electronic state with an energy more than 5 eV above
the ground level.

The telescope camera with large aperture and minimized chromatic and
spherical aberration developed by Bernhard Schmidt (1879–1935) in 1931
at the Hamburg-Bergedorf observatory, and other observational advances,
led to improved estimates of the relative abundances of the chemical
elements in the solar atmosphere and the extension of such estimates to
other stars and luminous celestial bodies. Improved laboratory measurements and calculations of atomic spetral line intensities afforded additional
refinements. The classical electromagnetic model envisaged an electron in
an atom set into oscillatory motion by the absorption of light with the
appropriate resonance frequency, giving a theoretical absorption intensity
of $\pi e^2/mc$ in terms of the charge and mass of the electron. Observed
intensities, measured by spectral line or band areas, differed from the
theoretical quantity by a factor termed the oscillator strength or f-value.
Observed f-values range up to the theoretical optimum ($f \sim 1$), but are
often substantially smaller.

For astrophysical applications, such as the determination of the sodium
abundance in the solar atmosphere from the depth of the Fraunhofer D_1
and D_2 line absorption (in the background black-body radiation continuum), the corresponding f-values measured in the laboratory served the
purpose. The f-values of some atoms and their ions were not accessible
experimentally at the time, and these f-values were calculated quantum
mechanically from the electric dipole moment due to the transition from
one stationary electronic state of the atom to another or, for the 'electric
dipole forbidden' transitions, from the corresponding magnetic dipole or
electric quadrupole transition moment. Lines appearing uniquely in the
spectra of low-pressure gaseous nebulae, like the green line observed in the
Orion nebula by Huggins and assigned to a terrestrially unknown element,
'nebulium', were shown in 1927 by I. S. Bowen at Pasadena to arise from
'forbidden' transitions between states within the ground electronic configuration of such species as O^{2+} and N^+.

During the 1950s it was established that hydrogen is the predominant element not only in the atmospheres of most stars, but also in the interstellar medium, especially in the spiral arm regions of the Galaxy. Possible sources of extraterrestrial radio signals were discussed by J. H. Oort's group at the Leiden observatory in 1944, and one of the group, H. C. van de Hulst, calculated that a radio line at 21 cm wavelength might be observable from neutral atomic hydrogen in interstellar space. The radio line is emitted by a hydrogen atom undergoing a spin–flip transition from a parallel to an antiparallel alignment of the spin axes of the proton and the electron. In 1946 I. S. Shklovsky of the USSR reported similar and more extended calculations, which included the radio lines expected from the deuterium atom and the OH and CH molecules.

The predicted 21 cm radio line of the hydrogen atom was detected by three groups simultaneously in 1951; by Oort, van de Hulst, and C. A. Muller in The Netherlands; by H. I. Ewen and E. M. Purcell at Harvard; and by W. N. Christiansen and J. V. Hindman in Australia. By 1959 measurements in the northern and the southern hemisphere of the intensity of the 21 cm radio line and its Doppler wavelength shifts had charted the overall galactic hydrogen distribution and also its motions. These measurements mapped out the spiral arms of the Galaxy with young stars embedded in relatively dense gas clouds rotating around the distant centre.

While the amount of hydrogen can be estimated from spectral line intensities for most stellar atmospheres, that of the next most abundant element, helium, is measurable only for the hotter stars, in which the spectra of the neutral atom and its ion He^+ are accessible. In the case of the Sun, hydrogen and helium make up 98 per cent of the solar atmosphere, and the heavier elements account for only 2 per cent by weight. Few stellar atmospheres contain a larger percentage of carbon and the heavier elements, technically termed 'the metals', and a number are significantly depleted.

In 1952 Walter Baade (1893–1960) at Pasadena divided stars into two main types on the basis of their metal content. Population I stars are located in the disc of our Galaxy, with only a small velocity directed perpendicular to the disc plane. Their heavy-element content by mass ranges from 2 per cent for stars near to the central region to 3 per cent for stars out in the spiral arms, and to a maximum of about 4 per cent for young stars forming in dense gas clouds. Population II stars belong to the globular clusters in the spherical halo of the Galaxy. They are far removed from the centre, moving with high velocity perpendicular to the disc plane, and with a metal content smaller by a factor of one-tenth to one-hundredth than that of the first type, the Population I stars. The *relative* abundances of the heavy elements, readily accessible spectroscopically from carbon $(Z = 6)$ to barium $(Z = 56)$, are similar for the stellar atmospheres of the two star population types.

4.2 The cosmic time-scale

The distinction between the metal-rich Population I stars of the disc in our
Galaxy and the metal-poor Population II stars of the halo was extended by
Baade to the stars of the nearby Andromeda galaxy in the local galaxy
cluster. This extension resolved a conflict between the several cosmic
time-scales proposed during the period of 1920–40. A small value for the
age of the universe, some 2 billion (10^9) years, had been estimated in 1929
by Edwin Hubble (1889–1953) at Mount Wilson from his relationship
between the velocity of recession (v) of the external galaxies and their
distance (r) from our Galaxy, $v = Hr$, where H is Hubble's constant. The
velocities of recession were obtained from the red-shift of atomic lines in
the galactic spectra, based on the decrease in frequency of waves from a
receding source, an effect discovered in 1842 by the Austrian physicist C. J.
Doppler (1803–53). The galactic distances were estimated by several
methods, dependent upon the particular distance range, the most crucial
being the use of the Cepheid variables.

In 1912 Henrietta Leavitt (1868–1921) at the Harvard College observatory found that the variation in the period of brightness of 1777 (variable)
stars in the Small Magellanic Cloud is proportional to their visual luminosity,
and thus to their relative magnitude, since these stars are approximately at
the same distance from the Earth. Harlow Shapley (1885–1972), as a
student with Russell at Princeton, established in 1913 that the Cepheid
variables are intrinsically puslating sources and not, as then generally
believed, eclipsing binary stars. At Mount Wilson in 1917, Shapley found
other sets of Cepheid variables in the globular clusters of stars in the
galactic halo, again with a brightness proportional to the length of the
period of variation, and placed the relation between period and magnitude
on an absolute basis, common to all the sets.

In 1924 Hubble found two Cepheid variables in the Andromeda galaxy,
and from their periods and absolute magnitudes, obtained through Shapley's
relation, he estimated the distance of the variable stars and their galaxy to
be 7.5×10^5 light-years. The distance of the Andromeda galaxy served as a
unit for the estimation of the separation of more distant galaxies, from the
apparent luminosity of the galaxy or of bright stars within them. This
distance unit governed the value of the Hubble constant H, and thence the
estimate of some 2 billion years since the beginning of the recession of the
galaxies.

Hubble's estimate was small in relation to Eddington's (1926) calcula-
tion of the lifetime of the Sun (~15 billion years) and, more significantly,
relative to the age of the Earth measured by the parent and daughter
element concentrations in radioactive minerals and the parental half-life. In
1921 Russell estimated the age of the Earth's crust at approximately 4
billion years from the analyses reported for uranium and thorium minerals.

Following his characterization of the minor isotope of uranium, Rutherford in 1929 gave 3.4 billion years as an upper limit to the age of the Earth from the present abundance ratio of ^{235}U to ^{238}U and their half-lives, on the assumption that originally the two isotopes were equally abundant.

An attempt was made to overcome these discrepancies by discarding all notions of an overall cosmic time-scale, in the theory of 'continuous creation' throughout an eternal universe, put forward in 1948 by Hermann Bondi, Thomas Gold, and Fred Hoyle at Cambridge. Einstein's cosmological principle (1917) that all observers in the universe are spatially equivalent was extended to the 'perfect cosmological principle' that such observers are equivalent also at all times. The extended principle implied that the average density of matter in the universe is constant over time, so that hydrogen must be continuously created at a rate of 10^{-43} g cm^{-3} s^{-1} in order to compensate for the thinning out of matter due to the recession of the external galaxies. The creation rate, while very small, corresponds to the birth of some 50 000 stars per second over the observable universe.

Evidence inconsistent with the continuous creation theory emerged early in the 1950s, when Martin Ryle (1918–84) at Cambridge and other radioastronomers discovered that the distribution of radio galaxies is not uniform: the radio sources become stronger and more numerous at greater distances. The radiation from the more distant radio galaxies was emitted in the remote past, so that the radio galaxies must have evolved from stronger to weaker radio sources on the cosmological time-scale. Stronger evidence for a singular origin of the universe came in 1965 from A. A. Penzias and R. W. Wilson at Bell Laboratories, New Jersey, with the discovery of the cosmic radiation background in the centimetre (microwave) region, characteristic of a black body at 2.7 K. Extrapolation backwards in time of the recession of the galaxies, producing a temperature elevation of the cosmic black-body radiation, leads to the high-temperature high-density origin of the universe—the Big Bang theory. Although it has become the standard theory, Big Bang creation raises a number of problems which are accommodated by continuous creation (Burbidge 1971).

The source of Hubble's low estimate for the age of the universe became apparent in the mid-1950s when it was found that there are two relationships between absolute magnitude and periodicity for Cepheid variable stars. There is one for the metal-rich Population I stars and another, approximately parallel but one and a half magnitudes lower at a given periodicity, for the metal-poor Population II stars. As a consequence, Hubble's estimate of the distance to the Andromeda galaxy was more than doubled, to 2×10^6 light-years, in line with other indications, such as the apparent brightness of novae in Andromeda and other external galaxies.

Revised values of Hubble's constant, still under discussion, give the universe an age between 10 and 20 billion years. A similar age range is estimated independently for the beginning of the nucleosynthesis of the chemical elements, and the median age is consistent with models of stellar evolution, which give the oldest stars in our Galaxy an age of approximately 15 billion years. These are the small-mass remnants of the original first-generation stars in the metal-poor Population II, belonging to the long-lived class of subdwarf stars near to the low-temperature end of the main sequence. The brighter, large-mass members of the first-generation stars have long since moved off the main sequence to become red giants and then spent dwarf stars, or have vanished in supernovae explosions to provide the heavy elements for the metal-rich second- and later-generation stars of Population I.

The Sun, in Population I, has an age of little more than 4.6 billion years, based upon the dating of meteorites and other components of the solar system by radioactive half-life methods. It is estimated that the Sun evolves by using up some 10 per cent of its hydrogen over about 7 billion years, and other stars of the main sequence, with mass M and luminosity L in units of the solar mass and luminosity, respectively, consume the same fraction of their hydrogen over about $7\,(M/L)$ billion years. The brightest stars of spectral class O, some 8×10^4 times as luminous as the Sun and containing approximately 25 solar masses, thus have an evolution time for 10 per cent hydrogen consumption of only approximately 2 million years, and all such stars are short-lived. In contrast, cooler stars of the K spectral class, with 0.4 of the luminosity and 0.8 of the mass of the Sun, have an evolution time of more than 10 billion years for 10 per cent hydrogen burning.

The difference in heavy-element content which distinguishes between the two stellar populations correlates both with the age and the galactic locations of the star types. The metal-poor Population II type form spherical systems of some 10^5–10^6 stars, the globular clusters, lying distant from the galactic centre in the near-spherical halo. The heavy-element content diminishes with an increase both in the age of the star and in its distance from the galactic centre. These old and distant stars do contain some heavy elements, which may have originated from an older pregalactic star population of short-lived supergiants, with no remaining representatives. The near-spherical distribution of globular clusters containing the oldest observed stars suggests that the original galactic nebula, with a minor heavy-element content from the pregalactic star population, was itself spherical. The development of the Galaxy to the form of a disc, containing most of the mass and with an increasing heavy-element content as second- and later-generation stars evolved, left the old globular clusters in the regions where they first formed (Fowler 1984; Tayler 1988).

4.3 Element and isotope abundances

The chemical compositions of the Population I main sequence stars largely represent the current cosmic abundance of the elements: the small content of heavy elements in the now less-abundant Population II stars refers to the general composition at a much earlier epoch. The observed spectra of only a few stars contain lines of elements heavier than barium ($Z = 56$), so that *cosmic* relative abundances are generally limited to elements lighter than the lanthanides. The case of the Sun is exceptional, and spectroscopic estimates of the *solar* relative abundances are available for virtually all of the elements up to thorium ($Z = 90$), with an upper limit for the abundance of uranium and a few of the lighter elements. In addition to the spectroscopic line-intensity method, studies of the cosmic-ray nuclei, characterized from the collision tracks produced by the particles, provide ancillary and complementary abundance data. The abundances of lithium, beryllium, and boron are relatively high among the cosmic-ray particles, indicating that these lighter elements originate principally from the high-energy collisions of heavier nuclei by spallation.

All of the bodies in orbit around the Sun are depleted in the volatile elements, so that none individually reflect the relative abundances of the elements in the solar system as a whole. Even the outer giant planets, rich in volatile elements, have atmospheres depleted in hydrogen. Further, the larger bodies, the planets and the Moon, are differentiated in chemical composition due to thermal processing during and after aggregation, and only the surface layers are generally accessible for elementary and isotopic abundance analysis. Some of the smaller bodies, the asteroids and comets, which cooled more rapidly during formation or aggregated from cold materials, appear to be less differentiated chemically, and fragments of these bodies are continuously available as recovered meteorites.

Each year the Earth acquires some 100 000 tonnes of extraterrestrial material, of which less than 1 per cent arrives in pieces large enough for recovery and analysis. Of the meteorites seen to fall and subsequently recovered, the majority are 'stones' (93 per cent), followed by the 'irons' (5.4 per cent), and a few 'stony-irons'. Most of the stones are chondrites, so termed from the millimetre-sized spheroids of once-molten silicates, the chondrules, set in a ground mass of sulphide minerals, metals and their oxides, and hydrated silicates. The most hydrated types, the carbonaceous chondrites, contain up to 6 per cent carbon, free and combined. The isotopic analysis of elements produced by radioactive decay in meteorites indicates that many of them originate from bodies which solidified some 4.55 billion years ago.

Both the age and the probable origin of meteorites from the smaller bodies of the solar system (asteroids and comets) suggest that they represent primitive material which condensed early in the formation of the solar system from a nebular of dust and gas. The carbonaceous chondrites have a composition generally regarded as the optimum representation of the primordial material from which the Sun and the planets were formed, except for the most volatile elements. The chemical analysis of recovered

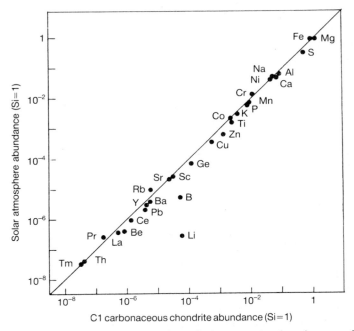

Fig. 4.3 The relation (1:1) between the relative atomic abundances of the less volatile chemical elements in the C1 carbonaceous chondrites and in the atmosphere of the Sun: each scale is normalized to a unit abundance for silicon (data listed by Wasson (1985)).

carbonaceous meteorites provides relative abundances of the non-volatile elements which correlate one to one with the corresponding solar abundances in most cases (Fig. 4.3).

The procedure of collating the elementary abundances on the Earth with those in meteorites and on the Sun was pioneered by the geochemist, Victor Goldschmidt (1888–1947) in Oslo, who introduced the classification of the chemical elements into siderophile chalcophile, and lithophile types, which accumulate in the metal, sulphide, and silicate phases, respectively. In 1937 Goldschmidt drew up a single scale of elementary abundances, combining the solar with the terrestrial and meteoritic data then available. Harold Urey (1893–1981) in Chicago revised Goldschmidt's estimate of the proportions of the metal, sulphide, and silicate phases in meteoritic material and, with Hans Suess in 1956, reported a more extended listing of the abundances of the elements and their individual isotopes, based primarily on solar system data. The Suess and Urey (1956) tabulation has been progressively updated with new data, listed by Wasson (1985) and by Suess (1987) (Fig. 4.4).

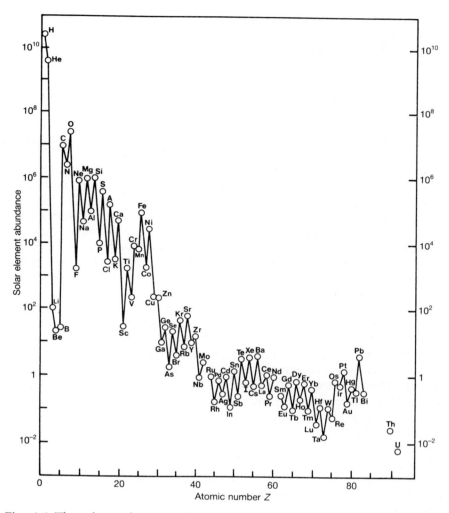

Fig. 4.4 The relation between the relative atomic abundances of the chemical elements in the solar system, normalized to an abundance of 10^6 for silicon, and the atomic number of the element (Z). The distribution illustrates that the abundance of a given even-Z element is generally larger than that of either of the adjacent odd-Z elements (data listed by Wasson (1985)).

Early observations on the relative abundances of the chemical elements were associated with surveys of mineralogical resources and with classifications of the elements based on theories of their constitution or origin. Johann Döbereiner (1780–1849) at Jena, in the course of classifying the elements into families of triads (1817–29), noted that the more widely distributed elements have relatively small equivalent weights. The first chief chemist of the US Geological Survey, Frank Clarke (1847–1931), published in 1889 a mineralogical listing of the relative abundances of the elements, which he attempted to correlate with the periodic classification and to explain in terms of the evolution of the elements from protyle. The production of helium in radioactive decay suggested that the α-particle might be another building block in the atomic nucleus, as proposed in 1904 by Rutherford and Soddy independently in their books on radioactivity, and elaborated by G. Oddo (1914) in his 'rule of four' to account for the marked abundance of elements, such as carbon and oxygen, with atomic weights approximating to an integral multiple of that of helium ($A = 4$).

In 1917 William Harkins (1873–1951) at Chicago, relying on the analysis of both minerals and meteorites, reported the new 'periodic law' that the elements with even atomic numbers (Z) are more abundant than their odd-numbered neighbours. He found a 70-fold abundance excess (on average) of the even-Z elements in meteorites, and noted that all of the five elements then unknown between hydrogen and uranium had odd atomic numbers, while the odd-Z radioactive elements had shorter half-lives than the corresponding even-Z set.

Harkins accounted for the new law in terms of the evolution of the elements from hydrogen and helium. He suggested that the elements with an odd atomic number have nuclei composed of both protons and α-particles to give the appropriate mass number (A), with electrons to adjust the charge, whereas the nuclei of even-Z elements, except for beryllium, are integral condensations of α-particles with charge-balancing electrons, giving these elements greater nuclear stability and abundance. When sets of isotopic relative abundances became available, Harkins (1931) extended the even–odd distinction, noting that isotopes with an even mass number (A) are systematically more abundant than neighbouring odd-A isotopes (Fig. 4.5). In 1934 J. Mattauch proposed the law, to which there are a few exceptions, that there are two or more stable isobars (isotopes of different elements with the same mass number) for A even but only one stable isobar for A odd.

Following the discovery of the neutron in 1932, and the measurement of the neutron capture cross-section of isotopes in neutron bombardment experiments, the individual number of neutrons N and protons Z in a given nucleus were found to be related to the isotopic abundance. Walter Elsasser at Paris in 1934 drew attention to the high relative abundance of isotopes containing 'magic numbers' of neutrons or protons: 2, 8, 20, 28, 40, 50, 82, and 126. 'Doubly magic' isotopes are particularly abundant, such as ^{40}Ca ($Z = 20$, $N = 20$) or ^{208}Pb ($Z = 82$, $N = 126$); while other abundant nuclei may be formed from such isotopes by facile β-decay, such as ^{56}Fe from ^{56}Ni ($Z = 28$, $N = 28$).

In 1948 Maria Goeppert Mayer (1906–72) in Chicago, and independently Hans Jensen (1907–73) at Hanover, accounted for the magic

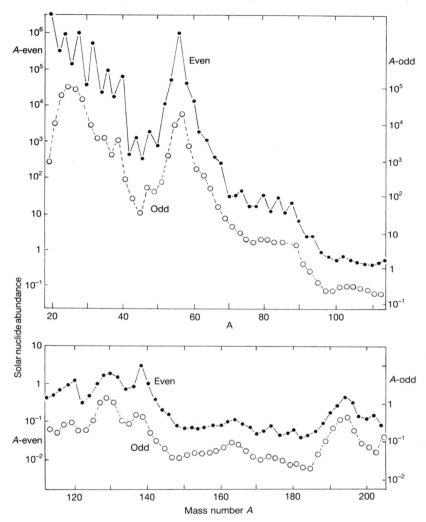

Fig. 4.5 The relation between the relative nuclide abundance of the elements in the solar system (Si = 10^6) and the mass number (sum of the proton number and the neutron number, $Z + N = A$) for the major isobars. The A-odd scale is offset 0.5 log units below the A-even scale, otherwise the relations overlap at mass numbers greater than about 160. The alternation of the A-even distribution from $A = 20$ to $A = 90$ illustrates the 'rule of four', from successive α-particle additions in helium burning (Section 5.3). The peak in the distributions around $A = 56$ corresponds to the stable nuclei of the iron group elements, formed by nuclear equilibration in the cores of massive stars (Section 5.1). The two sets of double peaks in the distribution at $A = 130$–8 and at $A = 194$–208 correspond to the closed-shell 'magic' neutron number of $N = 82$ and $N = 126$, respectively, the doubling of the peaks illustrates the two neutron capture processes (Section 5.4), the 'rapid' and the 'slow' (data listed by Cameron (1982)).

numbers in terms of closed nuclear shells of protons and neutrons, similar to the (extranuclear) closed shells of electrons in the noble gas elements. The analogy with the orbital electrons of an atom extends, in the nuclear shell theory, to the assumption that each nucleon (proton or neutron) spins on its axis and rotates around the centre of the atomic nucleus with strong coupling between the orbit and spin angular momentum. A parallel alignment of the two angular momentum axes corresponds to the lower energy.

The closed nuclear shell of 82 neutrons combines with different numbers of protons to form the stable nuclei of no less than seven different elements. A set of nuclei with a common number of neutrons, like the $N = 82$ case, Mayer proposed, 'for convenience, shall be called *isotones*' (a term coined by replacing the p for proton in *isotope* by n for neutron). The $N = 82$ closed shell includes the Z-odd nuclei of ^{139}La and ^{141}Pr, with an apparent 100 per cent isotopic abundance in each case, and an abundance respectively 27-fold and eight-fold larger than the average for the other Z-odd lanthanide elements. In contrast, the number of isotones for N even but non-magic is only three or four, while for N odd the number of isotones averages to less than one for a given value of N.

Mayer correlated the neutron absorption cross-section data available for the atomic nuclei over the mass number range $A = 51$ to $A = 209$ with the neutron number, finding minima at the closed shell values of $N = 50, 82, 126$. The minima suggested not only that closed shells confer a particular nuclear stability, but also that high abundances of nuclei with magic neutron numbers accumulate in the evolution of the elements by neutron capture, owing to the small capture cross-section and the consequent low probability of transmutation to nuclei with non-magic neutron numbers.

Pairing effects are important for nuclear stability, both for protons and neutrons individually. Suess (1987) lists data for 283 stable or long-lived isotopes, showing that the mass numbers are even for 173 and odd for 110. The A-odd isotopes divide into the approximately equivalent sets of Z even with N odd (53) and Z odd with N even (57). In contrast the A-even isotopes are predominantly Z even with N even (166). The few A-even isotopes (7) having Z odd with N odd are mainly the light nuclei, e.g. ^{2}H, ^{6}Li, ^{10}B, ^{14}N, with no stable even–even isobars. Heavier odd–odd nuclei are unstable through β-decay with respect to an even–even isobar. Thus ^{40}K ($Z = 19, N = 21$), widely used for dating solar system minerals, decays with a half-life of 1.25 billion years to doubly magic ^{40}Ca ($Z = 20, N = 20$) by β-emission (90 per cent) and to ^{40}Ar ($Z = 18, N = 22$) by K-electron capture.

A 'charged liquid-drop model' of the atomic nucleus, developed from 1936 by Niels Bohr and others, had been well supported by the discovery of nuclear fission in 1939, and the liquid-drop and nucleon-shell models appeared to conflict initially. During the 1950s the two models were unified by Aage Bohr, Ben Mottelson, and Leo Rainwater who investigated the pairing effects of the nucleons and their collective motions in an atomic nucleus, showing how the nuclear shape, charge distribution, and energy levels change with variations in the number of both neutrons and protons. The atomic nucleus has an average spherical shape only for the closed shell of the magic nucleon numbers. For the intermediate nucleon numbers, the nucleus becomes an oblate or prolate spheroid, or else triaxial, or even less symmetrical (Hamilton and Maruhn 1986).

4.4 The age of the elements

The discovery and characterization of radioactivity promised a solution to two major problems: the age of the Earth, and the source of the energy of the Earth, Sun, and other celestial bodies. By the 1890s, geologists had arrived at an age of some 600 million years for the oldest sedimentary rocks containing macrofossils (shells and bones). This estimate was based on the rate of sedimentation in river deltas and the total thickness of the sequence of sedimentary strata. William Thomson, Lord Kelvin (1824– 1907), disputed this age estimate on the grounds that the transformation of gravitational potential energy into heat by the contraction of the Sun could last no more than 24 million years, and gradual cooling similarly limited the geologically active lifetime of the Earth.

Another source of solar and terrestrial energy was needed for a longer lifetime. Both geologists and astrophysicists looked to radioactive decay as a further source of such energy, but Kelvin regarded radioactivity as merely a byproduct of cosmic radiation, making a negligible contribution to the energy resources of the solar system. In 1903 Pierre Curie and Laborde measured calorimetrically the energy liberated continuously by the radioactive decay of radium salts. They estimated that the thermal energy from the decay of a given mass of radium suffices to raise the temperature of the same mass of water from melting point to boiling point in about 45 minutes. From these measurements and the approximate abundances of the radioactive elements then available, Rutherford and Soddy concluded in the following year that the Earth might well be warming up rather than cooling down. The presence of helium in the spectrum of solar prominences appeared to provide evidence that radioactivity contributed to the energy of the Sun as well.

In addition, the constancy of the rate of disintegration of a radioactive element provided a clock which measured the time interval since the crystallization of the minerals, or even from the initial creation of the elements. Rutherford proposed methods for the radioactive dating of mineral formation and of the synthesis of the elements. The He/U or He/Th ratio in a uranium or thorium ore and the measured rate of helium production from the parent radioactive element dated the mineral, as did other radioactive parent/daughter ratios with a known rate of parental radiodecay.

Rutherford (1906) estimated the age of several uranium ores as more than 400 million years from the He/U ratio, the value being a minimum because of the possible loss of some helium from the mineral over time.

Following the detection by Aston of a new lead isotope ^{207}Pb in the ore broeggerite, Rutherford identified in 1929 the parent as the minor isotope of uranium, ^{235}U. From the present isotope ratio, ^{235}U/^{238}U, and the half-lives of the isotopes, Rutherford estimated that the stellar synthesis of the radioactive elements had terminated some 3.4 billion years ago, on the assumption that the abundances of the two uranium isotopes were then equal. The estimate was taken as an upper limit to the age of the Earth, which came out at twice the age of the oldest radioactive minerals then known. Rutherford concluded that uranium and other radioactive elements were certainly produced in the Sun 4 billion years ago and are probably synthesized there at present.

The development and extension of Rutherford's radioactive dating methods have given a cosmological time-scale to the formation of the solar system and an epoch for the beginning of heavy-element synthesis in the Galaxy. Many minerals in the meteorites condensed and became closed systems some 4.55 billion years ago, the age given by a number of the radioactive parent/daughter ratios. But minor inclusions in some meteoritic minerals contain an excess abundance of isotopes derived from short-lived radioactive ancestors, indicating that the synthesis of several radioisotopes incorporated into the presolar nebular terminated about 200 million years earlier.

There are a number of anomalies in the elemental abundances of meteorites compared with terrestrial abundances. The isotopes of xenon with mass numbers 129 and 131 to 136 are overabundant in some minerals from meteorites. Atoms of the isotope ^{129}Xe produced by β-decay from ^{129}I (which has a half-life of only 17 million years) occupy iodide lattice sites in the meteoritic minerals. Neutron bombardment of the mineral converts the naturally occurring ^{127}I in the iodide lattice sites to ^{128}Xe, and this isotope is released as a spike output at the same temperature as the excess ^{129}Xe on heating. The xenon isotopes of mass number 131 to 136, also overabundant in meteoritic materials, arise from the spontaneous fission of the plutonium isotope ^{244}Pu, which has a half-life of 75 million years (Rowe 1986).

Other isotope anomalies in meteoritic minerals add further evidence for the inclusion of short-lived radioisotopes during the condensation of the meteorites. The β-decay of ^{107}Pd with a half-lfe of 6.5 million years has left an excess abundance of ^{107}Ag in meteoritic material, and an excess of ^{26}Mg remains from the β-decay of ^{26}Al, which has a half-life limited to 0.73 million years. The heavier extinct radioisotopes which have left stable daughter isotopes in overabundance were probably generated in a nearby supernova explosion just before, and possibly triggering, the condensation of the presolar nebula into the Sun, planets, and other bodies of the solar system. The light radioisotope ^{26}Al is detected by γ-photon spectroscopy in the galactic disc at present, however, and it must be produced continuously, or at intervals which are short relative to the half-life. The repeated nova outbursts of unstable binary stars are a likely source, some 30–50 such outbursts being observed in our Galaxy each year.

A supernova is observed every 30 years or so in a galaxy external to our own. Each one ejects stable and radioactive elements into the interstellar medium, which contains the dense dust and gas clouds of prestellar nebulae. Repeated supernova explosions over the history of our Galaxy with inter-stellar mixing of the elements expelled gives rise to a quasi-steady-state abundance of the long-lived heavy radioisotopes, particularly the thorium isotope ^{232}Th (half-life 13.9 billion years) and the uranium isotopes ^{238}U (half-life 4.51 billion years) and ^{235}U (half-life 0.713 billion years). The present radioisotope abundance ratios, ^{232}Th/^{238}U = 3.75 and ^{235}U/^{238}U = 7.26×10^{-3}, extrapolate to the values of 2.32 and 0.317, respectively, at the epoch of the origin of the solar system. The latter values represent the quasi-steady-state abundance ratios, attained around the time of the inception of the solar system some 4.8 billion years ago, after extended aeons of production offset by continuous spontaneous radioactive decay.

Calculations of the relative rates of production of the heavy long-lived radioisotopes provide estimates of the length of time needed to attain the immediate presolar abundance ratios 4.8 billion years ago. The calculated production ratios for ^{232}Th/^{238}U and for ^{238}U/^{235}U of 1.80 and 1.42, respectively (Fowler 1978), or of 1.39 and 1.24, respectively (Thielemann *et al.* 1983), indicate that heavy-element production began in the Galaxy between 12 and 18 billion years ago. These values are wholly independent of other estimates for the age of the universe, based on the relation between the spectral red-shift and the distance of the remote external galaxies, or on the ages of the oldest stars, although the separate estimates are in remark-able agreement (Fowler 1984). The uncertainties of the nuclear cosmochro-nology estimates are mainly those of the heavy radioisotope production rates, the magnitude of the immediate presolar supernova nucleosynthesis, and the time interval between that event and the condensation of closed solid systems in the solar nebula. The uncertainties of the Hubble red-shift estimates are principally those of the constancy, or the increase or the decrease, of the recession rate of the distant external galaxies.

5
Stellar nucleosynthesis

5.1 Stellar energy

Rutherford's discovery of artificial radioactivity in 1919, and Aston's characterization of stable light isotopes in the following year, apparently verifying Prout's hypothesis, shifted interest in stellar energy production from the disintegration of the heavy radioactive elements to the nucleosynthetic reactions of the light elements. The 'anomalously high' mass of the hydrogen atom, with $M = 1.008$, implied the enormous loss of energy in the condensation of hydrogen to heavier atoms with, as it then appeared, an integral mass. Jean Perrin (1870–1942) at Paris, who had counted Avogadro's number (1909) from the Brownian motion in suspensions of microscopic particles, suggested in 1919 that the conversion of hydrogen to heavier elements might be the primary source of solar and stellar energies. Eddington agreed with the suggestion, observing in his address to the Cardiff meeting of the British Association for the Advancement of Science in 1920, 'what is possible in the Cavendish Laboratory may not be too difficult in the Sun'.

During the 1920s the elementary composition of the Sun and stars was not yet known quantitatively, and Eddington's assumption of a mean particle mass of approximately 2 for the bare nuclei and electrons of the stellar interiors, rather than about 0.5 for hydrogen alone, had led to his overestimate by a factor of about 100 for the solar luminosity. The demonstration by Russell in 1929 that hydrogen is the major elementary constituent of the Sun and the discovery of the neutron in 1932 provided a firmer basis for the transformation of earlier speculations into specific mechanisms for the origin of the chemical elements and the stellar energies. The particular mechanisms proposed for nucleosynthesis and energy production followed two main lines: first and earliest, those based upon a high-temperature equilibrium between the atomic nuclei, giving a relative abundance distribution which is frozen on cooling; and second, kinetic mechanisms whereby the elements evolve sequentially from protons and neutrons.

Richard Tolman (1881–1948) at Pasadena investigated the equilibrium between hydrogen and helium in 1922, and concluded that high temperatures ($> 10^6$ K) would be needed, the value being subsequently adjusted to approximately 10^9 K after the stellar H/He abundance ratio of about 4 by mass was established. Urey and Bradley (1931) examined the temperature and density conditions under which the observed isotopic abundances of

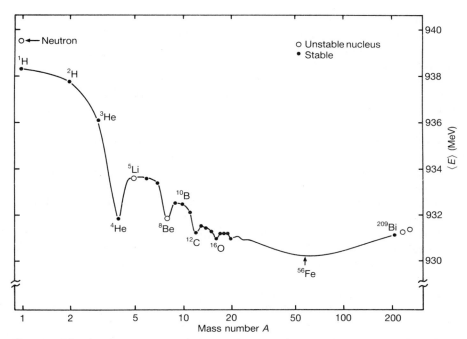

Fig. 5.1 The binding energy of the nucleons in the atomic nuclei: the relation between the mass number and the mean binding energy per nucleon $<E>$, Mc^2/A in MeV (10^6 electron volts), for the most stable isobar (atomic mass M) of each atomic mass number (A).

the lighter elements might be produced in thermodynamic equilibrium, but found no single set of conditions. A similar conclusion emerged in all subsequent studies of nuclear equilibration. Thermodynamic equilibrium leads to a predominance of the iron group elements since they are the most stable; they lie at the minimum of Aston's packing-fraction relation or, what is equivalent, of the relation between atomic number and binding energy per nucleon (Fig. 5.1). The relative abundances of the lighter and the heavier elements are deficient on all equilibrium mechanisms.

Non-equilibrium mechanisms of nucleosynthesis envisage the evolution of the heavier elements from hydrogen by neutron or proton capture followed by β-decays. The sequential neutron capture theory based upon a hot primordial *ylem* (*hyle*, first matter), put forward by George Gamow (1904–68) at Washington in 1935, encountered the early objection that nucleogenesis from hydrogen at stellar temperatures would stop at helium. In a study of energy production in the stars, Hans Bethe at Cornell showed in 1939 that there are no stable isotopes with an atomic mass of 5 or 8 and that the stable nuclei of lithium, beryllium, and boron would soon 'burn'

away at stellar temperatures. Carbon is the lightest element after helium
to maintain an appreciable steady abundance in stellar interiors since
its isotopes regenerate, along with those of nitrogen and oxygen, in a
catalytic cycle which converts four protons to an α-particle. The direct
combination of two protons to form a deuteron, followed by further
proton captures and β-decays to helium the PP chain, is a competitive
reaction sequence, dependent upon the composition and core temperature
of the star.

While Bethe's two processes for the conversion of four protons to an
α-particle, the PP chain and the CNO cycle, accounted for the luminosity
of the majority of stars, i.e. those of the main sequence of Russell and
Hertzsprung, the problem of the origin of carbon and the heavier elements
remained unexplained. By the 1950s it was clear that no single mechanism,
of either the equilibrium or the kinetic type, could account for the relative
abundances of all the known chemical elements. In 1957 Margaret and
Geoffrey Burbidge, William Fowler, and Fred Hoyle compared several
kinetic and equilibrium mechanisms with the main features of the empirical
isotopic abundances tabulated by Suess and Urey (1956). They based their
analysis on the view 'that the stars are the seat of origin of the elements', as
opposed to 'other theories which demand matter in a particular primordial
state for which we have no evidence'. They concluded: 'We have found it
possible to explain, in a general way, the abundances of practically all the
isotopes of the elements from hydrogen through uranium by synthesis in
stars and supernovae', a conclusion which remains largely unchallenged,
although substantially supplemented.

The Burbidges, Fowler, and Hoyle proposed eight main nucleosynthetic
processes in star interiors, all starting from hydrogen:

1. The conversion of hydrogen to helium at temperatures of $1-5 \times 10^7$ K
   and densities of $\rho \sim 10^2$ g cm^{-3}.

2. The burning of helium to carbon, oxygen, and neon at $T \sim 1-3 \times 10^8$ K
   and $\rho \sim 10^5$ g cm^{-3}.

3. The capture of successive α-particles by ^{16}O and ^{20}Ne producing
   ^{24}Mg, ^{28}Si, ^{36}Ar, and ^{40}Ca at $T \sim 10^9$ K.

4. The *equilibrium* e-process at $T \sim 4 \times 10^9$ K, required to account for the
   high abundance of iron group elements.

5. The *slow* s-process of neutron capture by iron group and lighter elements, with time intervals of 10^2-10^5 years between successive captures,
   allowing for intervening short-lived β-decays.

6. The *rapid* r-process of neutron capture in supernovae explosions, with
   intervals of $0.01-1$ s between successive captures, bypassing the limiting element, bismuth, of the s-process to generate thorium, uranium,
   and the transuranic elements.

7. The *proton capture* p-process, producing the rare light isotopes of the heavy elements in a hydrogen-rich medium at about 3×10^9 K.

8. An unknown *x*-process invoked to explain the production of the light temperature-vulnerable nuclei deuterium, lithium, beryllium and boron.

Further nucleosynthetic processes were added after the detection of the 2.7 K microwave radiation background in 1965. This discovery led to the further development of the Big Bang model, earlier suggested by Alpher, Bethe, and Gamow (1948) in a celebrated paper by Alpher and Gamow to which Gamow jocularly added Bethe's name in order to make up the Greek alphabetic sequence, α, β, γ. The Big Bang theory proposes that a major nucleosynthesis of the lighter elements took place between 1.5 and 5 min after the primary creation (the 'initial singularity'), long before the formation of the galaxies and their stars. The temperature-fragile nuclei that were not incorporated into stars and destroyed still survive in the interstellar and intergalactic medium. The observed present-day interstellar mass fractions of the lighter nuclei are reported as: ^{1}H = 0.70, D = 3×10^{-5}, ^{3}He = 6×10^{-5}, ^{4}He = 0.28, and ^{7}Li ~ 5×10^{-9} (Pagel 1982, 1987).

Mechanisms additional to the Big Bang nucleosynthesis gave rise to the lighter elements and identify the *x*-process, at least in part, as the *light-atom* or l-process of spallation in cosmic-ray particles and in stellar atmospheres where high-energy collisions break up heavier nuclei. The abundance of lithium, beryllium, and boron, relative to oxygen or hydrogen, is some 10^5 times larger among the cosmic-ray nuclei than in the solar system. Further major modifications of the general nucleosynthetic scheme are the subdivision of the α-capture process into several steady-state stages, the burning of carbon, neon, oxygen, and silicon, and the addition of explosive deflagration processes at these stages.

According to the Big Bang theory, the galaxies and then first-generation stars began to form from the materials of the primary nucleosynthesis a few hundred million years after the initial singularity. In the process of star formation, the contracting gas cloud of the proto-star heats up, through the transformation of gravitational potential energy into atomic kinetic energy. The light elements of the primary nucleosynthesis, other than hydrogen and helium, are destroyed through nuclear reactions before hydrogen burning begins at temperatures above 5×10^6 K. Contraction of the stellar material ceases when a steady state is reached, in which the gravitational attraction is balanced by the outwardly directed thermal pressure. The rate of energy production from the nuclear reactions in the core then becomes equal to the rate of radiant energy emitted from the stellar surface. At this stage the star remains relatively stable for the longest period of its active history, converting hydrogen to helium on the main sequence (Fig. 5.2).

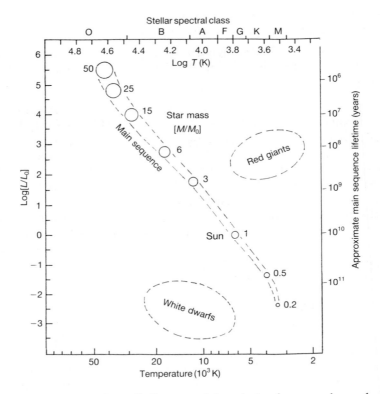

Fig. 5.2 A Hertzsprung–Russell diagram of the relation between the surface temperature of a star (the stellar spectral class) and its luminosity relative to that of the Sun, $[L/L_0]$. The majority of stars lie on the main sequence, burning hydrogen to helium, with a lifetime t inversely proportional to the stellar mass, $t \sim 10^{10}[M/M_0]^{-2.5}$ years. After using up much of their hydrogen, stars evolve into helium-burning red giants with a shorter lifetime. Thereafter, in the larger stars, carbon and oxygen burning may lead to a deflagration with the ejection of planetary nebulae, or to a supernova explosion. Smaller stars, and the smaller-mass remnants of spent larger stars, contract to white dwarfs and cool down after the expenditure of their nuclear fuel.

5.2 Main sequence stellar evolution

The hydrogen-burning stage of stellar evolution is long-lived because the conversion of four protons to an α-particle has the highest energy yield per nucleon (6.68 MeV) of any stellar nuclear reaction, while the energy production is buffered at a moderate level by a feedback mechanism ensuring a near-constant luminosity for a given stellar mass. Any excess energy production leads to a thermal expansion, and the consequent density reduction and cooling diminishes the hydrogen-burning rate. The earliest hydrogen to helium conversion, the proton–proton chain, consists of three sets of reactions termed PP I, PP II, and PP III.

These and other nuclear reactions are conventionally written in a form devised by Bothe during the 1930s, whereby a reaction between fundamental particles of the type $A + B \rightarrow C + D$ is expressed compactly as $A(B,C)D$, with the reactants to the left of the comma and the products to the right. The terms in parenthesis denote a general reaction type, applicable to a different reactant A' and product D'. Two or more successive reactions are expressed as a single sequence; thus $A(B,C)D \rightarrow D(E,F)G$ in the abbreviated form becomes $A(B,C)D(E,F)G$.

All three PP processes have two reactions in common. Two protons p combine in the first reaction to give a neutrino ν and a deuteron d, either with positron e^+ emission, accounting for 99.75 per cent of the conversion, or by electron e^- capture, limited to 0.25 per cent of the conversion:

$$p + p \rightarrow d + e^+ + \nu \,(99.75 \text{ per cent}) \quad \text{or} \quad p + e^- + p \rightarrow d + \nu \,(0.25 \text{ per cent}).$$

In the second reaction the deuteron combines with another proton to yield a photon γ and ^{3}He:

$$d + p \rightarrow {}^3\text{He} + \gamma$$

or $d(p,\gamma)^3$He in Bothe's notation.

The PP I process, accounting for 85 per cent of the overall conversion of hydrogen to helium in the Sun, is completed by the combination of two ^{3}He nuclei to give two protons and ^{4}He. The PP II and PP III processes, responsible for the remaining 15 per cent of the hydrogen to helium conversion, diverge after the combination of ^{3}He with ^{4}He to give ^{7}Be and a photon. In Bothe's compact notation, these reactions are expressed in the form

PP I ^{3}He(^{3}He, p + p)^{4}He (85 per cent)
PP II/III   or ^{3}He(^{4}He, γ)^{7}Be (15 per cent)
PP II ^{7}Be(e$^-$, ν)^{7}Li (99.9 per cent) $\rightarrow$ ^{7}Li(p, γ)2^{4}He
PP III   or ^{7}Be(p, γ)^{8}B (0.1 per cent) $\rightarrow$ ^{8}B(, e$^+$ + ν)2^{4}He.

The PP processes are important for the small-mass stars in the lower-temperature range of the main sequence, such as the Sun, and the relative contributions of the three chains depend on the particular stellar composition and temperature. As in the case of the Sun, the PP I chain is generally the most important, but the other chains become significant at higher stellar temperatures.

In the more massive second-generation stars, and possibly in the older stars of the first generation, the catalytic carbon–nitrogen–oxygen (CNO) sequence converts hydrogen to helium through a main cycle of six reactions. Starting from ^{12}C, four proton captures and two β^+-decays give rise to a helium nucleus (α-particle) with the regeneration of the ^{12}C nucleus. A minor subsidiary cycle branches off from the major reaction sequence at ^{15}N, with a branching ratio in the Sun of about 2000:1 in favour of the final (p,α) reaction of the main cycle, to which the subsidiary reaction sequence returns at ^{14}N. The main cycle consists of the sequence

$$^{12}C(p, \gamma)\,^{13}N(,e^+\nu)\,^{13}C(p,\gamma)\,^{14}N(p,\gamma)\,^{15}O(,e^+\nu)\,^{15}N(p,\alpha)\,^{12}C$$

and the following reactions make up the subsidiary cycle:

$$^{15}N(p,\gamma)\,^{16}O(p,\gamma)\,^{17}F(,e^+\nu)\,^{17}O(p,\alpha)\,^{14}N.$$

The conversion of four protons to an α-particle involves the loss of 0.0287 atomic mass units, which corresponds to an energy of 26.73 MeV from Einstein's (1905) relation, $\triangle E = \triangle mc^2$. Not all of the energy is emitted as thermal radiation photons at the stellar surface, since the neutrinos produced by the PP chain or the CNO cycle escape from the stellar interior with virtually no energy loss through collisions. The disintegration of the unstable isotope ^8B in the PP III process produces high-energy neutrinos (7.2 MeV), which are detected by the reaction $^{37}Cl(\nu,e^-)^{37}Ar$, followed by the β-decay of ^{37}Ar with a half-life of 35 days. Only a small fraction of the expected solar neutrino flux has been observed in an experiment, continuous since 1967, using a tank of approximately 5×10^5 litres of tetrachloroethylene as a detector and sited more than a kilometre deep in a derelict gold mine. To account for the discrepancy, Bethe and others suggest that the electron neutrino, ν_e, may be transformed into the muon neutrino, ν_μ, in the Sun.

The relative importance of the PP chain and the CNO cycle for stellar hydrogen-burning depends upon the heavy-element abundance and the temperature. For main sequence stars with a typical central density of 100 g cm^{-3}, hydrogen is converted to helium at the same rate by the two mechanisms at a core temperature of approximately 21×10^6 K for Population I stars or at approximately 27×10^6 K for those of Population II with the smaller abundance of carbon, nitrogen, and oxygen. With a core temperature of approximately 13×10^6 K the Sun produces only about 1 per cent of its energy by the CNO cycle, which becomes dominant only for the more massive (> 1.5 solar mass) and hotter stars of the Population I main sequence.

The conversion of hydrogen to helium by the main sequence stars does not greatly change the original H/He ratio of the primordial nucleosynthesis. It is estimated that each generation of stars contributes only a little more helium than heavier elements to the interstellar region. Much of the helium produced by main sequence hydrogen burning is consumed in the production of the heavier elements in the next stage of stellar nucleosynthesis, helium burning. The observed relative abundances of the isotopes of carbon, nitrogen, oxygen, and fluorine are in fair agreement with the corresponding theoretical steady-state abundances of the CNO cycles, except for ^{12}C, ^{16}O, and ^{18}O. These nuclei are generated additionally at the next stage of nucleosynthesis and are included among the direct products of helium burning.

5.3 Red giant nucleosynthesis

The Sun and other G stars have a main sequence lifetime of some 10 billion years. The lifetime changes approximately by an order of magnitude for each successive spectral class: 100 billion years for K stars and 1 billion years for F stars, down to only 1 million years for the bright O stars of the Harvard classification (O, B, A, F, G, K, M). Near the end of the main sequence lifetime the hydrogen of the stellar core becomes exhausted, although hydrogen burning continues in a shell adjacent to the central regions. The stellar core itself contracts and heats up through the conversion of gravitational potential energy as the core becomes more dense. The outer layers of the star expand and cool down to the redder spectral classes, K and M, but the energy radiated is substantial, owing to the large surface area.

Eventually the core helium reaches temperatures of the order of 10^8 K and densities of about 10^5 g cm^{-3} where collisions of two α-particles maintain a small steady-state abundance of the unstable isotope ^{8}Be ($\sim 10^{-16}$ s lifetime), which rapidly captures a third α-particle to form ^{12}C. While the second step is exothermic, the first is endothermic, and the overall energy yield is only 2.46 MeV per helium atom or some 10 per cent of the energy produced in the formation of a helium atom from four hydrogen atoms. Subsequent α-particle captures are rather more energetic: 7.15 MeV for the production of ^{16}O from ^{12}C by the ^{12}C$(\alpha,\gamma)^{16}$O process; 4.75 MeV at the next stage, ^{16}O$(\alpha,\gamma)^{20}$Ne; and 9.31 MeV for the reaction ^{20}Ne$(\alpha,\gamma)^{24}$Mg.

Isotopes left behind from the earlier CNO cycles, particularly ^{14}N, are processed during the helium-burning stage by reaction sequences of the type ^{14}N$(\alpha,\gamma)^{18}$F$(e^+,\nu)^{18}$O$(\alpha,\gamma)^{22}$Ne$(\alpha,n)^{25}$Mg. These reactions produce isotopes with an excess of neutrons over protons in the nucleus, such as ^{18}O and ^{25}Mg, which increasingly dominate the relative isotopic abundances of the elements following ^{40}Ca. Moreover, they supply the neutron flux required for the synthesis of the neutron-rich isotopes of the heavier elements.

When the helium supply of the central region of the red giant is near exhaustion, the core consists largely of carbon and oxygen. Gravitational contraction again increases the core density and raises the temperature to approximately 10^9 K. Under these conditions, ^{12}C nuclei combine with an energy release comparable with that of the previous stage through reactions such as

$$2\,^{12}\text{C} \rightarrow \alpha + \,^{20}\text{Ne} \ (4.62 \text{ MeV}) \quad \text{and} \quad 2\,^{12}\text{C} \rightarrow \text{p} + \,^{23}\text{Na} \ (2.24 \text{ MeV}).$$

Subsequently the ^{16}O nuclei react similarly to form mainly ^{28}Si and ^{31}P with some ^{31}Si and a little ^{32}S.

Thereafter, at stellar core densities of the order of $10^8 g\ cm^{-3}$ and temperatures of the order of $4 \times 10^9 K$, ^{28}Si burns by successive α-particle captures to form the $A = 4n$ nuclei in a quasi-equilibrium process over the range of mass numbers between 28 and 64. The isotope distribution centres on the iron group elements, which have the largest nuclear binding per nucleon of all nuclei. The equilibrium process accounts for the isotope abundance peak at ^{56}Fe, produced from the doubly magic nucleus ^{56}Ni by β-decay. The nuclear reactions of the iron group elements can provide no further energy for the star; but heavier elements are produced in stellar interiors. Atomic spectroscopic lines due to the unstable element, technetium, with a half-life of 2×10^5 years for the longest-lived isotope, ^{99}Tc, have been found in the spectra of a few stars.

5.4 Nucleosynthesis by neutron and proton capture

The production of elements heavier than iron in stars takes place in a shell around the heavy-element core. The main mechanism mediating the production consists of the capture of neutrons liberated by helium burning and by the subsequent reactions. Proton capture provides a subsidiary mechanism. In a collision with a positively charged atomic nucleus the neutron, unlike the proton, is not repelled, so that the capture of a neutron is relatively more facile. In the steady-state or slow neutron capture mechanism, the s-process, there are long intervals, from $10-10^5$ years, between successive captures of a neutron by a heavy nucleus. Neutron-rich isotopes build up slowly by the (n,γ) reaction from mass number A to $(A + 1)$ at a constant atomic number Z. Eventually a nucleus that is unstable to β-decay is built up and, unless the half-life is exceptionally long, it is transformed by the (e^-,ν) reaction to the isobar of mass number $(A + 1)$ and atomic number $(Z + 1)$. That isobar becomes, in turn, the target for a further chain of (n,γ) isotope-building reactions.

For a given element, the relative abundances of the isotopes formed by the s-process are determined by the cross-section of the nucleus for neutron capture. Nuclei with a small cross-section have a low probability of neutron capture, such as the nuclei with a closed shell, $N = 50$, 82, or 126, which accumulate and become the more abundant. The product of the measured cross-section σ_A and the relative abundance N_A for the nucleus of mass number A is found to be approximately constant for the isotopes of an element which are generated by the s-process. The constancy was first observed for the isotopes of samarium with $A = 148$, 150, and subsequently for the isotopes $A = 122$, 123, 124 of tellurium. The value of the product, $\sigma_A N_A$, for a range of elements gradually decreases as the mass number becomes larger, i.e. with the increasing number of neutrons captured between the parent nucleus from which the s-process started and the daughter isotope considered. The s-process terminates with the heaviest stable nucleus, ^{209}Bi. Conversion to ^{210}Bi is followed by β-decay to ^{210}Pb and subsequent α-decay to ^{206}Pb. In turn the ^{206}Pb is transformed by successive neutron captures to ^{209}Pb and thence by β-decay back to ^{209}Bi, completing the cycle.

Rapid neutron capture, the r-process, bypasses the terminal cycle of the s-process, adding successive neutrons at time intervals which are short (0.01–1.0 s) relative to the half-lives of the α- and β-decays to produce the long-lived heavy radioactive nuclides, ^{232}Th, ^{235}U, ^{238}U, and ^{244}Pu. In addition the r-process is required to account for the double peaks in the relation between isotopic abundance and mass number at $A = 130-138$ and 194–208, noted by Suess and Urey (1956). They associated the heavier peak in each case with the closed-shell neutron numbers (e.g. ^{138}Ba ($N = 82$) and ^{208}Pb ($N = 126$)), produced by the s-process, and attributed the lighter peaks to neutron capture nucleosynthesis under conditions of high neutron density (Fig. 5.3). Unstable nuclei with the magic neutron numbers $N = 82$ and 126 form by rapid neutron capture and then transform by β-decay to stable isotopes of lower neutron number at the respective lighter mass number peaks near ^{130}Xe ($N = 76$) and ^{196}Pt ($N = 118$).

A third pair of r-process and s-process abundance peaks is associated with the closed neutron shell, $N = 50$, over the mass number region $A = 80$ to 88 (e.g. ^{80}Br ($N = 45$) and ^{88}Sr ($N = 50$)), although these peaks are overlaid by the general fall-off in isotopic abundance after the equilibrium process peak at ^{56}Fe (Fig. 5.3). The isotope ^{87}Rb ($N = 50$) is unstable, undergoing β-decay to ^{87}Sr with a half-life of 4.7×10^{10} years. Measurements of the abundance ratio ^{87}Rb/^{87}Sr (in conjunction with the corresponding ^{87}Sr/^{86}Sr ratio, to provide an estimate of the zero time ^{87}Sr content of the sample) are widely employed for the geochemical dating of minerals.

A minority of the isotopes of the elements heavier than iron, some 30–36, all in low abundance, are proton rich: they cannot be synthesized by neutron capture reactions and they were 'bypassed' in both the s-process and the r-process. The distribution of these proton-rich nuclides (as a function of mass number) parallels at a lower level that of the more abundant isotopes produced by neutron capture, but with an abundance peak at the closed proton shell of $Z = 50$ (tin) as well as the neutron magic numbers $N = 50$ and 82 (Fig. 5.3). The distribution suggests that the bypassed isotopes are secondary products which originate from primary nuclei built up by neutron capture. These isotopes were formed either by the p-process of proton capture through (p,γ) reactions or by neutron ejection through (p,n) or (γ,n) reactions.

The proton capture reaction is the more probable, since the proton-rich nuclei are associated with the heavier-element peak of the abundance

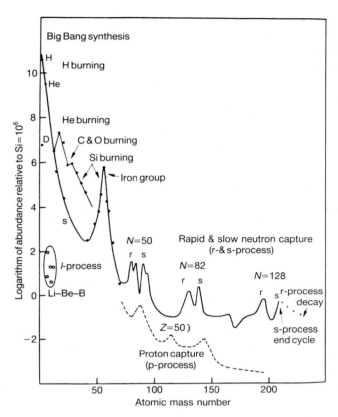

Fig. 5.3 The smoothed relation between the solar relative abundance of the elements ($Si = 10^6$) and the atomic mass number (A) resolved into sections characteristic of the main nucleosynthetic processes: (a) Big Bang synthesis of hydrogen, helium, and some deuterium and other light nuclides; (b) hydrogen burning to helium in the main sequence stars; (c) helium burning to carbon and oxygen in the red giant stars; (d) carbon and oxygen burning to second short period products; (e) silicon burning to $A = 4n$ products, which are equilibrated over the range $A = 28$ to $A = 64$ with optimum abundance at the iron group (peak at ^{56}Fe); (f) slow (s-process) and rapid (r-process) neutron capture, giving the neutron-rich nuclides and the double-peak abundance maxima at the closed-shell 'magic' neutron numbers, $N = 50$, 82, and 126; (g) the p-process of proton capture, giving an abundance maximum at the isotopes of tin ($Z = 50$, proton closed shell), as well as at $N = 50$ and $N = 82$; (h) the l-process generating light elements (Li, Be, B) by the spallation of heavier cosmic-ray nuclides (after Fowler (1984)).

maxima at the closed neutron shells, $N = 50$, 82, due to the s-process. In addition, the relative abundances of the proton-rich isotopes for a given element do not differ by more than a factor of three, consistent with proton capture but not with neutron ejection. The low abundance of the proton-rich nuclides is a consequence of the Coulomb charge barrier to the penetration of a nucleus by a proton. For mass numbers beyond the iron group, $A > 70$, the total abundances of the proton-rich isotopes, the r-process nuclei, and the s-process products lie in the approximate ratios 1:25:50.

There are some departures from the average abundance ratios, notably in the region near to the closed proton shell, $Z = 50$ (tin), where an additional subsidiary abundance peak is observed (Fig. 5.3). Tin has ten stable isotopes, three proton rich, two produced only by the r-process of neutron capture, one solely by the s-process, and four by both the s- and the r-processes:

Mass number, ^{A}Sn	112	114	115	116	117	118	119	120	122	124
Abundance, N_A (%)	1.0	0.7	0.4	14	7.6	24	8.5	33	4.8	6.1
Process origin	p	p	p	s	s,r	s,r	s,r	s,r	r	r
$\sigma_A N_A$ (10^{-30} m^2)				149	318	159	218	133	22	22.

Evidently the closed proton shell ($Z = 50$) of tin favours the formation of the element by proton capture. The product of the neutron capture cross-section σ_A (conventionally measured in the barn unit, 10^{-28} m^2) and the fractional abundance N_A of the tin isotopes clearly distinguishes the nuclides produced only by the r-process from the products of the s-process or those generated by both processes.

5.5 Explosive nucleosynthesis

Both the rapid neutron capture process and the proton capture process are features of explosive nucleosynthesis during supernova events. A star of less than 1.4 solar mass does not explode, but after some three billion years in the red giant phase the outer layers are ejected over about ten thousand years as a planetary nebula. The nebula has an abundance of elements similar to that of the solar system up to calcium, except for a helium enhancement by a factor of about two. The stellar core, containing most of the original mass, contracts to about one-thousandth of its former size and becomes a white dwarf star, which slowly radiates away its stored energy.

The finite small size of the white dwarf and the 1.4 solar mass limit are connected with the 'degeneracy pressure' of the electrons in the star, first investigated by R. H. Fowler (1889–1944) and developed by H. S. Chandrasekhar (1939). The degeneracy pressure arises from Pauli's exclusion principle, which prohibits the absolute degeneracy of any two electrons, i.e. an identical set of quantum numbers for each electron. The uncertainty principle connecting the electronic momentum range $\triangle p$ with the conjugate space coordinate range $\triangle q$ and Planck's constant $(\triangle p \triangle q \sim h)$ divides phase space into discrete cells, each of dimension h^3. No more than two electrons can simultaneously occupy any given cell, and those two electrons must then have opposite spin directions. A compression requires the electrons to occupy the lowest-energy cells in phase space. Thereby the Maxwellian energy distribution is distorted or even destroyed, so that the electron gas cannot respond to a decrease in temperature. The electron gas supports a velocity distribution up to high energies and exerts a correspondingly high pressure, maintained at large compressions (the degeneracy pressure), despite temperature changes.

The gravitational forces in stellar cores of more than 1.4 solar mass overcome the electronic degeneracy pressure and collapse the electrons into the protons of the atomic nuclei, forming neutrons. The energy required for the atomic decomposition to neutrons comes from the gravitational potential. If the neutron degeneracy pressure is sufficient to halt further collapse, a neutron star is formed; if not a 'black hole' results. A black hole is estimated to result from the collapse of a core four or more times larger than the solar mass in a star containing initially more than 50 solar masses.

Stars starting out in the 8–50 solar mass range ultimately generate a core of mainly iron group elements with approximately 1.5 solar mass. When the supply of nuclear reaction energy is exhausted, the core collapses to a neutron star residue. The core bounce from the compression produces a shock wave which ejects the outer layers in a supernova explosion on the time-scale of a few seconds. The high neutron density during the explosion generates the neutron-rich isotopes of the heavy elements by rapid neutron addition (the r-process) to iron group elements and the products from the previous slow neutron addition to those elements (the s-process). At the same time the minor neutron-poor isotopes of the heavy elements originate through the p-process from the previous products by proton capture in the hydrogen-rich shell of the exploding star, or through proton or photolytic neutron ejection by the (p,n) or (γ,n) reactions. The supernova remnant may become a pulsar, a rapidly spinning compact neutron star emitting a beam of synchrotron radiation at radio frequencies or over most of the electromagnetic spectrum. An example is the radiation source in the Crab nebula pulsating 30 times a second and produced by a supernova explosion in AD 1054.

Two types of supernovae (SN) were distinguished by Minkowski in 1940. Type I SN lack spectral lines at maximum light while type II SN show the Balmer series of hydrogen; and the light-decay curves of the two types differ in that the I SN radiation intensity falls off with a typical half-life of about 56 days whereas the II SN decay curve is less regular, often showing a stepped intensity decline. The type I SN explosions occur in all types of galaxies, ellipticals as well as spirals, and they may leave no compact remnant, while type II SN are confined mainly to the gas-rich arms of spiral galaxies and often leave behind a neutron star. The two main types of supernovae differ in the stage of nucleosynthesis achieved before the final explosion, dependent upon the initial stellar mass. The type I SN derive from stars beginning with 6–8 solar masses, which undergo a final explosive collapse with a carbon–oxygen deflagration. A major nucleosynthesis of the iron group elements takes place during the explosion. With less than 6 solar masses originally, the star may leave a white dwarf remnant of about 1.4 solar mass, consisting chiefly of carbon group elements, by exploding with the onset of carbon ignition. The more massive and highly evolved stars of metal-rich Population I, with a core of iron group and derived s-process elements, end more typically in a type II SN explosion.

The transition from one nuclear reaction sequence to another gives rise to some instability, since the particular feedback conditions of gravitational contraction and outward thermal pressure, dependent upon the density and temperature of the reaction zones, usually differ widely for the old and the new reaction processes. The instability may lead to an oscillatory 'hunting' of the feedback mechanisms, as in the variable stars exhibiting regular pulsations in size and luminosity. The feedback mechanisms may fail completely if the core pressure is sustained by electronic degeneracy. Unlike ideal gas pressure, degeneracy pressure is insensitive to a range of temperature changes, and the nuclear reactions, with strongly temperature-sensitive rates, produce a runaway escalation of the temperature. At a sufficiently high temperature, the degeneracy is lifted and the pressure increase is then catastrophically rapid. The ensuing detonation, with carbon, oxygen, and silicon explosive nucleosynthesis in successive concentric shells, produces the less abundant nuclei of odd mass and odd atomic number and the rarer neutron-rich isotopes, expelling them into the interstellar medium, along with the products of steady-state nucleosynthesis.

In a type I SN explosion silicon deflagration produces the doubly magic isotopes ^{56}Ni in abundance. The typical type I SN light-decay curve with a 56 day half-life corresponds to the rate of β-decay of ^{56}Ni, first, with a half-life of 6 days to ^{56}Co and, subsequently, with a half-life of 77 days to ^{56}Fe, liberating energetic positrons and photons at each stage. Absorption and emission lines of cobalt ions have been observed in the spectra of type I SN events (Trimble 1983). Earlier, the detection in hydrogen bomb tests of the isotope ^{254}Cf ($Z = 98$), produced from ^{238}U by the rapid neutron capture process, led to the suggestion that ^{254}Cf, which has a half-life of 56 days for spontaneous fission, might be responsible for the particular rate of the type I SN light decay. Although the energy released by the spontaneous fission of ^{254}Cf is large (220 MeV per decay), compared with α-decay ($\sim$5 MeV) or β-decay ($\sim$1 MeV), the estimated production of the californium isotope in a supernova event is small compared with the prolific generation of ^{56}Ni and other iron group radioisotopes, such as ^{59}Fe (β-decay half-life of 45 days).

The nuclei expelled into interstellar space in a supernova explosion are ejected with high energies, some with velocities close to the speed of light, forming high-energy cosmic rays. Some cosmic-ray particles have been detected with energies as high as 10^{20} eV. The escape of the cosmic-ray nuclei from a galaxy is retarded by the galactic magnetic field, so that they circulate throughout the galactic disc with a mean lifetime of about 10 million years to provide a constant isotropic background of cosmic radiation. Some low-energy cosmic radiation has a local origin, with an intensity that grows and decays with the onset and decline of solar flares, superimposed upon the more constant background of the solar wind. The solar wind and flare particles are mainly protons, with some 5 per cent of α-particles and some 1 per cent of heavier nuclei, together with electrons, some free and others orbitally bound to nuclei in ions, such as Li$^+$.

The high-energy cosmic rays were first detected from their secondary ionization products at the Earth's surface in 1900, when the Curie–Becquerel view that radioactivity had a cosmic-ray origin was current. Charles Wilson, finding tracks in his cloud chamber set up in a railway tunnel, supposed the rays might have a terrestrial origin, but Victor Hess (1883–1964) found, in a series of balloon ascents in 1911, that the secondary atmospheric ionization became larger at greater heights and remained constant in time at a given altitude, suggesting an extraterrestrial source of the primary cosmic radiation. Later high-altitude balloon experiments with stacks of photographic emulsions, particularly by Wilson's one-time student, Cecil Powell (1903–69), and subsequent satellite measurements with scintillation and Cerenkov counters, showed that the primary cosmic rays consist of a substantial range of the atomic nuclei moving with relativistic velocities.

As a function of atomic number, the distribution of the cosmic-ray atomic nuclei differ from the relative abundances of the elements for the solar system in two major respects. First, the thermally fragile nuclei of lithium, beryllium, and boron are overabundant in the cosmic rays by a factor of over 10^5, owing to the spallation of heavier nuclei. Cosmic rays are therefore a significant source of these relatively rare light nuclei. Second, the heavier elements produced by the rapid neutron capture process are overabundant. Of the two abundance peaks associated with the closed neutron shell at $N = 126$, the lighter r-process peak at platinum is larger than the heavy s-process peak at lead for the cosmic-ray nuclei. Moreover, the neutron-rich isotope ^{58}Fe is as abundant as ^{56}Fe among the cosmic-ray nuclei, whereas ^{56}Fe/^{58}Fe = 300 on average in solar system materials (Trimble 1975).

In an external galaxy, supernovae occur at intervals between 15 and 60 years, dependent upon the Hubble type. From his surveys of the galactic nebulae, Hubble (1936) divided the galaxies into two main types, the ellipticals and the spirals, envisaging an evolution from the near-spherical large star clusters, through progressively flattened elliptical forms, to rotating disc-shaped stellar assemblies which develop tenuous and then prominent spiral arms, often with a central 'bar' or elongated aggregate of stars. Although no longer regarded as an evolutionary sequence, Hubble's classification corresponds to a distinction between the galaxies containing many older metal-poor Population II stars and little interstellar gas—the ellipticals—and those with numerous young stars of Population I—the spirals—particularly in the gas-rich spiral arms. Type II supernovae occur only in the spirals, mainly in the spiral arms, whereas type I are observed in both elliptical and spiral galaxies, but with greater frequency in the spirals.

Supernovae are rarely observed visually in our own Galaxy of the Milky Way, owing to the optical opacity of the interstellar dust in the galactic disc. The dust does not obscure the radio-frequency emissions of supernovae remnants (SNR) on the remote side of the Galaxy. Less than a quarter of the 120 radio SNR catalogued have been detected optically. Chinese and Japanese chronicles record the appearance of some eight naked-eye 'guest stars', luminous for several months or even a few years, over the pretelescopic period of observation, prior to the publication of Galileo's *Starry messenger* (1610). The recorded positions of these 'guest stars' correlate with those of present-day SNR radio sources at a confidence level from the 'possible' for the earlier events to 'certain' for the more recent (Stephenson and Clark 1978).

A supernova in AD 1006 was particularly spectacular and the event was recorded not only by Chinese and Japanese astronomers, but also by Arabian and European observers. The appearance of new stars ran counter to the Aristotelian doctrine of the unchanging perfection of the heavens, and the 1006 event was the sole pre-Renaissance supernova recorded in the west. The more recently observed supernovae were the extraordinary 'New Star' of 1572, studied by Tycho Brahe (1546–1601) at his Uraniborg observatory on the island of Hveen, near Copenhagen, and the brilliant nova of 1604, investigated by Johannes Kepler (1571–1630), who succeeded Tycho Brahe as Imperial Astronomer at the Prague court of the Holy Roman Emperor, Rudolph II. The earliest telescopic observation of a supernova was made inadvertently by the first Astronomer Royal at Greenwich, John Flamsteed (1646–1719), who catalogued in 1680 a star which was subsequently characterized as a supernova, although Flamsteed himself did not appreciate that it was a new star.

6

The interstellar medium

6.1 The domains of interstellar matter

It is estimated that interstellar matter makes up some 10 per cent of the total mass of the Galaxy, equivalent to the mass of about 10^{10} stars. In the denser regions of the interstellar medium, the dark molecular clouds, stars are generated and evolve through the nucleosynthetic sequences. The heavier chemical elements produced by the transmutation of hydrogen and helium escape from the stars to the interstellar medium, where their relative abundance slowly increases. These heavier elements are liberated at a continuous and low rate through the stellar winds, and episodically at a higher rate in novae and ejected circumstellar shells, or finally and catastrophically in supernovae.

The presence of metallic atomic species in interstellar space was first suggested by the German astronomer J. F. Hartmann (1865–1936), who observed in 1904 that the Fraunhofer H and K absorption lines of Ca^+ in the spectrum from the nearby star pair, delta-Orionis, did not undergo Doppler wavelength shifts during the orbital motion of the two stars around one another, unlike other atomic lines. Initially the 'stationary lines' were attributed to clouds of calcium around binary systems, but O. Struve (1897–1963) at the Yerkes observatory found in 1928 that the apparent absorption intensity of the stationary Ca^+ lines is approximately proportional to the distance of the binary star, implying that the absorbing atoms are distributed throughout the intervening space between the stars and the Earth. Other neutral and charged species similarly detected in the interstellar medium include Na, K, Ca, Fe, Ti^+, CN, CH, CH^+.

Further evidence for obscuring matter in interstellar space came from the Swiss astronomer R. J. Truempler (1886–1956), who showed in 1930 that the progressive relative enhancement of the red region in the star's spectrum, with increasing distance of the star source, is consistent with the scattering of the corresponding blue spectral range by small particles of cosmic dust, thinly dispersed throughout space. Denser clouds of the particles or grains accounted for the long-known dark regions which appear all along the Milky Way, and for the 'reflection nebulae', or luminous clouds reflecting the continuum of radiation from nearby stars, detected by V. M. Slipher (1875–1969) at the Lowell observatory in 1912.

Radioastronomy, developed from the late 1940s, proved capable of probing the dark regions and opened up new perspectives of the extension and the composition of interstellar matter. The detection in 1951 of the

21 cm radio-frequency line of the neutral hydrogen atom, due to the emission of a radio photon in the spin–flip transition from a parallel to an antiparallel orientation of the proton and the electron spin axes, led to the mapping of the distribution of atomic hydrogen throughout the Galaxy, unobscured by the dark clouds opaque to visible radiation (Fig. 6.1). Neutral hydrogen atoms are found in two main domains: first, in aggregates of transparent diffuse clouds with a mean number density $<n>$ (particles/m^3) of 10^7–10^8, associated with the spiral arms of the Galaxy; and, second, in the intercloud gas of the galactic disc between the spiral arms, where the number density falls to approximately 10^5 m^{-3}. The diffuse clouds account for some 40 per cent of the mass of the interstellar medium but only some 5 per cent of its volume, whereas the intercloud gas fills some 40 per cent of the volume and makes up some 20 per cent of the mass (Table 6.1).

Table 6.1 Domains of the interstellar medium. For each domain the percentages refer to the fraction of the total galactic volume and to the total mass of the interstellar matter in the Galaxy, with the mean number density $<n>$ (m^{-3}) of the main component and the effective temperature T (K)

Domain	Main component	Volume (per cent)	Mass (per cent)	$<n>$	T
Dense clouds	Molecules (H_2)	~0.5	~40	10^9–10^{12}	10–50
Diffuse clouds	Atoms (H)	~5	~40	10^7–10^8	10–100
Intercloud gas	Atoms > ions	~40	~20	~10^5	~5×10^3
High-temperature gas	Ions (H^+, e^-)	>50	~0.1	~10^4	~5×10^5

Some of the hydrogen atoms of the intercloud gas, about 10–20 per cent, are dissociated to protons and electrons by the stellar ultraviolet radiation with photon energies equal to or greater than the ionization energy (13.6 eV) of hydrogen, giving the intercloud gas an effective temperature of approximately 5000 K. While some hydrogen atoms of the diffuse clouds are similarly ionized, the number density of atoms is so much larger that the fractional ionization is very small and the effective temperature averages to only 10–100 K. Above and below the mean plane of the Galaxy the intercloud gas thins out towards the halo of globular star clusters to an average number density of some 10^4 m^{-3}, forming a region of high-temperature gas. Here the hydrogen is largely ionized to protons and electrons at an effective temperature of approximately 5×10^5 K. The high-temperature gas domain occupies more than 50 per cent of the

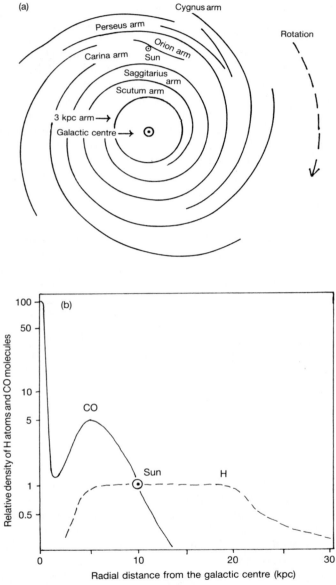

Fig. 6.1 The distribution of neutral hydrogen atoms and carbon monoxide molecules throughout the Galaxy, based on the intensity of the respective radio-frequency emission lines at 21 cm and 2.6 mm wavelength: the density distribution of molecular hydrogen parallels that of carbon monoxide with the ratio $[CO]/[H_2] \sim 10^{-4}$ in the dense molecular clouds near to the solar system, falling to about 10^{-6} in the lower-density diffuse clouds. (a) Schematic distribution of the dense molecular

clouds (carbon monoxide) and diffuse clouds (hydrogen) forming the spiral arms of the Galaxy; the arms embed additionally most of the stars observable by optical spectroscopy. (b) The radial distribution in the Galaxy of the smoothed mean density of hydrogen atoms and carbon monoxide molecules relative to the respective local density in the region of the solar system. The Sun lies some 10 kiloparsec (30 000 light-years) from the galactic centre, where the carbon monoxide density is a maximum in giant molecular clouds of up to 10^7 solar masses. The subsidiary carbon monoxide density peak near 6 kiloparsec (kpc) corresponds to the more active region of star formation in dense molecular clouds between 2 and 10 kpc (after Bally 1986; Blitz *et al.* 1983).

---

galactic volume but contributes only about 0.1 per cent to the mass of the interstellar medium.

Inside some of the transparent diffuse clouds lie dense opaque molecular clouds, composed mainly of hydrogen at a number density of $10^9 - 10^{12}\,m^{-3}$ with CO, OH, HCN, and other molecules at much lower relative abundance. The dense molecular clouds have a low average temperature (10–50 K) and occupy only about 0.5 per cent of the galactic volume, but they make up some 40 per cent of the mass of the interstellar medium. The mass of a dense cloud lies typically in the range from about $10^2 - 10^6$ solar masses. The opacity of the dense clouds arises from their relatively high number density, which preserves the molecules within them from photolysis by stellar radiation. The radiation flux of ultraviolet photons with energies required to dissociate the hydrogen molecule into atoms (>4.48 eV) is largely filtered out through absorption by the minor abundance of hydrogen molecules (about 10 per cent) in the diffuse clouds surrounding the denser molecular clouds.

Although opaque at visible and ultraviolet wavelengths, the dense clouds are transparent to a degree over the infrared region and more completely so at microwave and radio frequencies. The rotational transitions of molecules in the dense clouds give microwave lines which characterize the emitting species, such as the CO line at a wavelength of 2.6 mm (frequency 115 GHz). Other molecular energy modes have a quantization corresponding to transitions in a similar frequency region, like the umbrella inversion mode of NH_3 at 24 GHz (1.25 cm wavelength), or the spin–reorientation transition of the unpaired electron relative to the orbital electronic angular momentum around the bond axis in linear free radicals, as in the case of the 1.66 GHz (18 cm) line of the OH radical. In addition, electronic transitions between the higher excited states of atoms and ions have energies corresponding to radio frequencies, allowing the characterization of the more highly ionized domains of the interstellar medium. The recombination of an electron and a proton to form a hydrogen atom cascades through the main quantum number change from $n = 110$ to $n = 109$ with the emission of a line at 5.01 GHz (6 cm), and similar radio-frequency lines are observed for the helium recombination cascade through the quantum number changes $n = 160 \rightarrow 159 \rightarrow 158$ and $157 \rightarrow 156$.

6.2 The interstellar dust

Dust grains make up about 1 per cent of the mass of the interstellar medium. The dust particle number density generally parallels that of the gaseous molecules, atoms, and ions in the domains of the interstellar medium; but it is especially large in the circumstellar regions around either an old cooling star which has ejected much of the material from its outer shell or a young hot star embedded in the dense gas cloud from which it has formed. The circumstellar gas and dust clouds have temperatures in the 100–1000 K region, so that the dust particles give rise to infrared emission features, while absorbing and scattering the visible and ultraviolet radiation of the star they surround.

The scattering of starlight due to the dust particles is approximately proportional to the light frequency over the visible range, but it rises steeply in the ultraviolet region, with an absorption feature near 220 nm. The relation between the scattering extinction and the frequency of starlight indicates that the grain particles have a dimension of the order of 0.1 μm or less, and the 220 nm absorption suggests a possible graphite or metal oxide composition. The infrared absorption and emission spectra of circumstellar dust clouds demonstrate the presence of silicate grains and the silicon oxide molecule in the gas phase around oxygen-rich stars, or acetylenic molecules with graphite and silicon carbide particles around carbon-rich stars.

Visible and ultraviolet starlight transmitted through interstellar dust becomes linearly and circularly polarized in proportion to the scattering of the light. The polarization indicates that the size of the dust particles is of the order of the wavelength of the light, and that the particles have a preferred alignment. It is suggested that the galactic magnetic field produces the statistical particle orientation, acting upon the magnetite (Fe_3O_4) or other magnetic components of the interstellar grains.

Grain formation in an interstellar gas cloud depletes the cloud of its less volatile constituents. The relative depletion of the species in the interstellar clouds, compared with cosmic abundancies, thus provides an indication of the elementary composition of the dust particles. Oxygen, with the highest cosmic abundance after hydrogen and helium, affords a virtually unlimited source for the formation of refractory oxides, which are the first to condense upon cooling. Aluminium oxide, which condenses from the gas to the solid phase at 1530 K under a partial pressure of 10^{-1} Pa (10^{-6} atmospheres), or 1650 K at 10^{-4} atmospheres, is expected to form the first solid material on cooling, followed by the oxides of calcium, titanium,

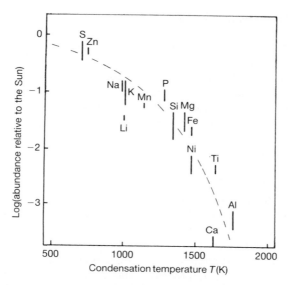

Fig. 6.2 The relation between the condensation temperature of the chemical elements and their depletion, relative to the corresponding solar abundances, in a diffuse interstellar cloud located in the direction towards the star zeta-Ophiuchi (after Morton 1974).

nickel, iron, magnesium, and silicon. In a diffuse cloud located in the direction towards the star, zeta-Ophiuchi, aluminium and calcium are depleted by more than three orders of magnitude, compared with their cosmic abundances, while elements with oxides condensing at lower temperatures are less depleted (Fig. 6.2).

The interstellar dust particles have a lower effective temperature than the surrounding gaseous atoms and molecules since the energy of an absorbed photon is shared out among the numerous bonded atoms within a grain, and re-emitted over a range of low-energy quanta. In the dense clouds, ices of water, methane, ammonia, and carbon monoxide condense into a mantle around the silicate core of a grain and undergo photochemical reactions in the stellar flux of ultraviolet radiation. Laboratory simulations of the photolysis of the ice mantles around the grain cores at 10 K indicate the formation of a range of organic molecules and radicals, such as HCO, HCHO, HOCO, $HCONH_2$, and HNCO (Greenberg 1984). The products undergo subsequent photochemical and 'thermal' reactions, although the latter may become temperature independent and non-Arrhenius in character, due to quantum-mechanical tunnelling, as in the formation of polyoxymethylene $(-CH_2O-)_n$ from formaldehyde HCHO (Goldanskii 1986).

6.3 The dense molecular clouds

The dense interstellar clouds emit a large number of microwave and radio-frequency lines, owing primarily to the rotational transitions of diatomic and polyatomic molecules and, in part, to electron-spin reorientation in free radical species. A comparison of the line frequencies with the microwave spectra of known species identifies the emitting molecule; an agreement for at least three lines serves as a general criterion for a positive identification. The comparison microwave spectrum may be directly measured, or calculated if the molecule remains to be synthesized at the time of the identification, as in the case of the higher polyacetylenenitriles, $H(C{\equiv}C)_n CN$, $n = 4, 5$. The intensity of the microwave line emitted from a dense cloud is proportional to the column density of the emitting species, i.e. the number of the molecules in a column with unit cross-sectional area between the source and the detector.

By 1985 some 90 interstellar molecular species had been identified, and several more are added each year (Winnewisser and Herbst 1987; Smith 1988). A number of the microwave and radio-frequency lines from the interstellar sources remain unidentified. Eddington once compared scientific activity to that of a fisherman who catchs only those fish that cannot evade the mesh size and construction of his particular net. The microwave spectroscopy of interstellar species illustrates Eddington's parable. Strong microwave absorption or emission requires a molecule with a permanent electric dipole moment, giving a periodically varying dipolar electric field as the molecule rotates. Non-polar molecules, such as hydrogen, acetylene or methane, give only an electric quadrupole or higher electric multipole field which varies periodically as they rotate: the same restriction holds for the vibrational motion of hydrogen and other diatomic elementary gases. The absorption or emission in the microwave region for the rotations, or the infrared range for the vibrations, is then substantially weaker, by a factor of about 10^{-6} in the electric quadrupole case. The weak emission is compensated by the high abundances of these molecules, and they have been identified; but other non-dipolar molecules, such as carbon dioxide, which is produced in grain–mantle photolysis simulations, as yet remain undetected by radio-frequency spectroscopy in the interstellar sources.

The interstellar molecules that are the most readily characterized by microwave spectroscopy are the linear heteronuclear systems, which have a single value for the moments of inertia about the two axes mutually perpendicular to one another and to the bond direction, giving a series of strong and nearly equally spaced lines. The largest molecules as yet discovered, tetra- and penta-acetylenenitrile, are of this type. So too are ^{12}CO and ^{13}CO, employed in surveys of the location, dimensions, and column density of the dark clouds in the spiral arms and at the centre of the Galaxy (Fig. 6.1): ^{13}CO is monitored when the high abundance of ^{12}CO gives a saturated signal. The interstellar molecules detected with more difficulty are the asymmetric tops, with three unequal moments of inertia, particularly ring compounds with a small electric dipole moment, such as cyclopropenylidene (C_3H_2), the first cyclic molecule characterized, and identified in 1985 (Table 6.2).

The interstellar molecules are produced both homogenously, in the gas phase, and heterogeneously, on the surface or within the ice mantle of the interstellar grains. In the gas phase, ion–molecule reactions are more important than neutral–neutral reactions, owing to the larger effective collision and reaction cross-section of the ion–molecule type. The charge of an ion induces an electric dipole moment in a neutral molecule, and the Coulombic attraction between the charge and the induced dipole promotes the collision and reaction of the two species, producing rate enhancements of about 10^2 relative to analogous radical–molecule or radical–radical reactions. The combination of atoms to form a diatomic molecule generally requires a third body to take up the energy released on bond formation. Three-body collisions are highly improbable in the interstellar clouds, and atom recombinations are largely confined to grain surfaces. Such energetic reactions on the surface or within the ice mantle of a grain evaporate the mantle locally and release the constituent molecules into the gas phase.

The chemical composition of an interstellar cloud lies far from the state of thermodynamic equilibrium. Essentially a dark cloud is an open flow reactor system with a steady-state composition governed by kinetic factors, fed by the input of high-energy photons and cosmic-ray nuclei from the stars, and drained by the output of lower-energy quanta and molecules produced by the chemical and physical processing of the cloud species. The two most abundant molecules in a dense cloud, hydrogen and carbon monoxide, are expected to equilibrate thermodynamically to form predominantly methane and water at low temperature, even at very low pressure. But many kinetically facile reactions lead to the production of carbon monoxide, which is consumed by relatively slow reactions to give products that are readily transformed back again to carbon monoxide. The steady-state composition of an interstellar cloud has become a problem in computer modelling, based upon the measured or estimated rate coefficients of many ($>10^3$) thermal reactions of various types, mostly ion–neutral with some ion–ion or neutral–neutral, together with numerous radiolytic and photochemical processes (Duley and Williams 1984).

6.4 Star formation from dense clouds

In a study of star formation (1919), James Jeans (1877–1946) at Cambridge investigated the kinetic and gravitational stability conditions of a sphere of gas as a function of the temperature, the molecular mass of the constituent particles, and their number density. He demonstrated that,

Table 6.2 Molecules detected in interstellar clouds and the species ([ ]) identified in circumstellar shells (Winnewisser and Herbst 1987)

Interstellar molecule		Isotopic species
H_2	hydrogen molecule	HD
H_2D^+	dihydrogen deuterium ion	
HO	hydroxyl radical	^{17}OH, ^{18}OH
H_2O	water	HDO, $^{17}OH_2$, $^{18}OH_2$
H_3N	ammonia	H_2ND, $^{15}NH_3$
HNO	nitroxyl hydride	
H_2S	hydrogen sulphide	
HCl	hydrogen chloride	
NO	nitric oxide	
NS	nitric sulphide	
NaOH(?)	sodium hydroxide	
[SiH_4]	silane	
SiO	silicon monoxide	^{29}SiO, ^{30}SiO
SiS	silicon monosulphide	^{29}SiS, ^{30}SiS, $Si^{34}S$
PN(?)	phosphorus nitride	
SO	sulphur monoxide	^{34}SO, ^{33}SO, $S^{18}O$
SO_2	sulphur dioxide	$^{34}SO_2$
CH, (CH⁺)	methylidine radical (ion)	$^{13}CH^+$
[CH_4]	methane	
[HC≡CH]	acetylene	
[$H_2C=CH_2$]	ethylene	
$H(C)_n$	poly-yne radicals, $n = 2$–5	DC_2
$CH_3(C≡C)_nH$	methyl poly-ynes, $n = 1, 2$	
C_3H_2	cyclopropenylidene	
[C_2Si]	silacyclopropyne	
HCN	hydrogen cyanide	DCN, $H^{13}CN$, $HC^{15}N$
HNC	hydrogen isocyanide	DNC, $H^{15}NC$, $HN^{13}C$
HNCO	isocyanic acid	
HNCS	thioisocyanic acid	
H_2CO	formaldehyde	DHCO, $O^{13}CH_2$
H_2CS	thioformaldehyde	
$H_2C=NH$	methylenimine	$^{13}CH_2 = NH$
CH_3OH	methanol	
CH_3NH_2	methylamine	
CH_3SH	methyl mercaptan	
HCOOH	formic acid	
H_2NCHO	formamide	
H_2NCN	cyanamide	
CH_3CN	acetonitrile	$N^{13}CCH_3$
CH_3CHO	acetaldehyde	
$H_2C=C=O$	ketene	
$HCOOCH_3$	methyl formate	
CH_3OCH_3	dimethyl ether	
C_2H_5OH	ethanol	
C_2H_5CN	propionitrile	
$H_2C=CCN$	acrylonitrile	
$H(C≡C)_nCN$	poly-yne nitriles, $n = 1$–5	D- and ^{13}C-analogues
$CH_3(C≡C)_nCN$	methylpoly-yne nitriles, $n = 1, 2$	

beyond a critical total mass value, namely the Jeans mass for a given temperature and density, the sphere of gas must collapse gravitationally. A less massive sphere of gas, on the contrary, would disperse into space. For the conditions of a dense interstellar cloud of hydrogen molecules at 20 K the Jeans mass amounts to some 20 solar masses, or a larger mass at a higher temperature, in proportion to $T^{3/2}$. During the gravitational collapse of such a gaseous sphere, the average density ρ increases and, since the Jeans mass is proportional to $\rho^{-1/2}$, the large sphere of gas breaks up into smaller contracting spherical bodies.

While idealized, the model of Jeans suggests that stars, and clusters of stars, form from the large dense clouds. The inference is supported by the observation that the mean density of the total matter in the volume of a star cluster approximately equals the average density in the dark molecular clouds of the Galaxy. Some dense interstellar clouds have the mass and size of a large star cluster, such as the giant molecular cloud in Sagittarius, towards the centre of the Galaxy, which is more than 100 light-years across and with a content of more than 10^6 solar masses.

The general initiation of the gravitational collapse of a dense cloud is attributed to a spiral density wave which sweeps through each arm of the Galaxy about once in every 100 million years. The density wave is required to account for the structure of the spiral galaxies. Owing to the larger angular velocity near to the centre of a galaxy, the spiral arms would become stretched and tightly wound up without the shepherding effect of the density wave. During the period of about 10 million years required for the density wave to pass through a given spiral arm of the Galaxy, a supernova explosion not too far removed from the dense molecular cloud produces a shock wave that precipitates the gravitational collapse of the cloud. If close enough, the supernova explosion injects into the dense cloud short-lived radioactive isotopes which remain as stable daughter isotopes in condensed materials.

The collapse of a dense cloud into a star on the main sequence occupies only a brief period in the overall lifetime of the star, so that observational evidence of this stage is sparse. A knowledge of star formation from cloud collapse derives mainly from computer simulations of various theoretical models which, differing widely in detail, present a common general view. The models envisage an initial collapse of low-density gas with the conversion of gravitational potential into molecular kinetic energy, thereby heating up the mass, which cools by the emission of black-body radiation. The build-up of the matter density in the central regions of the contracting mass leads to progressive photon scattering and absorption, and the increasing opacity of the mass prevents the escape of energy as radiation. The consequent rise in temperature dissociates the larger molecules in the cloud mass and, at about 1800 K, the hydrogen molecules are disrupted into hydrogen atoms. Further contraction and heating ionizes the hydrogen atoms, and then the helium atoms, to produce a plasma of protons, electrons, and helium nuclei in the core region of the proto-star.

Energy is supplied continuously to the proto-star mass not only by its own gravitational contraction but also by the infall of additional matter from the surrounding cold dense molecule cloud. The temperature rise in the hot contracting core increases the temperature gradient between the centre of the proto-star and its outermost shell, producing the convective transport of heat in the form of bubbles of hot gas rising from one layer to the next, with the result that the surface layer becomes hot and luminous. The surface area of the proto-star is large at this stage and, for a given energy emission per unit area (proportional to T^4 by Stefan's law), the total luminosity of the emission is very large. Gravitational contraction, while supplying energy for the radiant emission, reduces the surface area and subsequently the total luminosity progressively falls, despite the increase in core temperature.

At a sufficiently elevated temperature radiative diffusion takes over from convection as the major form of energy transport from the core to the surface. In radiative diffusion, photons are successively radiated, absorbed, emitted, reabsorbed and re-emitted progressively through the core and beyond to the outer layers, giving a steady state with no large-scale mixing of the materials in the core and surrounding layers. If the total mass is large enough, nuclear reactions in the core begin with hydrogen burning and the proto-star matures into a main sequence star.

Condensed masses smaller than about 0.1 solar mass never begin nucleosynthesis because they form an electronically degenerate core before the temperature required for hydrogen burning is attained. The degeneracy pressure withstands further gravitational contraction and thus prevents any further increase in temperature. Such a small condensed mass gradually cools and fades to a dark dwarf, which may be detected by its perturbation of the motion of a bright companion in a binary system. A small nearby star (0.15 solar mass, 5.5 light-years distant), named after its discoverer, E. E. Barnard (1857–1923) at the Yerkes Observatory, has an even smaller dark companion of only 1.5×10^{-3} solar mass, or about 1.6 times the mass of the planet Jupiter.

Proto-stars larger than 0.1 solar mass are expected to become observable before the onset of nuclear energy production, at the luminous convective stage under gravitational contraction, given sufficient attenuation of the surrounding opaque dense cloud. The stage is estimated to occupy only a brief period, about 1 million years for a star of solar mass, so that only a small population of convective proto-stars exists at any one time. Clusters of faint stars associated with dense clouds, the T Tauri stars, named after the initial case detected, appear to be at the convective stage. The luminosity of T Tauri stars varies irregularly and is consistent with convective energy transport from the core.

Several dense molecular clouds contain compact infrared sources. While the dense clouds are opaque in the visible and ultraviolet region, they transmit the longer infrared wavelengths, corresponding to the peak black-body emission from the surface of proto-stars in the visually opaque stage and the emission from the surrounding envelope of dust and gas. A relatively close large molecular cloud in Orion has a maximum infrared flux density between 50 and 100 μm wavelength, corresponding to a mean temperature of approximately 50 K in the central regions of the cloud itself. In addition the Orion molecular cloud contains compact infrared sources with peak emission between 5 and 15 μm, corresponding to black-body temperatures in the region of 500 K.

Another large molecular cloud, in the direction of the star rho-Ophiuchi, contains as many as 67 similar compact infrared sources which are thought to be a cluster of proto-stars at the visually opaque stage of stellar evolution. Nearly half of the 95 nearby smaller dense clouds contain a compact infrared source, and almost half of these are visible optically too. The visible sources are of the convective T Tauri type. The invisible infrared sources are either similar stars obscured by dust, or earlier opaque-stage proto-stars of about a solar mass, still deriving a significant part of their energy of the infall of further matter from the surrounding dense cloud (Beichman *et al.* 1986).

7

The solar system

7.1 Origins of planetary systems

Historically two general models have been put forward to account for the origin of the solar system. One type proposes the formation of the Sun and the planets from a single body of matter, as in the cosmic vortex theory (1644) of René Descartes (1596–1650), or the solar nebular hypothesis (1755) of Immanuel Kant (1724–1804), developed into a planetary ring theory (1796) by Pierre Laplace (1749–1827). The other type envisages a near collision between two bodies. In the theory (1745) of Georges Buffon (1707–88), the approach of a comet with solar mass produced a tidal disruption from the Sun and the planets formed from the ejected material.

Towards the end of the nineteenth century, objections to the Laplacian solar nebula hypothesis were raised on the grounds that, while the Sun contains over 99.8 per cent of the mass of the solar system, it carries only a minor fraction of the total angular momentum, rotating slowly with a period of about 26 days. The orbital motions of Jupiter and Saturn, the most massive of the planets, account for some 80 per cent of the angular momentum in the solar system. Subsequent modelling (1916–19) by James Jeans and Harold Jeffreys (1891–1989) turned accordingly to the tidal disruption theory. It turned out, however, that the tidal removal of hot material from the Sun would result in the explosive dispersal of the ejected matter into cosmic dust and gas, rather than the formation of a planetary system. But the two-body tidal theory may be saved if the material for planet formation is tidally removed, not from the Sun, but from a cool low-density proto-star in a near collision (Woolfson 1984).

The angular momentum problem of the nebula hypothesis was solved by the discovery of the solar wind of charged particles and by magneto-hydrodynamic theories which demonstrate that the more intense stellar wind of a proto-star in a condensing cloud would carry away angular momentum from the central body to the dense regions of the surrounding nebula disc. Recent observations of very young stars, less than 1 million years old, show from the spectral Doppler shifts that matter is continuously emitted from such stars in an apparent bipolar outflow. The equatorial regions of the young star are obscured by a disc of dust and gas, and the bipolar jets of matter are observable only in the directions perpendicular to the plane of the disc.

Discs of dust or gravel-sized particles have been detected around several dozen nearby young stars by the infrared astronomical satellite (IRAS), in orbit 1983–4 (Habing and Neugebauer 1984), and the infrared telescope facility (IRTF) on Hawaii (Wolstencroft and Walker 1988). The presence of dust discs or shells around many more stars is inferred from the detection of an infrared excess in their radiant emission, i.e. an infrared output over and above that expected by fitting the visible and ultraviolet radiation of the central star to a black-body radiation curve.

The equatorial discs around pre-main-sequence stars consist of silicate or carbon dust with a particle size of less than 1 μm at temperatures of 160–500 K. The discs surrounding young main sequence stars have a larger grain size of less than 1 mm and a lower mean temperature of about 85 K. Some discs have an inner dust-free zone, forming a ring. The rings may be due to a planet at the inner edge, sweeping up infalling grains, or to grain ejection from the inner region by radiation pressure from the central star. The equatorial discs or rings have a radius of 85–500 AU, where the astronomical unit (AU) is measured by the semi-major axis (radius) of the Earth's orbit round the Sun (1.496×10^8 km). The solar system itself extends out to the planet Pluto at 39.7 AU with a cloud of some 10^{10} comets in a spherical envelope at 10^4–10^5 AU from the Sun.

The process of planetary system formation from a dense molecular cloud has two observational gaps, both spanned by computer modelling. The first gap lies between the radio-frequency and optical evidence for the character of the molecules and grains of the dense clouds and the infrared observations of the equatorial gravel belts surrounding young stars. The second and more singular gap is temporal, separating the extensive physical and chemical observations of the only planetary system for which we currently have direct knowledge—the solar system—and the inferred proto-Sun of some 5 billion years ago, surrounded by the gas and dust of the solar nebula.

Models for the formation of the solar system generally agree that the pre-main-sequence Sun passed through a hot T Tauri stage, with a convective interior, intense ultraviolet radiation emission, and a strong stellar wind. The radiation pressure and stellar wind swept much of the gas and the finer grains first from the inner region of the solar nebular, occupied now by the terrestrial planets, and then from the solar system as a whole. Possibly giant gaseous proto-planets had aggregated gravitationally before this stage, but the inner members were disrupted by the tidal removal of their gaseous atmospheres which were then swept to the outer reaches of the solar system. On this view the terrestrial planets were left as the collapsed involatile cores of the original set of giant gaseous proto-planets, now represented only by Jupiter and the other major outer planets.

Alternatively, the solid grains of the original cloud spiralled down to the central plane of the rotating nebula, forming a flat equatorial dust layer which soon aggregated to gravel belts. Cohesion of the gravel first to metre-sized and then kilometre-sized bodies gave circumsolar rings of

Table 7.1 The mass, size, and orbit of the planets, the Moon, and the Sun, in units of the Earth's mass $(M_E = 6 \times 10^{24}$ kg), equatorial radius $(R_E = 6380$ km), and orbital separation from the Sun $(1 \text{ AU} = 1.5 \times 10^8$ km), together with the density of the body

Body	Mass (M_E)	Radius (R_E)	Solar distance (AU)	Density (g cm^{-3})
Sun	3.3×10^5	1.1×10^2	—	1.41
Mercury	0.056	0.382	0.387	5.44
Venus	0.815	0.949	0.723	5.25
Earth	1.000	1.000	1.000	5.52
Moon	0.012	0.272	—	3.34
Mars	0.107	0.533	1.524	3.94
Asteroids	7×10^{-4}	—	~2.8	—
Jupiter	317.9	11.12	5.204	1.32
Saturn	95.1	9.41	9.55	0.69
Uranus	14.5	3.98	19.21	1.31
Neptune	17.2	3.81	30.11	1.66
Pluto	0.11	0.5	39.44	—

planetesimals, moving in near-circular orbits of varying eccentricity and thus prone to collision. During the removal of gas from the solar system by radiation pressure and the stellar wind, adjacent planetesimals with low relative velocities cohered on collision, forming a larger body that continued to grow until the local supply of planetesimals was exhausted. The final size of a planet depended upon the supply of materials and the competition of its neighbours (Table 7.1). The giant volatile-rich planet, Jupiter, must have formed largely before the gas was swept from the solar nebula, and its competition for the nearby planetesimals accounts for the gross depletion of mass in the asteroid region of the solar system and for the small mass of Mars relative to the Earth and Venus (Wetherill 1980).

7.2 The solar nebular condensation

Models for the condensation of the solar nebula aim to account for the large density difference between the terrestrial planets orbiting between the Sun and the Asteroid belt and the giant outer planets (Table 7.1), and for what is known of the chemical composition of the bodies within the solar system. The models assume that the initial solar nebula was made up of the chemical elements with their present relative abundances. The elements were available as the molecules and the dust of the parent dense molecular cloud in the outer reaches, but mainly as the dissociated free atoms and

Table 7.2 The relative molecular composition ($[H_2] = 1$) of the atmospheres of the outer planets (Owen 1985) and their black-body radiation temperature, measured from the infrared emission (IR $T(K)$)

	Jupiter	Saturn	Uranus
IR $T(K)$	128	98	56
Hydrogen H_2	1	1	1
Methane CH_4	$\sim 2 \times 10^{-3}$	10^{-3}	0.03
Ammonia NH_3	2×10^{-4}	2×10^{-5}	—
Water H_2O	10^{-6}	—	—
Phosphine PH_3	4×10^{-7}	3×10^{-6}	—
Carbon monoxide CO	3×10^{-9}	trace	$<2 \times 10^{-4}$
Hydrogen cyanide HCN	2×10^{-9}	$<7 \times 10^{-9}$	—

ions in the hotter central region, corresponding approximately to that of the terrestrial planets.

Early models, beginning notably with Urey (1952c), were based on the assumption that the initial nebula was homogeneous and in thermodynamic equilibrium, with a gradient of decreasing temperature and pressure away from the Sun in the plane of the nebula disc and in the outward directions perpendicular to the disc. In the outer regions the dust, gas, and ices of the nebula condensed to form the giant planets with their atmospheres of molecular hydrogen and the hydrides of the lighter elements (Table 7.2). The chemical composition of the outer giant planets, although variably depleted in the light elements to some degree, lies closer to the solar system abundance of the elements than that of the terrestrial planets.

The major chemical segregation of the inner region begins with the expulsion by the intense proto-solar wind and radiation pressure of hydrogen, helium, and the other inert gases in approximate inverse proportion to their atomic weights. Other elements become depleted in proportion to their volatility, but this trend is offset by compound formation, so that they are retained as oxides, nitrides, sulphides, and phosphides, and by the survival of undissociated particles of graphite, carbides, silica, and other materials in the cosmic dust.

As the inner nebula cooled down from around 2000 K materials slowly condensed out at equilibrium with the gases of the nebula in the following sequence (Lewis and Prinn 1984):

1. Above 1600 K, refractory metal oxides, e.g. CaO and Al_2O_3, forming mineral assemblies such as $CaTiO_3$, $Ca_2Al_2SiO_7$, $MgAl_2O_4$, and the rare earth oxides Ln_2O_3 (where Ln is a lanthanide element).
2. Near 1400 K, metallic Fe–Ni alloy, containing small amounts of the

siderophile elements carbon, nitrogen, phosphorus, cobalt, and heavier transition metal elements.

3. Around 1350 K, the reaction of SiO with magnesium and water gases to give $MgSiO_3$ (enstatite).

4. Near 1000 K, the reaction of atomic alkali metal vapours with the calcium aluminosilicate formed at higher temperature to produce the alkali aluminosilicates.

5. Below 680 K, the reaction of metallic iron with the gaseous hydrogen sulphide of the nebula to give FeS (troilite) and with the water vapour to form FeO-containing silicates, such as pyroxene $(Mg,Fe)SiO_3)$, from the enstatite produced at higher temperatures.

6. The hydration of the higher-temperature minerals by water vapour to produce first tremolite $(Ca_2(Mg,Fe)_5(Si_2O_{11})_2(OH)_2)$ near 400 K and then the more hydrated serpentine $((Mg,Fe)_3Si_2O_5(OH)_4)$ around 350 K.

According to the equilibrium condensation model, the overall composition of a given planet is an approximate reflection of the materials that condensed out of the solar nebular at the specified temperature and pressure prevailing at the particular distance of the planet from the Sun. The initially condensed grains were in equilibrium with the gas of the nebula and had the composition required by the particular temperature position in the condensation sequence, which was different for the several terrestrial planets. Once the condensation grains had accreted and were buried inside planetesimals, they were no longer in reactive contact with the gas of the nebula and their overall elementary and mineralogical composition became frozen and remained unaffected by subsequent falls in the temperature of the nebula.

The aggregation of material formed at different heliocentric distances and at different times modifies the general relationship expected between the solar distance and the planetary composition. Subsequent segregation processes additionally affect the relationship, such as the separation of the compositionally inequivalent Moon and the Earth during or after their accretion from a common swarm of homogeneous planetesimals. But the equilibrium condensation model successfully accounts for the general trend of decreasing mass density from Mercury to Mars in the series of terrestrial planets, particularly after correcting the densities to 1 atmosphere pressure, where both the Earth and Venus have uncompressed densities of approximately 4.2 g cm^{-3} with smaller changes for Mercury (5.4 g cm^{-3}) and Mars (3.3 g cm^{-3}), compared with the measured densities (Table 7.1).

Thus, according to the equilibrium model, the materials condensing out at approximately 1400 K to form Mercury were rich in metallic iron–nickel (density about 7.8 g cm^{-3}) and contained most of the available refractory calcium and aluminium compounds but not much magnesium silicate, even though magnesium and silicon have a ten-fold larger solar abundance than calcium and aluminium. On this basis Mercury is expected to have a high density from the large metallic core, the small silicate mantle, and the aluminium-rich crust. A more complete condensation of the magnesium silicates (densities 3.2–3.8 g cm^{-3}) occurred in the region of Venus (~900 K) and the Earth (~600 K) so that these planets have a higher silicate/metal ratio than Mercury and a lower density in consequence. At the position of Mars (~450 K) and the asteroids (~400 K) much of the iron condensed out as troilite (FeS, density ~4.8 g cm^{-3}) or as FeO in the silicates, giving these bodies the smallest mass densities of the terrestrial planets. For each planet the slowly accreting material is taken to be uniform in composition. The subsequent differentiation, by internal melting from the heat of radioactive decay, into a metal core, silicate mantle, and aluminosilicate crust, follows without a change in the overall composition of the once-homogeneous condensed material.

While the equilibrium homogeneous condensation model satisfactorily explains the relative densities of the terrestrial planets, it does not account for the composition of their atmospheres, nor for the hydrosphere in the case of the Earth. At the proposed formation temperature of the uniform materials of the Earth (~600 K), and more particularly Venus (~900 K), neither hydrated minerals nor condensed carbon and nitrogen compounds are expected to be stable enough to provide sufficient water, nitrogen, and carbon dioxide. Hydrogen, already lost from the inner nebula, would be required for the generation of water, and water would be needed for the production of gaseous carbon and nitrogen compounds from carbides and nitrides. Moreover, the oxidized ferric iron Fe(III) of the Earth's crust and upper mantle could not condense homogeneously with the metallic iron Fe(0) of the core, and much of the iron would assume the intermediate ferrous Fe(II) oxidation state.

An alternative model, depending upon a fast heterogeneous accretion, was designed to meet these problems. Condensates formed in the nebula at a given distance from the Sun have different compositions because both the temperature and the pressure decrease with an increasing separation from the mean plane of the disc in the perpendicular directions. At distances remote from the plane of the disc low-temperature materials separate out with a composition resembling the oxidized, volatile-rich, carbonaceous meteorites. These aggregates were at equilibrium with the local gas of the nebula, which was continuously replenished by the infall of relatively unprocessed compounds and dust from the parent molecular cloud. Once the low-temperature grains aggregated into first-generation planetesimals a few centimetres in size, the internal material became chemically isolated from the gas phase, which changed in local composition as the planetesimals fell through the gas to the plane of the disc.

Condensates formed in the plane of the disc at the same heliocentric distance within the inner solar system had a high-temperature composition, metallic iron–nickel, SiO_2, MgO, Al_2O_3, CaO, and other oxides, resembling the stony-iron meteorites. At intermediate distances from the plane the condensates had compositions corresponding to mixtures of the reduced high-temperature component (A) and the oxidized low-temperature component (B), and resembled the stony meteorites, particularly the ordinary chondrites with their millimetre-sized chondrules in a matrix of fine mineral and metal grains, which serve as analogues, if not examples, of the primary condensates. (The constituents of components A and B are listed in Table 9.3.)

The mixture of first-generation planetesimals in the plane of the disc varied from largely component A type at the heliocentric distance of Mercury to mainly component B type in the further limits of the Asteroid belt adjacent to Jupiter. The aggregation to second- and subsequent-generation planetesimals gave the terrestrial planets an overall composition determined by the initial mix. It is estimated that the Earth formed from a mixture of 85–90 per cent of the reduced high-temperature component A and 10–15 per cent of the oxidized low-temperature component B (Ringwood 1984). The model accounts not only for the relative mass densities of the inner planets, but also for the water and the gases of the hydrosphere and atmosphere, and the coexistence of ferric iron in the crust and upper mantle with metallic iron in the core.

The secondary segregation of a heterogeneously accreting body into a dense metallic core, silicate mantle, and lighter aluminosilicate crust took place at an earlier stage than envisaged by the homogeneous equilibrium model. The initial heterogeneous accretion of planetesimals was relatively fast, and the conversion of kinetic energy into heat by collisions melted the interior sufficiently to allow a continuous density segregation of the metallic core during the early accretion. In the case of the Earth at least, the core and mantle are not in equilibrium. The mantle and crust are richer in the highly siderophile elements of component B, such as the platinum metals, than expected from the large liquid iron/silicate partition coefficients of these elements (>50 000). Accordingly the accretion was heterogeneous over both the time of successive aggregations and the spatial region of the aggregation at a given heliocentric distance, with a larger fraction of the component B materials accreting at the later stages.

7.3 Asteroids and comets

The asteroids and their collision fragments, the meteoroids, are the smallest bodies moving in near-circular orbits in the plane of the solar system between the outermost terrestrial planet, Mars, and the innermost giant planet, Jupiter. It is considered that they approximate to surviving representatives of the original volatile-depleted planetesimals of the inner solar system. The volatile-rich comets serve as possible representatives of the planetesimals of the outer solar system, and their degassed extinct cores are a probable source of some carbonaceous meteorites.

The asteroids were the first of the inner planets to cool down, to attain a frozen chemical and mineralogical composition that set the radioisotope decay clocks of recovered meteorites at zero time about 4.55 billion years ago. Chemical segregation and mineralogical differentiation in the chondritic meteorites are only small scale, confined within millimetre-sized chondrules enclosed in a matrix of finer grains, suggesting an origin from small parent bodies. It appears that there never was an original giant asteroid with a fully differentiated core, mantle, and crust, which fragmented into the present-day Asteroid belt, as was once supposed. The largest asteroid, Ceres, discovered by Giuseppe Piazzi (1746–1826) in 1801, has a diameter of only 1020 km. Another 2000 asteroids with diameters greater than 10 km have been catalogued, and it is estimated that there are a further 5×10^5 asteroids or meteoroids with diameters greater than 1 km, together with many more of smaller size. The IRAS survey of infrared sources over 98 per cent of the sky in 1983–4 detected some 2×10^4 minor bodies in the Asteroid belt from their infrared emission of degraded solar energy.

The total mass of all the known asteroids amounts to only a small fraction ($\sim 7 \times 10^{-4}$) of the Earth's mass, supporting the suggested competition from Jupiter for the original planetesimals. The asteroids orbit at distances between 2.1 and 3.5 AU from the Sun and have a distribution of orbital periods of 3–7 years with gaps that indicate the continuing influence of Jupiter. The gaps occur for periods which would be integral fractions, 1/2, 3/7, 2/5, and 1/3, of Jupiter's period. Close approach between the asteroid and Jupiter at these critical values of the orbital period, repeating regularly, would ultimately perturb the asteroid gravitationally into a non-resonant orbit.

The orbits of asteroids with non-resonant periods remain affected to a degree, and the gravitational perturbations of Jupiter continuously provide a supply of asteroids with Earth-crossing orbits, namely the Apollo bodies. The class takes its name from the first one discovered, the Apollo asteroid about 2 km in diameter, detected photographically just inside the Earth's orbit at closest approach to the Sun (perihelion) by Karl Reinmuth at Heidelberg in 1932. The estimated rate of supply of Apollo asteroids and meteoroids, giving a steady-state number of some 2000 with diameters over 1 km, accounts for a substantial fraction of the observed meteorite falls on the Earth.

Meteorite falls observed with an array of two or more telescopic cameras, equipped with a rotating shutter to record a succession of timed locations, show that the orbits of these meteorites are Apollo like, extrapolating back from the Earth into the Asteroid belt at furthest separation from the Sun (aphelion). Analyses of the isotopes produced in the surface layers of meteorites by cosmic-ray exposure give lifetimes of a few million years in Apollo-like orbits after the fragmentation of the parent body.

The reflectance spectra of the asteroids over the ultraviolet, visible, and near-infrared regions, by comparison with the corresponding spectra of recovered meteorites, show similar spectral classes in the two sets of objects. The spectra of the asteroids fall into two main classes, S-type and C-type, resembling the spectra of stony-iron and carbonaceous meteorites, respectively, and a minor M-type, with a metallic reflectance spectrum like the iron meteorites. Asteroids with the C-type spectrum become more frequent as the mean distance of these bodies from the Sun increases. The observation is consistent with the expected fall in the solar nebula temperature with an increasing separation from the proto-Sun, since the chemical composition of the carbonaceous meteorites suggests formation temperatures of only 360–400 K.

Not all meteorites and Apollo bodies have an asteroid origin. Meteor showers of shooting stars originate from cometary material in a stream following the orbit of an active or extinct periodic comet, the showers having the same period. The Orionid meteor stream follows the orbit of the comet characterized as periodic in 1705 by Edmund Halley (1656–1742), and the Taurid streams follow that of the near-extinct comet similarly characterized in 1819 by Johann Encke (1791–1865). The parent body of the Geminid meteor stream, an extinct comet nucleus in an Apollo-type orbit, was discovered in 1983 by the IRAS survey.

The majority of comets are not periodic. They come from regions well beyond the outermost planet Pluto, at the rate of about four per year, in both direct and retrograde parabolic orbits at all angles of inclination to the mean plane of the planetary motions and they do not reappear. This distribution led J. H. Oort and E. J. Opik to propose in 1950 the presence of a spherical shell of some 10^9–10^{11} cometary bodies in circular orbits between 10^4 and 10^5 AU from the Sun. The comets of the cloud are occasionally perturbed by passing stars into new orbits that bring them into the inner solar system. Some comets may be further perturbed by close approach to the giant planets and become periodic with elliptical orbits either elongated, like Halley's comet with an average period of 77 years and an aphelion beyond Neptune, or Apollo type, like Encke's comet with a mean period of 3.3 years and an aphelion inside the orbit of Jupiter.

8

Meteorite and comet constitutions

8.1 The characterization of meteorites

Although recovered meteorites were long regarded as remarkable objects, their extraterrestrial origin was not widely accepted much before 1800 (Burke 1986). Meteorites appear alongside fossils as curios in the mineral collections of the eighteenth century, and recovered specimens from observed falls, like that of 1794 at Sienna in Italy, were attributed to essentially terrestrial agencies, following the tradition of two millenia. Such agencies might be the activity of a distant volcano, or concretions in the atmosphere of effluvia from subterranean sources, due to lightning or a hurricane.

Aristotle (384–322 BC) had argued that since meteors and comets are subject to generation and decay, like all mundane entities, they must be sublunary phenomena, originating from fire, the lightest of the four terrestrial elements. The heavenly bodies, composed of the superior fifth element the quintessence, moved eternally with uniform motion in the perfection of circular orbits impelled by spirit-movers. The main tenets of the hierarchical two-tier world system of Aristotle, taken over and amplified by the Scholastic philosophers from the thirteenth century on, were replaced during the scientific revolution of the sixteenth and seventeenth centuries, but details survived. Tycho Brahe tracked the orbit of a comet in 1577, showing that it cut across the supposedly crystalline quintessential spherical shells of the heavens, and his observations of the supernova of 1572, like those of Kepler on the supernova of 1604, established that the heavens too are subject to generation and decay. While the comets now took their place in the heavens, the faster-moving meteors and fireballs were still regarded as essentially terrestrial phenomena.

The democratization and mechanization of the universe were prerequisites for the unification of celestial and terrestrial mechanics. The unification relieved the universe of its angelic motors, needed to sustain the motions of the medieval cosmos, and they were transmogrified into the mundane creatures inhabiting many other worlds, each equivalent in all essential respects to our own (Dick 1982). During the nineteenth century the extraterrestrial creatures in turn fell from grace in serious scientific opinion to become merely microbes, although they regained something of their angelic ancestry as the seeds of life, and the recognition from early in the century that meteorites are of extraterrestrial origin provided these seeds with vehicles for their panspermatic journeys throughout the depths of space and aeons of time.

The question of the origin of meteorites was raised influentially by the physicist Ernst Florenz Chladni (1756–1827), noted for his work on acoustics, who travelled widely collecting meteorite specimens. In 1794 Chladni published a book arguing that the recovered falls and the several known masses of 'native iron' had a cosmic origin: these must be the residues of extraterrestrial fireballs since they are coated with a vitreous slag-like skin, indicative of intense heating.

The then-recent chemical revolution of Lavoisier stimulated the extension of chemical analysis to all accessible substances, and it was soon shown that the meteorites had a distinctive and common composition, unlike that of any known terrestrial materials, supporting Chladni's view. The first important analyses were carried out in 1799–1802 by Edward Charles Howard (1774–1816), a junior scion of Norfolk lineage who turned to chemistry and made a fortune from his sugar-refining innovations. His study, with the collaboration of a French émigré, Jacques-Louis Bournon (1751–1825), was organized by the president of the Royal Society of London, Joseph Banks (1743–1820), who was himself a noted collector of minerals, fossils, and meteorites.

Howard's analytical survey with Bournon's mineralogical report of 1802 covered two irons, two stony-irons, and four stones (ordinary chondrites, the most common form of meteorite). From ground-up specimens of the stones and stony-irons, four main mineral constituents were separated and individually analysed. These were: first, metal particles, separable by means of a magnet; second, red–yellow iron sulphide, FeS, later named troilite; third, millimetre-sized glassy brown globules, the chondrules, recognized for the first time; and last, a white-to-grey earthy matrix 'which serves as a kind of cement to unite the others'. The most striking result came from the analysis of the magnetic iron particles and the bulk material of the iron meteorites, which turned out to contain a large amount of nickel, from 9 per cent to as much as 28 per cent (at the time nickel was regarded as a rare element). The troilite, FeS, was singular too, the most common terrestrial analogue being 'fool's gold', pyrite, FeS_2.

Howard's analyses, which included the silica (SiO_2, 45–50 per cent), magnesia (MgO, 17–24 per cent), and iron oxide (FeO, 32–42 per cent) content of the chondrules and matrix, were soon confirmed. M. H. Klaproth (1743–1817) in Berlin and N. L. Vauquelin (1763–1829) with A. F. Fourcroy (1755–1809) in Paris extended the analyses to other meteorite specimens, particularly those from the fall of 1803 at L'Aigle in France, observed by a member of the Academy of Sciences. These analyses, showing a substantially higher magnesia and iron oxide content in stony meteorites than in crustal terrestrial rocks, together with the nickel in the metallic iron and the unusual sulphide, FeS, established the extraterrestrial origin of the meteorites during the period 1800–10 (Sears 1975).

The further characterization of the meteorite components followed from advances in chemistry, mineralogy, and metallurgy. Polished sections of iron meteorites etched with nitric acid show a surface repeat pattern of more and of less reflective bands, observed by Howard and other early workers but named after Aloys von Widmanstatten (1754–1849) who investigated the structures in 1808. The bands consist of two different iron–nickel alloys, kamacite (<7.5 per cent nickel) and taenite (>25 per cent nickel). The latter is the stable form for all compositions at high temperature (>1200 K). Since the composition of most iron meteorites lies between the two low-temperature stability limits, layers of kamacite, typically about 1 mm thick, separate out parallel to the faces of the cubic taenite crystal as the alloy mass cools. Two sets of bands appear on the etched polished surface of an iron meteorite, intersecting at a right angle if the section is cut parallel to the (100) face of the cubic taenite, or at 60° if parallel to the (111) face, or at intermediate angles for arbitrary sections, as shown in 1885 by Gustav Tschermak (1836–1927) at Vienna.

The kamacite bands in the iron meteorites are not uniform in composition across the thickness of the layer, exhibiting a minimum nickel content at the centre with maxima at each layer boundary. The variation of the nickel content across the M-shaped composition profile depends upon the rate of cooling of the nickel–iron mass. In recent years the composition variation over the surface bands in the Widmanstatten pattern of iron meteorites has been measured by electron microprobe analysis, in which an electron beam approximately 1 μm (10^{-6} m) wide scans across a specimen to excite the characteristic X-ray fluorescence lines of the elements at each micrometre-diameter spot selected. The nickel-content profiles across the kamacite layers of many iron meteorites correspond to cooling rates of about 1 K each million years. This estimate corresponds in turn to the cooling rate of an iron–nickel core in a parent body of some 300 km radius. Only two present-day asteroids, Ceres and Pallas, have larger radii and 10 or more parent bodies are required to account for the diversity of iron meteorite types, probably with radii nearer 100 than 300 km (Wasson 1985).

The diversity of minerals, element composition, and isotope ratios in the major group of meteorites, the stones, requires some 20–70 parent bodies of a small size (Dodd 1986). The small size follows from the observation that the meteoritic minerals are all low-pressure forms, e.g. graphite rather than diamond in the case of carbon, apart from some evident cases of high-pressure shock from collisions. Most of the stones, the ordinary chondrites, have an average composition of some 40 per cent of olivine $(Mg,Fe)_2SiO_4$; 30 per cent pyroxene, $(Mg,Fe)SiO_3$; 5–20 per cent of nickel–iron alloy; 6 per cent troilite, FeS; and about 10 per cent of the layered aluminosilicate, plagioclase (Table 8.1). The chondrules of these meteorites consist mainly of olivine and/or pyroxene. Both olivine and pyroxene are low-pressure forms, transformed by pressure into the respective higher-density forms of ringwoodite and majorite.

Table 8.1 The common alloys and minerals of meteorites

Kamacite	α-iron, with less than 7.5 per cent nickel; body-centred cubic structure
Taenite	γ-iron, with more than 25 per cent nickel; face-centred cubic structure
Troilite	stoichiometric iron sulphide, FeS; approximate hexagonal close-packed structure; dimorphic with hexagonal and cubic crystal forms
Magnetite	magnetic iron oxide, Fe_3O_4; cubic crystals
Olivine	orthorhombic crystal series, $(Mg,Fe)_2SiO_4$, containing individual orthosilicate tetrahedra, intermediate between Fe_2SiO_4 (fayalite) and Mg_2SiO_4 (forsterite), typically 60–85 per cent of the latter in the solid solution
Pyroxene	$(Mg,Fe)SiO_3$, intermediate between $MgSiO_3$ (enstatite) and $FeSiO_3$ (ferrosilite); made up of continuous single chains of silcate tetrahedra, each sharing two oxygen atoms. Dimorphic with orthorhombic (orthopyroxene) and monoclinic (clinopyroxene) crystal forms. Cation substitution common, e.g. 10 per cent calcium in pigeonite (monoclinic)
Plagioclase	triclinic crystal series intermediate in composition between $Na[AlSi_3O_8]$ (albite) and $Ca[Al_2Si_2O_8]$ (anorthite); aluminosilicate sheet structure in which each tetrahedral anion shares three oxygen atoms. Hydrolysed to clay minerals, e.g. kaolinite, $Al_2Si_2O_5(OH)_4$
Serpentine	water-bearing layer silicate from the hydrolysis of olivine or pyroxene; composed of continuous sheets of silicate tetrahedra each sharing three oxygen atoms; e.g. $(Mg,Fe)[Si_4O_{10}(OH)_8]$.

The minerals of the meteorites, like their terrestrial analogues, were characterized following initial elementary analysis by the external morphology of the crystal form, i.e. the symmetry class determined from the interfacial angles. Many of the minerals turned out to have a variable composition, forming an isomorphous series of crystals between end members with the same crystal symmetry; olivine between Mg_2SiO_4 (forsterite) and Fe_2SiO_4 (fayalite); and pyroxene between $MgSiO_3$ (enstatite) and $FeSiO_3$ (ferrosilite). In both series cation substitutions are common, particularly by calcium, and dimorphism in the case of pyroxene gives two subseries, one of orthorhombic crystal form (orthopyroxene) and the other monoclinic (clinopyroxene).

The crystal structures of the silicate minerals were pioneered in the 1920s by the X-ray diffraction studies of William Lawrence Bragg (1890–1971) and Linus Pauling, who distinguished the silicate classes by the O/Si ratio: 4/1 for the orthosilicate or neosilicates, made up of independent $[SiO_4]^{4-}$ tetrahedra, as in the olivines; 3/1 for the metasilicates or inosilicates, made up of continuous single chains of tetrahedra each sharing two oxygens, such as the pyroxenes; 5/2 for the phyllosilicates, consisting of continuous sheets of tetrahedra each sharing three oxygens, as in the serpentines; and 2/1 for the tektosilicates where the tetrahedra each share all four oxygen atoms, forming a continuous three-dimensional framework, e.g. quartz.

8.2 Meteorite distribution

Many recovered meteorites are 'finds', discovered many years after their fall, as opposed to 'falls' which are sought out and found shortly after an observed fall. The stones are not particularly conspicuous on the ground to the casual observer and a substantial fraction (47 per cent) of recovered stony meteorites are 'falls', searched for after the visual display. The irons are so unusual, however, that most of them recovered (96 per cent) are 'finds' (Table 8.2). Only a very small fraction of meteorite falls are ever recovered, since the majority fall into the oceans, deserts, and other un-inhabited regions. The extrapolation of data from the area of sky scanned by the Canadian telescopic camera network indicates that about 19 000 meteorites greater than 100 g in weight fall on the Earth annually: of these some 4100 are estimated to weigh over 1 kg and 830 over 10 kg.

Until recently the rate of recovery of meteorites from a given area has shown an approximate proportionally to the population density of the region, although influencd by the local scientific interest. Historically speci-mens from 63 falls have been recovered in France over the past two and a half centuries, the highest rate worldwide, although the rate corresponds to only about one in a million of the falls potentially recoverable. Of the 63 falls recovered, 27 were made in the particular period 1800–50 when the chemical analysis of meteorites was of especial interest because of the

Table 8.2 Summary of classified meteorites (Graham *et al.* 1985)

Type	Total	Falls	Finds
Irons (14 classes)	725	42	683
Stony-irons (3 classes)	73	10	63
Achondrite stones (5 classes)	132	69	63
Ordinary chondrites (5 groups)	1681	784	897
E group (enstatite, $MgSiO_3$)	31	16	15
H group (high iron, 4 types)	681	276	405
L group (lower iron, 4 types)	669	319	350
LL group (lowest iron, 4 types)	96	66	30
C group (carbonaceous, 6 types)	67	35	32
Total authenticated meteorites	2611	905	1706

increasing certainty of their extraterrestrial origin and the widespread belief in a plurality of inhabited worlds, if only of primitive lifeforms.

In 1969 a Japanese expedition discovered that the Yamato Mountains in the Enderby Land region of Antarctica is a rich repository of meteorites which had fallen in the past. Subsequently other localities, notably the Allan Hills of Victoria Land on the opposite side of Antarctica, were found similarly to be ample sources of ancient meteorites. The fourth edition of the British Museum *Catalogue of Meteorites* (Graham *et al.* 1985) lists a grand total of 2784 meteorites, with the comment that the meteorites found in Antarctica since 1977 are too numerous for inclusion, apart from exceptional specimens (>500 g). During the 1977–8 season 300 specimens were recovered from the Allan Hills, and in the following season more than 3300 meteorite samples were found in the Yamato Mountains area alone.

Meteorites falling in Antarctica are soon covered in snow and preserved in the ice continuously formed and augmented in the central areas, where the rate of snowfall exceeds the rate of sublimation. The accumulation of mass produces a glacial outward flow of ice under its own weight, carrying the meteorites enclosed and other debris, until a rock barrier is reached near to the coastal regions. The upwelling of the glacial ice at the barrier exposes the surface ice to the continuous winds that ablate the ice sheet, not only by promoting sublimation but also by the scouring action of airborne ice particles. The enclosed meteorites and any terrestrial rocks carried by the glacier are then deposited in the zone of ablation at the rock barrier.

Meteoroids in orbit are exposed to cosmic-ray particles which continuously produce radioactive and stable isotopes in the surface layers. Production ceases with the fall of the meteorite, and the decay of the radioisotopes measures the interval between the fall and the recovery of the specimen. The Antarctic meteorites so far dated fell during the past million years. Earlier falls probably lie deeper in the ice, possibly going back as far as 16 million years, the estimated age of the Antarctic ice-cap.

The Antarctic meteorites for the most part correspond to the classes earlier established from falls recovered elsewhere. One specimen has an unusual composition, corresponding to that of the lunar rocks brought back from the Apollo and Luna landings on the Moon. The specimen bears evidence of impact shock, and it may have been ejected from the Moon as a result of a meteoroid impact. Another small group of meteorites collected outside Antarctica, the shergottites, nakhlites, and the Chassigny fall (the SNC achondrite stones), came from a larger parent body active volcanically 2 billion years ago, and they are considered to have originated from Mars following a similar meteoroid collision. The analytical data obtained by the Viking landers show a close match between the composition and gas content of the SNC achondrite stones and of the Martian soil and atmosphere.

While most of the stony meteorites are chondrites, characterized by small silicate spheroids in a fine-grained matrix that was never molten, a small group of achondrite stones were once liquid and crystallized from the melt, like the irons. The achondrites include the meteorites of lunar and Martian origin, and they resemble the basalt rocks of the Earth and the Moon. The more numerous chondrites are classified by their fractional iron content. Studies during the nineteenth century suggested that the total iron fraction of a chondrite is approximately constant, a smaller amount of metallic iron being compensated by a larger amount of ferrous oxide combined in the silicates. In 1916 George Prior (1862–1936) at the Natural History Museum in London proposed the rules that 'the less the amount of nickel–iron in chondritic stones, the richer it is in nickel and the richer in iron are the magnesium silicates'.

Subsequent investigation of Prior's rules, principally by Urey and Craig in 1953, showed that the total iron-to-silicon ratio is not constant for the chondrites but falls into two main groups of high iron (Fe/Si 80 at.%) and low iron (Fe/Si 57 at.%). Later a subsidiary LL group with even lower iron content was added (Fe/Si 53 at.%), together with the E group where the iron is present largely as metal with virtually no FeO, the silicate being enstatite, $MgSiO_3$. In contrast the C group of carbonaceous chondrites contain little metal, and most of the iron is oxidized in the form of ferrous silicate or magnetite, Fe_3O_4.

8.3 The carbonaceous meteorites

With the rise of organic chemistry, following the introduction of new analytical methods by Justus Liebig (1803–73) at Giessen from about 1825, the black friable meteorites containing organic material attracted particular attention, due not least to their bearing on the possibility of extraterrestrial life. Four of the 12 carbonaceous chondrites recovered during the nineteenth century fell in France, at Alais (1806), Orgueil (1864), Ornans (1868), and Lancé (1872), and specimens were much in demand for analysis. Thenard and Vauquelin immediately analysed the Alais meteorite, reporting the presence of substantial carbon (2.5 per cent). Berzelius at Stockholm, examining Alais specimens in 1834, was surprised to find internal water and hydrated silicates analogous to clay minerals, together with a carbonaceous material resembling the humus of terrestrial soils; but, Berzelius concluded, the organic matter probably did not have a biological origin.

An opposite view was taken by his former student, Friedrich Wöhler (1800–82) at Göttingen, who delighted in meteorite analysis and obtained specimens of the carbonaceous stones that fell at Cold Bokkeveld (1838) in South Africa and at Kaba (1857) in Hungary. In 1859 Wöhler and his student Moritz Hörnes (1815–68) found soluble organic material in the specimens, as well as insoluble polymer, and they suggested that these heat-sensitive substances in the interior of the meteorite were insulated by the outer layers and remained cold during the fall. The Orgueil meteorite was examined soon after its fall in 1864 by Stanislas Cloez (1817–83), who reported a carbon–hydrogen analysis for the peat-like material the meteorite contained. Reduction of the material with hydrogen, described by Marcellin Berthelot (1827–1907) in 1868, afforded a series of paraffin hydrocarbons, similar to petroleum oil.

Interest in the carbon compounds of meteorites declined in the late nineteenth century when it became clear that a wide range of organic compounds could be synthesized in the laboratory without the intervention of vital biological agencies. For some years the synthesis of urea from ammonium cyanate by Wöhler in 1828 was regarded only as a singular curiosity without general significance. But the systematic syntheses of simple organic compounds from carbon itself by Berthelot (1860) convincingly generalized the abiotic approach to organic synthesis, and by the end of the century many chemists had adopted the view that even petroleum oil has an inorganic origin, from the action of water on metal carbides in the Earth.

During the second half of the twentieth century, a particular interest in the carbonaceous meteorites revived, with planned and then realized space exploration projects, geochemical studies of the Earth's early biosphere, and laboratory simulations of prebiotic chemistry. These studies were all underpinned by a range of new or much-refined analytical techniques in chromatography, mass spectrometry, radiochemistry, with X-ray and electron microscopy and electron microprobe analysis. A new period began in 1969 with the return of specimens for analysis from the first landing on the Moon (Apollo 11), two large falls of carbonaceous chondrites, one over 2000 kg at Allende in Mexico and the other of about 500 kg at Murchison in Victoria, Australia, and the discovery of the rich accumulation of ancient meteorites in Antarctica.

Initially carbonaceous meteorites were classified according to their bulk water and carbon content into three types, C1 ($\sim$20 per cent H_2O; 3–5 per cent C), C2 ($\sim$10 per cent H_2O; 1–3 per cent C), and C3 ($\sim$2 per cent H_2O; <1 per cent C). Subsequently these were subdivided into six types based upon petrological properties, such as the chondrule form, and named after a representative example with the original numerical affix indicating the water and carbon content. Thus the refractory-rich CV types (e.g. Allende, CV3) are named after the Vigarno fall in Italy (1910); the CM2 type (e.g. Murchison) after the C2 Mighei stone which fell in the Ukraine (1889); and the volatile-rich CI type, equivalent to C1, after the Ivuna fall in Tanzania (1938). The CV types are distinguished by large chondrules ($\sim$1 mm) from the CM and CO (Ornans fall) types, where the chondrules are small ($\sim$0.2 mm), and the CI, which virtually lack chondrules (<1 per cent).

The C1 carbonaceous chondrites are the least depleted in the volatile chemical elements of all the meteorite classes. They are taken to be the most primitive material known in the solar system from the observation of a 1:1 relation between the relative abundances of the less-volatile elements in the C1 meteorites and the corresponding solar abundances (Fig. 4.3). In addition the relation suggests that the C1 carbonaceous meteorites originally condensed from a homogeneous solar nebula. The analyses of the oxygen isotope ratio $^{18}O/^{16}O$ in the dolomite CaMg $(CO_3)_2$ and in the hydrated silicate (serpentine) of C1 meteorites indicate that they were formed at a temperature of approximately 360 K from comparisons with the isotope fractionation produced by equilibration of these minerals with water as a function of temperature in the laboratory. Other 'cosmothermometers', such as the reaction of olivine with water to form serpentine, give a similar formation temperature of around 360 K for C1 chondrites, compared with approximately 380 K for the C2 and about 420 K for the C3 type.

Subsequent extensions of oxygen isotope analysis to the anhydrous minerals of the C2 and C3 carbonaceous meteorites show that the original solar nebula could not have been wholly homogeneous, owing to the survival of a small presolar ^{16}O-rich reservoir (Clayton *et al.* 1988). The three stable isotopes of oxygen have the mean abundance ratios (R_i, for the heavier isotope, *i*) of $^{17}O/^{16}O = 3.91 \times 10^{-4}$ and $^{18}O/^{16}O = 2.055 \times 10^{-3}$. As isotopic variations are small, isotope abundance data are usually expressed as δ-values in parts per thousand (per mil, $^o/_{oo}$) relative to a standard which, for hydrogen and oxygen isotopes, is standard mean ocean water (SMOW). In general the $\delta(^iZ)$ for the heavier isotope *i* of the element with atomic number Z is defined by

$$\delta(^iZ) = [R_i(\text{sample})/R_i(\text{standard}) - 1] \times 1000.$$

In the case of oxygen, reaction processes under given conditions produce physico-chemical mass fractionations which are approximately twice as large for $\delta(^{18}O)$ as $\delta(^{17}O)$, so that a plot of $\delta(^{17}O)$ versus $\delta(^{18}O)$ for a set of abundance data is expected to be linear with a slope of 0.5. Such plots for the oxygen isotope ratios of all terrestrial and lunar minerals and the large majority of meteoritic specimens conform to the expectation, giving a linear relationship with a mean slope of 0.516 in the middle of the calculated range 0.50–0.53. A small fraction (1–2 per cent) of the anhydrous minerals contained in refractory white inclusions in C2 and C3 carbonaceous meteorites, up to 1 cm across in the CV3 type (Allende), are anomalous, however, and give the corresponding linear relationship a mean slope near unity (0.94). The minerals are standard materials, olivine $(Mg,Fe)_2SiO_4$ and pyroxene $(Mg,Fe)SiO_3$ or spinel $MgAl_2O_4$, and the oxygen isotope anomalies are confined to the refractory white inclusions, rich in the calcium and aluminium oxide minerals. Outside the inclusions, in the chondrules and the matrix, the olivine and pyroxene have their normal oxygen isotope composition.

The mean slope near unity in the relationship between $\delta(^{17}O)$ and $\delta(^{18}O)$ indicates that the anomalous minerals of the refractory inclusions are enriched in ^{16}O relative to the other two isotopes, marking an exception to the general solar system oxygen isotope abundance distribution. The refractory inclusions could not have been vaporized or even molten in the solar nebula, and they represent presolar material, generated in an oxygen reservoir enriched in ^{16}O and never mixed in with the bulk of the solar nebula matter to give isotopic uniformity. If the formation of the solar system was triggered by a supernova explosion, the ^{16}O-enriched reservoir and the refractory condensate were the product of a different and probably earlier event of explosive helium, carbon, or oxygen burning.

The hydrogen isotope ratios of the organic substances in the C1 and C2 carbonaceous meteorites additionally suggest the survival of presolar material, possibly organic compounds from the parent molecular cloud. Deuterium, with twice the mass of its isotope, hydrogen, readily gives large fractionation ratios, and the terrestrial standard mean ocean water (SMOW) is enriched four-fold in deuterium with respect to galactic hydrogen, owing to the more facile escape of hydrogen from the Earth's gravitational field. The insoluble carbonaceous polymer (kerogen) of some C2 meteorites is enriched in deuterium by a factor of four relative to SMOW, i.e. 16-fold with respect to galactic hydrogen. The microwave spectra of the dark interstellar clouds indicate that molecules such as hydrogen cyanide (HCN) and formaldehyde (HCHO) in the clouds are enriched in deuterium by factors as high as 10^5, owing to the slightly larger effective bond strength (smaller zero point vibrational energy) of C—D relative to C—H bonds. The survival of a small fraction of such deuterium-enriched molecules as polymeric reaction products accounts for the high D/H ratios of the kerogen in the carbonaceous meteorites (Lewis and Anders 1983).

The deuterium-enrichment of the soluble organic substances found in the C1 and C2 meteorites (Table 8.3) is smaller than that of the insoluble kerogen by generally less than one-half. Not only the D/H ratio but also the $^{13}C/^{12}C$ and the $^{15}N/^{14}N$ ratios are enhanced for the amino acid extracts from the Murchison meteorite, suggesting that the molecules of the interstellar clouds provided a feedstock for their production, at least in part (Epstein *et al.* 1987). Alternative evidence indicates that the amino acids and many of the other soluble organic compounds present in the C1 and C2 meteorites may have been produced on the surface of the meteorite parent body from the hydrogen, carbon monoxide, and ammonia of the solar nebula, catalysed by the magnetite Fe_3O_4 and the hydrated silicates (serpentine) found in these meteorites (Anders and Hayatsu 1981). The reactions involved are the Fischer–Tropsch type (FTT).

The FTT reaction was discovered in 1818 by Döbereiner, who passed

Table 8.3 Distribution of carbon in the Murchison CM2 meteorite (Wood and Chang 1985)

Substance	Abundance
Insoluble carbonaceous phase	1.3–1.8 per cent
Carbonate and CO_2	0.1–0.5 per cent
Aliphatic hydrocarbons	12–35 ppm
Aromatic hydrocarbons	15–28 ppm
Monocarboxylic acids (C_2—C_8)	~170 ppm
Hydroxy acids (C_2—C_5)	~6 ppm
Amino acids	10–20 ppm
Alcohols (C_1—C_4)	~6 ppm
Aldehydes (C_2—C_4)	~6 ppm
Ketones (C_3—C_5)	~10 ppm
Ureas	~20 ppm
Amines (C_1—C_4)	~2 ppm
Pyridines and quinolines	0.04–0.40 ppm
Pyrimidines	~0.05 ppm
Purines	~1 ppm
Polypyrroles	≪1 ppm
Sum	1.43–2.35 per cent
Total carbon	2.0–2.5 per cent

steam through a heated iron tube containing glowing charcoal to obtain, initially, carbon monoxide and hydrogen, together with magnetite, Fe_3O_4, which catalysed the subsequent formation of an oily product from the water–gas (CO/H_2) mixture. Studies of the reaction were taken up in 1902 by Paul Sabatier (1854–1941) and Jean-Baptiste Senderens (1856–1936), who found that the products varied with change of catalyst. From 1922, Franz Fischer (1877–1947) and Hans Tropsch (1889–1935) developed an industrial process for the production of synthetic petroleum from water–gas mixtures with a variety of metal oxide and silicate catalysts.

The FTT reactions are versatile, giving alcohols, acids, and other carbonyl compounds, including hydroxyacids, as well as aromatic and aliphatic hydrocarbons, mainly normal paraffins initially but reformed to branched chains and olefins on prolonged contact with the catalyst (Anderson 1984). The addition of ammonia to the water–gas feedstock results in the formation of amines, amino acids, and N-heterocyclic compounds, including the pyrimidine and purine nucleotide bases and porphyrin-like polypyrrole pigments. The range of amino acids afforded by the FTT reactions is not large, but it includes the aromatic amino acids which are not produced by the main alternative route for prebiotic syntheses on the surface of the

Earth or the parent bodies of the meteorites, namely the Miller–Urey electric discharge reactions (Section 13.1).

The passage of an electric discharge through CH_4–NH_3–H_2O or other reducing gas mixtures over an aqueous phase produces some 35 different amino acids, and the relative yields of 18 of these match the relative abundances of the corresponding amino acids isolated from the Murchison meteorite (Miller 1984). Some 55 amino acids have been found in C1 and C2 meteorites, but only 8 of the 20 protein amino acids are detected, and all of the chiral molecules isolated are racemic mixtures.

The isotopes of carbon in carbonaceous meteorites are substantially fractionated between the inorganic carbonate and the organic material overall, and additionally between the different classes of organic substances. The isotopic ratio ($^{13}C/^{12}C$) is higher in the carbonate than in the averaged organic carbon by some 60–80 parts per thousand ($\triangle\delta^{13}C \sim 60$–$80\%_{00}$). The FTT reaction carried out in the temperature range 375–400 K, indicated by the 'cosmothermometers' for the formation of the carbonaceous meteorites, gives a carbon isotope fractionation of the same sign and magnitude between the carbon dioxide and the hydrocarbons produced. The range of carbon isotope fractionations over the different classes of soluble organic compounds isolated from the Murchison meteorite is large, however, with a spread of $\triangle\delta^{13}C \sim 60\%_{00}$ between the 'heavy' amino acids and the 'light' benzene fraction.

Studies of the isotopic fractionations of the substances found in carbonaceous chondrites lead to the conclusions that unchanged presolar materials are a significant component of these meteorites, and that no single reaction mechanism accounts for the range of organic substances they contain. Much of the soluble organic matter is trapped in mineral matrices, probably between the layers of hydrated silicate minerals, where the original organic substances, including presolar materials, underwent hydrothermal change (like the minerals themselves) on the parent bodies of the carbonaceous chondrites (Mullie and Reisse 1987).

8.4 Composition of the comets

Comets usually become visible when they move into the inner solar system, at about 3 AU from the Sun. As a comet approaches the Sun, the visible components are seen as an approximately spherical head, the coma made up of dust and gas streaming out to 10^4–10^5 km, and two tails, each some 10^6–10^8 km long. These tails are always directed away from the Sun, and so point ahead when the comet is outward bound. One tail consists of ionized gas or plasma and follows a linear radial path from the Sun. The other tail is made up of dust particles in a path bent towards the Sun by the latter's gravitational attraction.

Early knowledge of the composition of the head and the tails of comets, studied spectroscopically first in the visible and ultraviolet range and then in the infrared and radio regions, was confirmed and greatly extended by the close encounter of five spacecraft with Halley's comet in March 1986, a month after it began its outward journey from the Sun (comet Halley, 1986). The composition of the comet was probed during the close approach of three of the spacecraft, Giotto approximately 600 km, Vega 1 around 9000 km, and Vega 2 about 8000 km. These three carried mass spectrometers for the analysis of both the gas and the dust, together with a range of other sensors. The other two, Suisei and Sakigaki, were designed to probe the outer atmosphere of the comet and its interaction with the solar wind and magnetic field, venturing to within 15×10^4 and 7×10^6 km of the nucleus, respectively.

The sensors aboard the spacecraft showed the nucleus of comet Halley to have an ellipsoidal form some $16 \times 8 \times 8$ km^3 with regular brightness variations of 2.2 and 7.4 day periodicity, ascribed to rotations around the minor and major axes. The nucleus has a black crust, probably carbonaceous, at a surface temperature of 330 K. The crust appeared to be penetrated by a dozen or more point craters and rifts (possibly lines of craters), emitting jets of gas and dust from the interior. The neutral molecules in the gas are largely water vapour (80 per cent by volume), ammonia (10 per cent), methane (up to 7 per cent), and carbon dioxide (up to 3.5 per cent). At the time of the Giotto encounter, the estimated output of water was some 15 000 kg s^{-1}. Close to the comet, in the head, ions derived from water predominate: $[H(H_2O)_n]^+$, $n = 0, 1, 2, 3$, together with H^+, H_2^+, C^+, CH^+, CC^+, CO^+, and smaller abundances of Na^+, S^+, and Fe^+.

The amount of dust in the jets from the craters in the nuclear crust of the comet varied with the time of measurement from 10 to 25 per cent of the mass of gas released. At peak jet activity, it is estimated that the dust output was some 10^4 kg s^{-1}. The abundance of dust particles increased with decreasing size, down to the 10^{-17} g limit of measurement, reflecting a progressive breakdown of the loosely bound larger particles. The first particles encountered, at 637 000 km by the inbound Vega 1, had the smallest of the masses measured, and the coarser grains were found only at closer approach. Giotto encountered particles up to 40 mg in mass and probably one even larger (100 mg).

Analysis of the dust particles by the mass spectrometers showed the grains to be composite, consisting of a loosely aggregated mineral core (density 1–2 g cm^{-3}) embedded in a fluffy organic ice-mantle (density 0.3–1 g cm^{-3}). The relative mean abundances of the elements in the mineral cores correspond to those of the C1 carbonaceous meteorites, i.e. to solar system abundances apart from hydrogen and other volatile elements. The mass spectra of the organic ice-mantles, compared with typical mass–ion breakdown patterns of polyatomic molecules, indicate the presence of linear and cyclic compounds belonging mainly to unsaturated series. The molecules detected include olefins and acetylenes, imines and nitriles, aldehydes and carboxylic acids, together with pyridines, pyrroles, pyrimidines, purines, and a range of benzene derivatives. While much water was detected as ice-cluster ions, e.g. $[H(H_2O)_n]^+$, $n = 1, 2, 3, \ldots$, no evidence was found for the presence of alcohols, amino acids, or saturated hydrocarbons (Kissel and Krueger 1987).

The 1986 spacecraft encounters with comet Halley confirm the view held for several decades that comets are loosely bound aggregates of interstellar dust and condensed hydrides of the lighter elements ('dirty snowballs'), formed at the outer edges of the solar nebular from materials of the parent molecular cloud. Additionally the spacecraft findings support the more recent view that the interstellar grains are composite, with a mineral core and organic mantle. The vast efflux of gas and dust from the nucleus of comet Halley illustrates the process of cometary decay at each passage through the inner solar system around the Sun, leading to successive stages of declining activity, like those displayed by comet Encke over the past century. Ultimately only the inert nuclear residue remains, like the body in the Apollo-type Earth-crossing orbit of the cometary debris responsible for the periodic Geminid meteor showers.

For the extinct core of a periodic comet the probabilities are high that it will collide ultimately with a planet or one of the smaller bodies in the solar system. Some of the carbonaceous meteorites probably originate from comets, particularly the C1 type, which are friable and break up when immersed in water. Relatively few carbonaceous meteorites are recovered as they often disintegrate and burn up in the atmosphere. A large carbonaceous meteorite fell in Canada at Revelstoke in 1965, judging by the fireball and detonation, but only a 1 g specimen was found, and that only because it was black and fell on a snowfield. No specimens were recovered from a larger fireball and detonation over the Tunguska river of central Siberia in 1908, seen and heard at distances more than 1000 km away. No specific crater was found, but the forests were flattened over an area of 70 km diameter. It is estimated that a comet with a nucleus of 40 m diameter, mass 5×10^7 kg, exploding 8.5 km high, was responsible for the Tunguska event (Hughes 1981).

9
The Earth and its formation

9.1 The Earth's core, mantle, and crust

The theory that the Earth contains much iron goes back to the Elizabethan court physician William Gilbert (1540–1603) of Colchester, or even to Pierre de Maricourt in the thirteenth century. Both were concerned with the magnetic compass in connection with navigation and, following Maricourt, Gilbert constructed a model of the Earth from a sphere of lodestone, Fe_3O_4 (magnetite), in order to map out the magnetic meridians with chalk lines, using a compass needle. During the nineteenth century the analysis of meteorites and the rock minerals of the terrestrial crust, together with the spectroscopic identification of chemical elements in the Sun, showed that there is much iron in the solar system present as the metallic nickel–iron alloy in the iron, stony-iron, and many of the stony meteorites, but oxidized in the accessible terrestrial rocks with apparently more of the ferric Fe(III) than the ferrous Fe(II) form.

Metallic iron in the Earth was required to account not only for the Earth's magnetic field but also for its overall mass density (5.52 g cm^{-3}). The crustal rocks are relatively light (~3 g cm^{-3}) and even the deeper mantle material is less dense (4.5 g cm^{-3}) than the Earth as a whole, and so metallic iron with its larger density (7.6 g cm^{-3}) was proposed as a major constituent of the Earth to compensate for the lightness of the terrestrial rocks. The hypothesis was supported by the established metal and mineral constitution of the meteorites, taken as representative of typical solar system material. The density of metallic iron, as well as its general absence from near-surface rocks, implied that the metal lay at the centre of the Earth, but not necessarily as a distinct core. A continuous increase of density from the surface to the centre of the Earth was usually envisaged, often through a plastic or liquid interior, required to account for volcanoes, or even through a supercritical gaseous interior, as advocated in 1905 by the physical chemist Svante Arrhenius (1859–1927) at Stockholm, with the support of geophysicist colleagues.

The Earth's layered structure was characterized from the 1890s principally by the detection and analysis of earthquake waves propagated through the bulk mass of the Earth (Brush 1980). An earthquake or surface detonation generates two types of acoustic wave in the Earth: one longitudinal along the direction of propagation, from pressure changes, the P-wave; and the other transverse to the transmission direction, from shearing variations, the S-wave. The P-waves are transmitted by both solids and liquids, with a

sharp change in velocity at the phase boundary, whereas the S-waves are
not transmitted by liquids and have a smaller velocity than P-waves in a
given solid. Both types of seismic wave undergo reflection and refraction at
an interface. Analyses of earthquake data from the seismic stations around
the world, now numbering more than a thousand, provide not only radial
images of successive layers but also tomographic scans of the three-
dimensional internal structure of the Earth (Anderson and Dziewonski
1984).

Seismology and density data identify a solid inner iron core, extending to
some 1300 km from the Earth's centre, surrounded by a thicker (2180 km)
liquid iron outer core containing lighter elements (S, O, Si) which reduce
the mass density and lower the melting point. The iron core occupies 16.2
per cent of the volume of the Earth and has 32.4 per cent of the mass, the
remainder being made up largely of the mantle, some 2880 km thick. The
crust accounts for only 0.4 per cent of the mass and 0.7 per cent of the
volume of the Earth (Table 9.1). Although there is no direct analytical
evidence for the composition of the core and the lower mantle, which is

Table 9.1 Radial layer thickness, volume, mass, and density of the Earth's
constituent shells

	Thickness (km)	Volume (10^{12} km^3)	Mass (10^{24} kg)	Density (g cm^{-3})
Whole Earth	6371	1.083	5.976	5.52
Core	3471	0.175	1.936	11.0
Mantle	2883	0.899	4.016	4.5
Crust	17*	0.008	0.024	2.8*

*Mean values for the continental and oceanic crust with respective averages in
thickness of 40 and 6 km and in density of 2.7 and 3.0 g cm^{-3}.

thought to consist mainly of high-pressure forms of silicates, $(Mg,Fe)SiO_3$
(perovskite), and oxides, $(Mg,Fe)O$ (magnesiowüstite), bulk sound velocity
measurements for a variety of metals and minerals under pressure support
the proposed core and lower mantle compositions (Ringwood 1984).

9.2 Geomagnetism and continental drift

The Earth's liquid iron core, long proposed, was established by analyses of
seismic data, particularly the studies from 1926 of Harold Jeffreys (1891–
1989) at Cambridge, and the solid core was inferred in 1946 by his former
student Keith Edward Bullen (1906–1976) from a P-wave velocity change
in the core, indicating a liquid–solid interface. Convection currents in the
liquid core due to the heat from the decay of radioisotopes and the latent

heat of crystallization of the iron explain the origin of the Earth's magnetic field in terms of the dynamo effect, suggested in 1919 by Joseph Larmor (1857–1942) to account for the magnetic field of the Sun and developed by Edward Bullard (1907–80) to interpret terrestrial magnetic data.

Much earlier, in 1691, Edmund Halley had suggested that the Earth had an inner solid core and a fluid outer core in order to explain the time variation of the direction of the terrestrial magnetic field. The variations with which Halley was concerned, although important for ocean navigation, were small over the century or so for which data were then available. After the work of Pierre Curie (1859–1906) on the temperature variation of the magnetic properties of crystals (1895), remanant permanent magnetism was found in igneous rocks such as lavas. The thermoremanent magnetization (TRM) registers the orientation of the Earth's magnetic field during the period when the rocks cooled down through 748 K, the Curie temperature point at which the magnetite Fe_3O_4 in the rocks becomes ferromagnetic, usually lowered to the 500–700 K region by the entry of titanium ions into the magnetite crystal lattice.

Early studies around 1900 showed that the TRM of recent lavas is orientated parallel to the Earth's present magnetic field, but in old lavas the TRM has a variety of directions, including the reverse antiparallel orientation. Following experience with magnetic mines in World War II, Blackett at the Imperial College in London developed a sensitive magnetometer which detected the weaker depositional remanent magnetism (DRM) of sedimentary iron-containing rocks, such as sandstones, as well as the TRM of igneous rocks. The collection of TRM and DRM data worldwide in the 1950s for rocks dated from the parent/daughter decay ratios of their radioactive contents showed that the direction of the Earth's magnetic field has continuously wandered historically from the parallel orientation to the terrestrial rotation axis, undergoing reversals every few million years.

Moreover, rocks of the same age from different continents, while agreeing on a common direction for the Earth's magnetic field from data for recent specimens, gave divergent orientations for older samples. North American and European rocks have the same geomagnetic orientation back to about 200 million years ago, but then diverge and are mutually displaced by approximately 30° of longitude for earlier specimens. The earlier geomagnetic rock orientations give a common direction for the Earth's magnetic field if North America and Europe formed a single land mass until about 200 million years ago when the Atlantic Ocean opened up and the two continents moved to their present relative positions on the Earth. Similar results followed for other continents: Africa and South America were once joined, and the subcontinent of India, Australia, and Antarctica once formed a single land mass.

After the coastline of Brazil and the west coast of Africa were first mapped out during the sixteenth century Francis Bacon (1561–1626) noted in 1620 the excellent fit of the two coastlines, and subsequently it was supposed that an intermediate continent of Atlantis had sunk under the Atlantic Ocean. During the nineteenth century similarities between different continents of coastal rocks and of flora and fauna, particularly of extinct creatures represented by the fossils of their bones, led to suggestions that the several continents were once connected by land bridges or were once a single land mass and had drifted apart. George Darwin (1845–1912) at Cambridge proposed in 1879 that the Moon had been born from the Earth, leaving the Pacific Ocean as a scar, and some geologists inferred a consequent readjustment of the remaining continental crust by fission and separation of the fragments.

The case for the theory of continental drift was argued forcibly from 1912 on by the German meteorologist and explorer Alfred Wegener (1880–1930), who suggested, from the geological evidence, that there had once been a single vast land mass, Pangea ('one land'), which broke up into the several continents that slowly drifted to their present positions. Wegener's theories enjoyed rather more support during the period of 1920–1950 from the geologists than from the geophysicists. The South African geologist Alexander du Toit (1878–1948) was a vigorous proponent of continental drift from 1927 on, advocating the origin of the present continents from two original land masses, Laurasia in the north and Gondwanaland in the south, separated by the ocean of Tethys.

The most formidable opponent of the theory was the classical geophysicist Harold Jeffreys, who argued from 1926 that, while the iron core of the Earth is liquid, the mantle and crust are far too rigid to allow any large-scale relative motion of the continents. The geophysical mantle of Kelvin fell upon Jeffreys, both of whom lacked first-hand experience of the seminal discoveries that revolutionized the earth sciences—Kelvin of radioactivity and Jeffreys of geomagnetism. Just as Kelvin had opposed the time-scale proposed by the nineteenth-century geologists and publically questioned Rutherford's claim that radioactive decay provided internal heat to the Earth, so Jeffreys took exception to the geological theory of the spatial drift of the continents and declined to accept the evidence for the drift from rock palaeomagnetism available from the 1950s (Hallam 1973, 1983).

The mechanism of continental drift, ocean floor spreading, and plate tectonics was worked out from further rock palaeomagnetic evidence and seismic data from 1960 on. Acoustic echo sounding mapped out submerged mountain chains in the oceans, the mid-ocean ridges (MOR), such as the Mid-Atlantic Ridge running from Iceland to the Antarctic. Seismic wave velocity changes in the MOR indicated that liquid magma lay not far below the surface, and jets of water heated up to 300°C and rich in metal sulphides, the hydrothermal vents, were detected on the MOR. Recovered ocean crust specimens showed a pattern of remanent magnetism in bands, symmetrical either side of the MOR, of a field direction alternately parallel and antiparallel (reversed) to the present orientation of the Earth's magnetic field. The dating of the ocean crust samples from radioisotope decay or the fossils in the sediments indicated a progression from recent to about 150 million years old with increasing distance from the MOR. Moreover, the succession of oceanic bands of remanent magnetism field direction in time tallied with the corresponding time sequence of TRM and DRM field directions for continental rocks.

The theory of ocean floor spreading accounted for these observations. Molten rock magma welling up from the mantle through the MOR records the current direction of the Earth's magnetic field on cooling down through the Curie point to give the solid basalt ocean floor crust. The basalt carrying its TRM record is pushed out symmetrically on either side of the MOR as fresh magma rises and is extruded. A given band of ocean crust moves outwards from the MOR over 100–200 million years, accumulating sediments above as it progresses, until a barrier is encountered, either a continental land mass or another block of oceanic crust moving outwards from different MOR. At the barrier the ocean crust with its accumulated sediments descends into the ocean trenches and returns back to the mantle in the process of subduction.

The pattern of the MOR, where new ocean crust is created by the ascent of magma from the mantle, and the deep ocean trenches, where the old crust returns back again, defines a worldwide dynamic system of six major plates and several smaller ones moving relative to one another over a few centimetres each year. Each major plate includes both oceanic and continental crust and an underlying supporting layer of solid mantle, forming the lithosphere some 100 km thick. The plates rest on a plastic layer of the mantle, the asthenosphere, from 100 to 600 km deep, which contains little molten material ($\sim$1 per cent) but sustains cycles of convective currents by thermal creep, ascending at the MOR and descending at the ocean trenches. The ascending and descending branches of a convection cycle are connected by one horizontal motion over the deeper mantle and another in the reverse direction at the top of the asthenosphere layer, moving the lithosphere plate immediately above.

Some ascending plumes of the convection currents in the asthenosphere come up from the deeper mantle to the surface where they mark localized hot spots, the ascending column being mapped out by three-dimensional seismic wave tomography. The hot spots continuously generate new volcanic islands, as near Iceland at the northern end of the Mid-Atlantic Ridge, or along the arc of the Hawaiian island chain. The Hawaiian islands increase in age from the recent and still volcanically active islands in the east to the old and now volcanically extinct islands in the west, owing to the westward spread of the ocean floor over the hot spot localized in the deeper mantle underneath.

9.3 Crust and mantle compositon

The crust and the mantle of the Earth were first clearly distinguished by the detection of an increase in seismic wave velocity by almost 1 km s^{-1} some 30–40 km below the surface of continents or about 6 km below the ocean floor from about 7 km s^{-1} for the P-waves and about 4 km s^{-1} for the S-waves in the topmost crustal layers. Andrija Mohorovičič (1857–1936) of the Zagreb observatory in Yugoslavia observed the velocity change during a local earthquake in 1909, and the Moho discontinuity now bears his name. The discontinuity is due in part to a change in chemical composition from a silica–alumina (sial) rich crust to the silica–magnesia (siam) rich mantle, and in part to the change from the lighter low-pressure mineral forms of the crust to the corresponding denser high-pressure forms of the mantle.

The two main components of the surface layer, oceanic crust and continental crust, while both rich in aluminosilicates, differ in their main igneous rock constituents, the basalts and the granites, respectively. After taking part in a scientific expedition to Iceland in 1846, Robert Bunsen, then at Marburg, proposed that there are two primary magmas, granitic and basaltic, and that other igneous rocks are mixtures of these two. The denser oceanic basalts contain more iron, titanium, magnesium, and calcium than the relatively light continental granites, which are richer in sodium and potassium and contain more silica. In 1924 F. W. Clarke and H. S. Washington of the US Geological Survey reported analytical data for more than 5000 different rock samples, from which it was shown that the frequency distribution of the percentage of silica in the rock samples is bimodal. The two frequency maxima at 52.5 per cent SiO_2 and 73.0 per cent SiO_2 correspond to the most common basalt and granite rock types. Each type contains an average 16 per cent of alumina Al_2O_3 as the next most dominant oxide, followed by the oxides of iron, calcium, sodium, magnesium, potassium, and titanium in order of decreasing percentage content.

These estimates of crustal composition were dependent upon the particular rock analyses available, and V. Goldschmidt at Oslo argued in 1933 that a natural average would be provided by the glacial clays. During the ice ages the glaciers had moved over large distances, eroding many different types of rocks in their progress, and had finally deposited the debris in the moraines. Goldschmidt's analyses of the Norwegian glacial clays agreed substantially with the estimates of Clark and Washington and with more recent analyses of the main elements making up the mean composition of the continental crust.

Initially the composition of the Earth's mantle was modelled, like that of the core, upon the composition of the meteorites. As the ordinary chondrites contain on average some 70 per cent of magnesium–iron silicates and

10 per cent of aluminosilicates, it was inferred that the mantle of the terrestrial planets has a similar composition. The inference is supported by the analysis of mantle xenoliths ('strange rocks'). These are nodules of mantle rock ejected in volcanic eruptions from depths of 100–200 km below the Earth's surface, distinct in their mineralogical and chemical composition from the surrounding igneous rock of solidifed lava. The minerals of the mantle nodules are high-pressure forms and they are substantially richer in magnesium (MgO, 39 per cent) than crustal rocks (MgO, 5 per cent) and depleted in aluminium (Al_2O_3, 3.5 per cent). The iron content of the nodules is comparable with that of the crust but it is present mainly in the ferrous form (FeO, 8 per cent), and both the sodium and potassium abundance are much reduced.

Mantle xenoliths are found in association with the high-pressure allotrope of carbon in the once-volcanic pipes of South African diamond mines. The diamond crystal is the stable form of carbon only at pressures corresponding to depths of 150 km or more below the surface of the Earth. Inclusions of minerals containing ^{40}K and ^{87}Rb in diamonds show, from the relative amounts of the decay products of these radioisotopes, that the diamonds crystallized and became closed systems some 3 billion years ago. Similarly the mantle xenoliths are older than their surrounding matrix; one found in New Mexico embedded in a basalt of solidified lava only a few million years old has a $^{87}Rb/^{87}Sr$ age of 1.27 billion years.

The analyses of the mantle xenoliths, on the assumption that they are representative of the mantle as a whole, provide an estimate of the mantle composition. The combination of the present mantle composition with the weighted (0.59 per cent) contribution of the crustal composition gives the relative abundances of the elements in the primitive mantle of the Earth surrounding the iron–nickel core before its differentiation into the present mantle and crust. A comparison of the relative abundances of the chemical elements in the primitive mantle with the corresponding values for the C1 carbonaceous meteorites (normalized to $Si = 1$) affords a set of depletion or enhancement factors for the elements in the primitive mantle of the Earth relative to the materials of the solar nebula, as represented by the C1 chondrites.

The general trends found in the abundance changes of the elements in the primitive mantle compared with the C1 meteorites (relative to unit silicon abundance) are as follows (Table 9.2; Wanke *et al.* 1984):

1. Elements forming refractory oxides (Mg, Al, Ca, Ti, Sc, and most refractory trace elements) are slightly enriched by a factor of about 1.3.

2. Moderately oxyphilic transition metals (V, Cr, Mn) are slightly depleted by a factor of 0.25–0.7.

3. Iron and the moderate siderophiles (Co, Ni, Cu, Ga, W) are depleted by a factor of 0.1–0.2, as are the moderate volatiles (Na, K, Rb, F, Zn).

Table 9.2 The abundance ratio of the chemical elements in the Earth's mantle and crust relative to the C1 carbonaceous meteorites (normalized to unit abundance for silicon) with the percentage of the element in the crust (crust percentage) (Wanke *et al.* 1984) and the nebula 50 per cent condensation temperature (K) of the element at approximately 10^{-4} atmospheric pressure (Wasson 1985)

Element	Abundance ratio	Crust percentage	Condensation $T(K)$	
Al	1.35	2.2	1650	
Ti	1.53	2.3	1549	
Ta	1.78	26	~1550	
Sr	1.6	6.3	—	
Ca	1.40	1.1	1518	
Ba	1.27	57	—	Refractory
Sc	1.43	0.74	1644	lithophile
Ln*	1.51*	8.9*	1533*	elements
Hf	1.16	7.3	1652	
U	1.78	24	1420	
Mg	1.18	<0.1	1340	
Si	1.00	0.77	1311	
V	0.73	1.0	1450	
Cr	0.56	<0.1	1277	
Li	0.74	3.8	1225	
Na	0.29	5.0	970	
Mn	0.28	0.50	1190	
K	0.22	45	1000	
Rb	0.18	63	~1080	Moderately
F	0.18	16	736	volatile
P	0.032	7.0	1151	lithophile
Cs	0.024	84	—	elements
Cl	8.7×10^{-3}	96	863	
Br	9.0×10^{-3}	90	~690	
I	0.012	68	—	
C	6.6×10^{-4}	48	—	
Re	3.2×10^{-3}	2.5	1819	
Os	3.2×10^{-3}	0.20	1814	Highly
Ir	2.9×10^{-3}	0.20	1610	siderophile
Au	1.9×10^{-3}	4.6	1225	elements
W	0.13	32	1802	
Fe	0.16	0.49	1336	
Co	0.10	0.13	1351	
Ni	0.097	<0.1	1354	
Cu	0.13	1.0	1037	Siderophile
Ag	6.9×10^{-3}	14	952	elements
Ge	0.021	0.59	825	
As	0.041	7.9	1157	
Sb	0.022	21	912	
Ga	0.21	2.9	918	
Zn	0.070	0.93	660	Chalcophile
In	0.115	2.2	456	elements
Cd	0.017	2.3	430	
Te	4.2×10^{-3}	<0.1	680	
Se	3.6×10^{-4}	6.7	684	
S	1.1×10^{-4}	39	648	

*Average values for the lanthanide series, covering La, Ce, Nd, Sm, Eu, Gd, Tb, Dy, Ho, Er, Yb, and Lu.

4. The highly siderophile elements (Ir, Os, Re, Au, etc.) are strongly depleted by a factor of about 2×10^{-3}, and so too are the strongly volatile elements (Cd, Ag, I, Br, Cl, Te, Se, C) but more variably (10^{-2}–10^{-4}).

9.4 The accretion of the Earth

The normalization of the relative abundances of the elements in the primitive mantle and in the C1 carbonaceous meteorites to unity for the standard element, silicon, is conventional. Alternatively, the small enrichment by a factor of about 1.3 of the primitive mantle in the elements forming refractory oxides could be taken as a corresponding slight depletion of the mantle in silicon if the refractory oxide elements were adopted as the standards for relative abundance normalization. According to the latter interpretation, the missing silicon went into the core in metallic form. In the liquid outer core metallic silicon lowers the melting point of the iron and produces a required reduction in mass density. A similar role is suggested additionally or alternatively for sulphur, as FeS, or oxygen, as FeO, in the outer core.

Metallic silicon in the liquid iron of the outer core could not be at chemical equilibrium with oxidized silicon of the main mantle constituents, the ferromagnesium silicates. Measurements of the distribution of the elements between liquid iron containing 10 per cent nickel and molten basalt (8 per cent FeO) at 1300°C indicate that most elements are far from equilibrium at the interface between the Earth's core and mantle. For iridium the partition coefficient, $D(Ir) = [Ir_{metal}]/[Ir_{silicate}]$, approximates to 10^5, and the corresponding values for rhenium and gold are both greater than 3×10^4, so that these highly siderophile elements would be further depleted by at least another order of magnitude from the abundance ratios found if the core and mantle were in equilibrium (Table 9.2). The absence of a core/mantle equilibrium is even more strikingly illustrated by the moderately siderophile elements which have mantle abundance ratios in the range 0.1–0.2 relative to the C1 meteorites, for their partition coefficients (iron/silicate) cover a wide range from $D(Ga) = 30$, through $D(Cu) = 80$, $D(W) = 260$, and $D(Co) = 200$, to $D(Ni) = 750$.

The strong depletion of the highly volatile elements in the Earth's mantle and the smaller depletion of those that are moderately volatile are consistent with the equilibrium homogeneous accretion model for the formation of the terrestrial planets, but the distribution of the siderophile elements requires a non-equilibrium heterogeneous accretion process. Accretion began with highly reduced material of chondrite composition, largely free of volatile elements, with iron and the siderophile elements present as metals and only the refractory oxyphile elements as oxides. This material corresponds to the high-temperature condensate, the component A of the two-stage heterogeneous accretion model. Owing to the elevated temperature

reached during the accretion, the metals segregate into a core throughout the first stage, but the mantle was never wholly molten, since the
refractory oxides have not segregated out on a large scale, as would be
expected for crystallization from the melt.

After the accretion of some two-thirds of the Earth, the accreting material
becomes progressively more oxidized and richer in the moderately volatile
elements (Zn, F, alkali metals), again in chondritic abundances. Siderophile
as well oxyphile elements are added as oxides, including iron as FeO, in a
low-temperature condensate, the component B of the second stage of
accretion. A large fraction of the highly siderophile elements (Ir, Os, Re,
Au) in the component B addition is extracted from the mantle into the core
by low-melting eutectic mixtures formed by the FeO with FeS, or with
metallic iron, but the extraction does not significantly affect the abundances of the moderate siderophiles (Ni, Co, Cu, W, Ga). At a late stage of
the component B accretion highly volatile elements, such as the halogens,
and molecular species like water and hydrocarbons, as in the recovered C1
chondrites, are added to the now cooler Earth as a surface layer.

The constituents of component A are modelled upon the estimated bulk
composition of the Earth without the volatile alkali metals and with all
oxidized iron reduced to the metallic state, while the materials of component B are taken to be equivalent to those of the Orgueil C1 carbonaceous meteorite (Table 9.3). Components A and B represent probable end
compositions of a progression from the initial high-temperature reduced
condensates, akin to the stony and stony-iron meteorites, to the final
low-temperature oxidized condensates, with aqueous and organic constituents like the C1 carbonaceous chondrites, owing to the removal of the
reducing gas, hydrogen, from the solar nebula by radiation pressure and
the solar wind.

9.5 The time sequence of the Earth's formation

The time at which the metal core of the Earth segregated from the silicate
mantle distinguishes between the equilibrium homogeneous model for the
accretion of the terrestrial planets and the non-equilibrium heterogeneous
model. Core formation was a slow and gradual process, according to the
equilibrium model, but the non-equilibrium model specifies a fast core
separation, concurrent with the accretion of component A. The closed
systems of the meteoritic minerals provide not only the starting point of
4.55 billion years ago for the first condensations from the radiometric
parent/daughter abundance ratios, but also the stable isotope ratios at a
zero time from which subsequent events, such as metal core formation,
may be dated.

Among the main isotopes of lead, ^{204}Pb is non-radiogenic, whereas
^{206}Pb and ^{207}Pb are produced by the radiodecay of ^{238}U and ^{235}U, respect-

Table 9.3 Composition of the reduced component A, devoid of volatiles, condensing from the solar nebular at elevated temperatures (>1300 K), and the oxidized component B, rich in volatiles, condensing below 380 K (Ringwood 1984)

Material	High-temperature component A (per cent)	Low-temperature component B (per cent)
Fe–Ni alloy	34.1	—
SiO_2	32.8	21.7
MgO	27.2	15.2
Al_2O_3	2.8	1.6
CaO	2.3	1.2
Cr_2O_3	0.5	0.35
TiO_2	0.2	0.1
FeO	—	22.9
NiO	—	1.2
MnO	—	0.2
Na_2O	—	0.7
K_2O	—	0.07
P_2O_5	—	0.3
Sulphur	—	5.7
Organics	—	9.7
Water	—	19.2

ively, and ^{208}Pb is generated similarly from ^{232}Th. Thus the non-radiogenic isotope ^{204}Pb serves as the standard denominator for the lead isotope ratios, with a radiogenic isotope as numerator, for radiometric dating. The radiometric age is derived from the exponential increase in the ratio with time from an initial level. Zero time on the radiometric clock was set when the mineral crystallized and became a closed system, isolated from chemical exchange with its environment.

The partitioning of the elements between molten iron and liquid silicate shows that uranium and thorium remain almost entirely in the silicate phase, whereas lead is mildly siderophile with a distribution coefficient, $D(Pb) = [Pb_{metal}]/[Pb_{silicate}]$, of approximately 2.5. At the time of core formation virtually all the uranium and thorium remained in the silicate mantle but a significant fraction of the lead entered the core, leaving the mantle depleted, particularly in the non-radiogenic isotope ^{204}Pb, the standard required for the lead isotope ratios used in age determinations.

An estimate of the time of core formation, when the mantle became a closed system, from the radioactive decay of the parent isotope ^{238}U rests on the value of the ratio $^{206}Pb/^{204}Pb$ for the whole Earth, prior to the segregation of metal from silicate. The required value of $^{206}Pb/^{204}Pb$ and of the analogous ratios for the other radiogenic isotopes of lead is provided by the lead content of the mineral troilite (FeS) in iron meteorites. The meteoritic troilite contains virtually no uranium nor thorium, so that the isotope ratios of the lead content of the mineral have remained primitive, referring to zero time for the solar system. The lead isotope ratios of meteoritic troilite and of the Earth's mantle indicate that core formation was largely complete around 4.4 billion years ago, within the first 150 million years of the Earth's history.

The accretion of planetesimals was still substantial at that time, since the craters due to meteoroid impacts on the older parts of the Moon's surface, the lunar highlands with radiometric ages of 4.4–4.0 billion years, have a much higher density (number per unit area) than the craters in the younger lowland basalts of the lunar mare, formed 4.0–3.2 billion years ago according to radiometric dating. The flux of meteoroids bombarding the Moon fell by a factor of about one-tenth from 4.4–4.0 billion years ago and then by another factor of about one-tenth over the following billion years, continuing on at a decreasing rate throughout lunar history to the low level of the current impacts, recorded by the seismometers placed on the Moon during the Apollo missions.

The lunar crater Giordano Bruno has a very sharp crater rim which has not been rounded off, like other lunar craters, by the 'gardening' process of small bolide impacts. Accordingly the Giordano Bruno crater must be recent and its size, some 20 km in diameter, indicates that it was produced by the impact of a body somewhat less than 1 km across. The position of the crater corresponds to that reported by five observers of the crescent of the new Moon on the evening of 18 June 1178 (Hartung 1976) when

suddenly the upper horn split into two. From the midpoint of this divison a flaming torch sprang up, spewing out over a considerable distance fire, hot coals and sparks. Meanwhile the body of the Moon which was below writhed as it were in anxiety ... and throbbed like a wounded snake.

The history of the bolide bombardment of the Earth probably followed a course similar to that of the Moon, but the craters from the early meteoroid impacts on the Earth have been entirely weathered away. It is estimated that a 20 km crater on the Earth's surface remains recognizable for some six hundred million years and a 10 km structure for half that time. The Barringer crater near Flagstaff, Arizona, with a diameter of 1.22 km is only about fifty thousand years old. Traces of around a hundred large craters up to 140 km in diameter are detected by aerial photography but they are mostly less than a billion years in age. Meteorite fragments are found in the vicinity of a dozen or so, indicating that they are impact craters.

Some 3×10^4 kg of meteoritic iron are strewn around the Barringer crater, which was generated by the impact of a mass estimated at 10^8 kg. The largest craters of 140 km in diameter, at Vredefort in South Africa and at Sudbury in Ontario, Canada, are the most ancient, with ages of 1.97 and 1.84 billion years, respectively. The Sudbury impact structure, produced by a bolide some 10 km in diameter, is rich in nickel as the sulphide ore. The ore probably segregated by crystallization from the molten crust produced by the impact, although it is possibly of meteoritic origin.

As the mantle and crust of the Earth are depleted in the highly siderophile elements (Ir, Os, Re, etc.) by about 2×10^{-3} relative to the C1 chondrites, the discovery of an iridium-rich layer of clay, 65 million years old, separating the different fossil types of the Cretaceous and the Tertiary periods led Luis Alvarez (1911–88) and his coworkers at Berkeley to propose that a bolide some 10 km in diameter had collided with the Earth at that time. The iridium-rich clay layer occurs throughout the world, extending to samples from the Pacific ocean floor some 65 million years old. The total amount of iridium in the layer worldwide ($\sim 5 \times 10^8$ kg) and the meteoritic abundance of iridium (0.5 ppm) require a 10 km diameter for the impacting body. The bolide may have been a comet, since non-protein amino acids (racemic isovaline and α-aminobutyric acid) have been detected at the Cretaceous/Tertiary boundary at a higher abundance than expected for a carbonaceous meteorite (Zhao and Bada 1989). As a result of the collision, the Earth was enveloped by an opaque blanket of dust which severely limited plant photosynthesis for several years and led to mass extinctions of a wide variety of species, including the dinosaurs and other giant reptiles of the Cretaceous period (136–65 million years ago).

Over the past 500 million years there have been five major mass extinctions, and the expected frequency of a collision beween a 10 km diameter bolide and the Earth is one in about 100 million years. Two further iridium-enriched layers have been found, one at 11 million and the other at 38 million years ago. The latter is associated with the major fossil changes that mark major extinctions at the boundary of the Eocene and Oligocene periods. There is evidence for a periodicity of about 26 million years in major extinctions, consistent with the iridum-enriched layers at 11, 38, and 65 million years before the present. It is suggested that the Oort–Opik cloud of comets is regularly perturbed by a solar companion star (Nemesis) with an orbital period of 26 million years. The perturbation sends a shower of some 10^9 comets into the inner solar system, and a few collide with the Earth over the course of a million years or so, producing mass extinctions (Alvarez 1987).

9.6 The atmospheres of the terrestrial planets

The inner planets, formed from volatile-depleted planetesimals, lacked any significant primitive atmosphere. Gases and volatile elements were occluded or chemically combined in the planetesimals, as they are in recovered meteorites, and the thermal processing of the planet released noble gas atoms and molecular gases to form a secondary atmosphere. In the materials of the terrestrial planets the most depleted of all the chemical elements are the noble gases, particularly their primordial non-radiogenic isotopes. The radiodecay of ^{40}K continuously generates ^{40}Ar, and ^4He is produced similarly from the actinides, but ^{36}Ar and ^3He, like ^{20}Ne, ^{84}Kr, and ^{132}Xe, are primordial. From a comparison of abundances in the atmospheres of the Earth and the Sun, the ratio of ^{36}Ar, terrestrial/solar, is approximately 10^{-6}, compared with a corresponding ratio of approximately 10^{-2} for ^{35}Cl or ^{37}Cl, which condensed in ionic or molecular combinations. But the radiogenic isotope ^{40}Ar, while less abundant than ^{36}Ar in the solar atmosphere, is predominant in the Earth's atmosphere.

The elements of the noble gas series have a characteristic abundance pattern, similar for the Earth, Venus, Mars, and the chondritic meteorites, termed the planetary rare gas distribution, which differs from that of the Sun in that the heavier elements of the series are relatively more abundant. The ratio $^{20}Ne/^{36}Ar$ approximates to 0.4 for the Earth, Venus, Mars, and most meteorites, compared with a value of 35 for the corresponding ratio of the two isotopes in the solar atmosphere. The planetary rare gas distribution, common to the terrestrial planets and the chondritic meteorites, supports the model of an accretion of chondrite-like planetesimals in the formation of the inner planets of the solar system.

While the pattern of the noble gas abundance distribution is common to the terrestrial planets, the absolute abundances, e.g. of ^{36}Ar, fall with an increase in the heliocentric distance by two orders of magnitude between Venus and the Earth, and by another two orders of magnitude from the Earth to Mars. The systematic abundance decrease outwards from the Sun suggests that the gas pressure in the solar nebula at the time of the planetesimal condensation was about 75 times higher at the heliocentric distance of Venus than at that of the Earth and lower by a similar factor in the region of Mars (Wayne 1985).

Soon after the accretion of the terrestrial planets, it is probable that their surfaces were covered by magma oceans, possibly 200 km deep in the case of the Earth, due to the heat liberated during accretion and core formation. The basalt maria of the Moon are the solidified remnants of seas of once-molten rock magma. Intense volcanic activity during crust formation led to the degassing of occluded and chemically combined volatiles, forming the early secondary atmosphere. Predominantly reducing gases are obtained by heating up to 1500 K meteoritic material modelling the terrestrial planetesimal composition (98 per cent of high-iron chondrite and 2 per cent of C1 chondrite). The main gases liberated up to about 1000 K are, in order of decreasing abundance, methane, nitrogen, hydrogen, ammonia, and water vapour, with minor amounts of hydrogen sulphide, carbon monoxide, and carbon dioxide. With full outgassing over the 1000–1500 K range, nitrogen and ammonia drop to the level of minor constituents and comparable

Table 9.4 The main constituents of the atmospheres of the terrestrial planets and their black-body radiation temperature, from the infrared emission (T_e (K)) and the mean surface temperature T_s (K)) with the pressure relative to that of the Earth (P_s (atm))

	Venus	Earth	Mars
T_e (K)	240	235	210
T_s (K)	745	280	225
P_s (atm)	90	1	0.01
CO_2 (per cent)	96.5	0.08	95.3
N_2 (per cent)	3.5	78	2.7
O_2 (per cent)	0.003	21	0.13
H_2O (per cent)	<0.05	~0.2	~0.03
Ar (per cent)	0.003	0.9	1.6

amounts of methane, carbon monoxide, carbon dioxide, water vapour, and hydrogen are released (Lewis and Prinn 1984).

The lifetime of a reducing atmosphere on the terrestrial planets is expected to be short, owing to the photolysis of the hydrides by the solar ultraviolet radiation and the escape of hydrogen atoms from the gravitational field of the planet in its upper atmosphere. The planetary atmosphere then becomes neutral, composed mainly of carbon dioxide and nitrogen, as observed for Venus and Mars at present (Table 9.4). On the Earth uniquely, temperature conditions over much of the surface have been restricted to the liquid range of water over the past 4 billion years or so, thereby facilitating the fixation of atmospheric carbon dioxide. First, in the weathering reaction, aqueous bicarbonate attacked silicate minerals to liberate silica and form carbonate sediments, which date back to about 3.8 billion years ago. Second, photosynthetic organisms reduced dissolved carbon dioxide with hydrogen split from water, liberating the molecular oxygen produced into the atmosphere at significant partial pressures from about 2 billion years ago.

It has been argued that the atmosphere of the Earth was never reducing and that it has evolved continuously over the whole history of the Earth by the volcanic emission of gases from the mantle through the crust. The view rests on the nineteenth-century uniformitarian principle that the history of the Earth must be explained only in terms of the geological forces observed in operation today. The principal components of present-day volcanic gases are steam and carbon dioxide, which are particularly soluble in silicate melts. At 30 kbar pressure and 1625°C melts of diopside ($CaMgSi_2O_6$) dissolve up to 25 per cent of water and 5 per cent of carbon dioxide by weight (Holland 1984). The gradual release of such quantities of water and carbon dioxide from the mantle of the Earth by volcanic action over 4 billion years suffices to account for the present volume of the oceans and mass of the carbonate deposits. But the argon isotope ratios indicate that the Earth was over 75% degassed by 4 billion years ago. The accumulation of ^{40}Ar in the mantle from the radiodecay of ^{40}K now gives a mantle ratio of $^{40}Ar/^{36}Ar > 10^4$, estimated from the gases entrained in the volcanic lavas at the mid-ocean ridges. The atmospheric ratio of $^{40}Ar/^{36}Ar$ is only 295.5, corresponding to the mantle value more than 4 billion years ago when the major degassing of the Earth must have occurred (Ozima 1987).

It is estimated that, if all the carbon dioxide fixed in the terrestrial sedimentary carbonates were liberated, the Earth's atmosphere would consist of 98 per cent carbon dioxide with a total pressure of about 60 atmospheres at the surface, resembling the atmosphere of Venus and, in relative composition, that of Mars (Table 9.4). The total amounts of carbon dioxide, free and combined, are comparable for the Earth and Venus, with a surface weight of 9×10^4 and 7×10^4 g cm^{-2}, respectively. The surface temperature of Venus ($\sim$745 K) is high enough to calcine limestone to calcium oxide and carbon dioxide and, like Mars, Venus now lacks the liquid water required for the weathering reaction of carbon dioxide with silicates to form carbonates and SiO_2. Only traces of water vapour remain in the atmosphere of Venus, and the deuterium content is high, with a D/H ratio of 1.5×10^{-2} compared with the cosmic ratio of 4.3×10^{-5} from the Big Bang nucleosynthesis.

The enrichment of deuterium in the residual water vapour on Venus, by a factor of about 360, is attributed to the photolysis of water vapour in the upper atmosphere and the preferential escape of hydrogen rather than deuterium atoms from the planet's gravitational field, owing to the larger mean velocity of the hydrogen atoms by the factor of about $(2)^{1/2}$. Terrestrial ocean water has a D/H ratio of 1.5×10^{-4} and the enrichment in deuterium over the cosmic ratio by a factor of 3.6 arises by the same mechanism of atmospheric photolysis followed by hydrogen escape, at least in part, for the Earth received, and continues to receive, deuterium-enriched organic material from the carbonaceous meteorites. Through exchange and oxidation reactions of the organic material, some of the deuterium enrichment ends up in the oceans.

The atomic kinetic energy required to escape from the gravitational field of Mars is only one-quarter of the corresponding escape energy from Venus or the Earth because the mass of the planet is smaller by an order of magnitude. Atoms of carbon, nitrogen, and oxygen, as well as hydrogen, escape from the gravitational field of Mars after liberation by photolysis from the atmospheric molecules. Thus the molecular nitrogen in the atmosphere of Mars is 65 per cent enriched in ^{15}N due to the preferential escape of ^{14}N atoms produced by the solar photolysis of nitrogen. The ^{15}N enrichment implies that only about 5 per cent of the original nitrogen content of Mars now remains and that the degassing occurred early in the history of the planet. Continuous outgassing would maintain or less profoundly change the isotope ratio for the solar system in general, as in the case of the $^{18}O/^{16}O$ ratio of Martian materials. The ^{18}O enrichment is small (<5 per cent) although Mars continuously loses oxygen atoms ($\sim 6 \times 10^7$ atoms s^{-1} cm^{-2}). The lack of a significant ^{18}O enrichment indicates that Mars has a large reservoir of condensed oxygen. Part of the reservoir is visible as the polar caps of ice and solid carbon dioxide, and there are probably large deposits of carbonates and hydrated minerals as well, for there is evidence of the aqueous weathering of the Martian crust in the past.

The channels resembling dry river beds on the surface of Mars suggest that water once flowed on a planetary surface warmed by the greenhouse effect, before the large-scale escape of hydrogen, carbon, nitrogen, and oxygen atoms from the relatively weak gravitational field. The importance of the thermal insulation provided by polyatomic molecules in planetary atmospheres, the greenhouse effect, is illustrated by the large difference between the effective black-body radiation temperature of Venus, estimated from the infrared emission from the planet, and the near-surface temperature, measured by the Venera probes (Table 9.4). The polyatomic and dipolar diatomic molecules in the atmosphere absorb much of the infrared radiation from the planetary surface and emit a substantial fraction back again to warm the surface crust.

10
The evolution of the Earth

10.1 Geological eras and periods

During the nineteenth century geologists and palaeontologists characterized a global order of successive rock strata in which depth correlated with
age on an extended but uncertain time-scale. The global sequence of strata
was defined by the mineral character of the rocks and by their fossil
content. The two criteria were often complementary, since different strata
of the same mineral type were found to contain contrasting fossils, as in the
case of the Old Red Sandstone and the New Red Sandstone with the
economically all-important Coal Measures in between. At the beginning of
the century the strata were classified into three main types: Primary,
Secondary, and Tertiary. The Primary rocks were primitive and apparently
devoid of fossils. They were often crystalline, like granite, and deep lying,
forming the cores of mountain ranges. The Secondary rocks were varied
mineralogically—limestones, sandstones, and shales—and they contained
numerous fossils of extinct species. The uppermost Tertiary rocks were
mineralogically less consolidated, such as chalk rather than limestone or
marble, with most of their fossils resembling the shells or bones of still-
living species.

The strata containing fossils were investigated in detail and, by the end
of the century, only the Tertiary remained to name a major geological
division or era, which was divided into six periods between the Pleistocene
('the most recent') and the Palaeocene ('the old recent'). The Secondary had
been separated into two eras, the Mesozoic ('middle life') and the Palaeozoic
('old life'), each divided into several periods or strata systems. The coal
strata formerly termed the Coal Measures were gentrified as the Carbon-
iferous system, and the flanking Red Sandstones became the Devonian
below and the Permian above (named after the locality of Perm in Russia).
The Palaeozoic era ended with the Permian period, having started with the
Cambrian period, when macroscopic fossils first appeared in the strata,
first as shells of calcium carbonate and then as bones of calcium phosphate.

The original Primary geological division was renamed the Precambrian,
which attracted less attention than the fossil-defined eras during the nineteenth century. The strata of the Precambrian aeon did not enter into the
early estimates of the duration of the Phanerozoic ('evident life') aeon,
covering the Palaeozoic, Mesozoic, and Tertiary (or Cainozoic) eras. These
time estimates were based upon the rates of erosion of land masses, or the
rates of deposition of sediments in the estuaries of rivers such as the Nile or

Table 10.1 The Precambrian and Phanerozoic time-scale in millions (10^6) of years before the present (bp)

Aeon	Era	Periods	Base date
Phanerozoic	Tertiary	Palaeocene to Pleistocene	63
	Mesozoic	Triassic to Cretaceous	240
	Palaeozoic	Cambrian to Permian	570
Precambrian	Proterozoic	Late	900
		Middle	1600
		Early	2500
	Archaean	Late	2900
		Middle	3300
		Early	3900
		Hadean	4500

the Ganges, compared with the summed total thickness of successive strata in the global geological column of sediments. Charles Darwin (1809–82) in the first edition of his *Origin of species* (1859) hazarded the conjecture that the erosion of the Weald in south-east England had occupied some 300 million years, but he reduced and then removed the estimate in later editions as inessential to his main purpose, following the criticisms of Kelvin from 1862 on. From model rates of cooling, Kelvin estimated the age of the Earth and the Sun at no more than 100 million years and probably only 20 million years.

The question of the age of the Earth remained unresolved at the end of the nineteenth century when many geologists supported the duration of some 600 million years for the Phanerozoic aeon of fossil-containing strata systems. A few geologists advocated a much shorter period and one, John Joly (1857–1933) at Trinity College, Dublin, devised a new dating method in 1899. Edmund Halley had suggested in 1690 that an estimate of the rate at which rivers carry salt to the seas, and of the total salt content of the oceans, would provide an age for the Earth's water-circulation system (its 'external alembics'). Joly applied the method and arrived at an age of 90–99 million years for the age of the Earth, supporting Kelvin's estimate. Subsequently Joly publicly accepted that radioactive decay provided the Earth with an internal heat source significant enough to undermine any age estimate based upon the rate of cooling alone, a conclusion Kelvin had conceded only in private to J. J. Thomson (Burchfield 1975). Joly, however, remained the only geologist of note in the 1920s who rejected the age estimates based upon the new technique of radiometric dating.

The geologists who pioneered the radiometric dating of th sedimentary strata and the igneous rocks were Joseph Barrell (1869–1919) at Yale and

Arthur Holmes (1890–1965) at Durham and then Edinburgh (from 1943). The radiometric age determinations of the geological periods over the 600 million year Phanerozoic aeon made by Barrell in 1917, and by Holmes from 1913 until 1960, brought to an end the major controversy on the age of the Earth and drew attention to the vast duration of Precambrian time. The span of the Precambrian remained uncertain until the 1950s, Holmes estimating the limits of 1.5–3.0 billion years in 1931, but the age of the Earth was fixed at 4.55 billion years by the time his last Phanerozoic time-scale appeared (1960).

What appeared to be fossil remains of micro-organisms were detected by optical microscopy in thin polished sections of Precambrian rocks from the 1880s, but little attention was paid to such finds until the 1950s. The supposed microfossils were often dismissed as inorganic artefacts, as some of them undoubtedly were. The first generally accepted Precambrian fossil structures were the stromatolites, columns up to 10 m high of sedimentary carbonates and compacted microcrystalline silica (chert) in successive layers containing microfossils of filamentary photobacteria. Present-day stromatolite columns, found in the coastal waters off the Caribbean islands and the Arabian Peninsula and in saline pools off the coasts of Western Australia and California, contain mats of filamentous blue–green cyano-bacteria which entrain suspended sediments to give a layered structure by upward seasonal growth.

The earliest known authentic stromatolite structures are 3.3–3.5 billion years old, found near Warrawoona in Western Australia, and similar structures of comparable antiquity are found in South Africa. Stromatolites became common in the Proterozoic era, 2.5–0.57 billion years bp (before the present), but their prevalence declined during the Cambrian period of the Palaeozoic era, 570–240 million years bp, due to grazing by burrowing animals, which confine stromatolites to limited habitats today. Authentic microfossils of coccoid unicellular and colonial organisms date back to the Proterozoic era with a few dubious earlier cases (Schopf 1983; Schopf and Packer 1987).

10.2 The Earth's early oceans and continents

An ocean of liquid water and an atmosphere containing carbon dioxide prevailed over the surface of the Earth by 3.8 billion years ago, the age of the earliest known sedimentary rocks. These sediments are found in West Greenland near the margin of the inland ice at Isua as an outcrop on an igneous crustal dome in the form of an arc or belt over 40 km long and variably 1–3 km wide. The contents of the Isua sediments are typical products of the action of water and carbon dioxide on igneous silicate and aluminosilicate rocks: calcite ($CaCO_3$), microcrystalline quartz SiO_2 (chert), and hydrated aluminosilicate clay minerals. The sediments contain iron as magnetite (Fe_3O_4) and silicates ($FeSiO_3$), together with crystals of galena (PbS) and zircon ($ZrSiO_4$), and small amounts of organic carbon as graphitic kerogen. There is little radiogenic lead and no uranium in the galena; its lead isotope ratios, $^{206}Pb/^{204}Pb$ and $^{207}Pb/^{204}Pb$, are the lowest so far reported for any terrestrial material, giving the galena an age of 3.74 billion years. The zircons contain uranium and radiogenic lead from which an age of 3.77 billion years is estimated, in agreement with the age of other materials in the sediments, dated by the Rb/Sr radiometric method and from the decay of ^{147}Sm with a half-life of 106 billion years to ^{143}Nd (Moorbath 1983).

Few traces remain of the earlier igneous rocks from which the Isua sediments derived by weathering. The earliest known igneous crustal minerals are individual crystals of zircon which, from their $^{206}Pb/^{238}U$, $^{207}Pb/^{235}U$, and $^{208}Pb/^{232}Th$ ratios, crystallized between 4.3 and 4.1 billion years ago. The zircons are found in conglomerates of metamorphosed silica sediments of more recent age (3.5–3.1 billion years) at sites consisting of narrow rock belts about 70 km long near Mount Narryer and the adjacent Jack Hills in Western Australia (Compston and Pidgeon 1986). Like the Isua rocks, these zircon crystals have remained intact since the time of their formation, preserved from recirculation in the mantle through plate tectonics.

At the present time plate tectonics maintains the conservation of crustal mass in a steady state, with the creation of new ocean crust from mantle material at the mid-ocean ridges and its return to the mantle in the deep ocean trenches, carrying with it continental crust in the subduction zone and the accumulated sediments from continental erosion. How far back in time the plate tectonic process goes remains an open question. The nineteenth-century uniformitarian view, that the past geological history of the Earth can be explained solely in terms of the forces prevailing today, would require the production of the entire mass of continental crust at a very early time to be recycled continuously through the mantle by the plate tectonic mechanism ever since. A more probable geophysical view takes crust formation to have been cumulative, beginning with the irreversible chemical differentiation of the early mantle, from which alumina separated to form the lighter aluminosilicate crust (sial), carrying with it the lithophilic elements (the alkali metals and the lanthanides and actinides), followed by the development of the plate tectonic mechanism on a minor and then a major scale.

The use of radioactive lithophiles to date the continental crust and mantle materials, by the Rb/Sr and the Sm/Nd methods, gives a *mean* age of 1.8–1.5 billion years for the continental crust. The young mean age implies that either very little early crust was formed or that the massive crust formed was rapidly recycled. The radiometric dating methods indicate major continent formation events 3.0–2.5 billion years ago on a worldwide basis and again 2.0–1.5 billion years ago. It is estimated that at least 50 per cent and possibly 75 per cent of the present-day mass of continental crust was already in existence 2.5 billion years ago. Granites dated to the period 3.3–2.5 billion years bp were produced by the partial melting of earlier continental crust, but the 3.8 billion year old Isua rocks were deposited directly on a proto-oceanic crust, rich in magnesium silicate (siam), before there was any continental crust in the region, and perhaps on the Earth as a whole.

In the Hadean period 4.5–3.9 billion years ago the heat production from radioactive decay in the Earth was some four times larger than it is today, with larger convection currents in the mantle as a consequence. These currents effectively recycled the thin proto-crust which resembled the skin on a lava lake or slag furnace. The skin became sufficiently thick, extensive, and rigid enough to allow an early form of the plate tectonic process, and with it the formation of the first continental crust on a modest scale, only about the time of the Isua rocks some 3.8 billion years ago (Moorbath 1985).

The output of solar radiation energy around 4 billion years ago is estimated to have been some 25 per cent lower than it is today, sufficiently low to have produced Earth surface temperatures well below the freezing point of water in the absence of compensating effects. The four-fold larger heat production from radioactive decay in the Earth provided a minor thermal contribution, but the main compensation came from energy conservation in a major greenhouse effect, due to the large partial pressure of carbon dioxide before much of the gas became fixed as sedimentary carbonate by the silicate weathering reaction.

It is estimated that if all of the carbon dioxide fixed as carbonates today were gaseous 4 billion years ago, at a pressure of up to 60 atmospheres (much carbon dioxide would have remained in the mantle, not yet released volcanically), a runaway greenhouse effect would not have resulted, unlike the case of Venus. Water would have remained liquid, and a fast silicate weathering reaction to give carbonate would ensue at the relatively high temperatures. Surface silicates, water, and a volcanic carbon dioxide supply provide a buffered system. The rate of silicate weathering decreased through the consumption of carbon dioxide and the temperature fell due to the decline of the greenhouse effect; but when the solar luminosity increased and temperatures rose, the weathering reaction became thermally accelerated (Walker *et al.* 1983).

A clement climate in the Precambrian aeon is indicated by the limited glaciations of the period, compared with the extensive and frequent glaciations during the Phanerozoic aeon. There are only two sets of Precambrian glacial deposits: one dated to 2.5–2.0 billion years ago, found in Australia, Canada, and South Africa; and the other, found worldwide (except Antarctica), dated to the late Proterozoic (1.0–0.57 billion years bp). The earlier set may be related by a mechanism as yet uncertain (possibly an

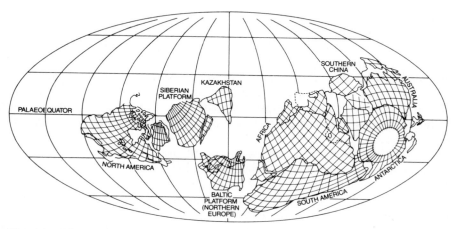

Fig. 10.1 The geography of the world during the transition from the Precambrian to the Cambrian period. In the late Precambrian much of the Earth's land mass is considered to have been concentrated in a single supercontinent (Pangea) between 30°N and 60°S of the palaeoequator. The break up of the supercontinent near the beginning of the Cambrian period produced extensive stretches of new coastline in warm, equable regions near the palaeoequator. The conditions favoured the proliferation of new lifeforms, marking the transition from mainly unicellular organisms, leaving microfossils, to structured multicellular organisms, with macroscopic fossil remains. Grid lines drawn on the continents mark present-day lines of latitude and longitude. (Reproduced from the illustration by Tom Prentiss on page 98 of 'The emergence of animals', by Mark A. S. McMenamin (1987). *Scientific American*, **256**, No. 4, April.)

increased consumption of carbon dioxide in photosynthesis) to the development of a significant partial pressure of oxygen in the Earth's atmosphere, since the Canadian glacial deposits are sandwiched between sedimentary strata consisting of reduced material underneath with oxidized substances above.

The reasons for the late Proterozoic glaciations are obscure, particularly in view of the distribution of the large land masses of the Earth close to the equator of the period (Fig. 10.1). The palaeomagnetic evidence for about 0.6 billion years ago indicates that much of the supercontinent of Gondwanaland, made up of present-day Antarctica, Africa, South America, India, China, and Australia, and the units of Laurasia, which are North America, Europe, and Siberia today, lay between the latitudes of 30°N and 60°S, with the palaeoequator passing through the regions that became Antarctica, India, Africa, Siberia, and North America.

More quantitative evidence for the temperatures prevailing over the surface of the Precambrian Earth comes from deposits of gypsum

($CaSO_4.2H_2O$) dating from 3.5 to 0.7 billion years ago. Gypsum separates from aqueous solution below 58°C or from saturated sodium chloride solution below 18°C. Above these temperatures under the respective conditions the calcium sulphate crystallizes without water of hydration (anhydrite, $CaSO_4$). The coexistence of gypsum with halite ($NaCl$) in deposits found in Canada (~2 billion years old) and Australia (~1.6 billion years old) suggests approximately 18°C as an upper limit for the water temperature at these times and places, probably then nearer to the palaeoequator.

A minor contribution to the warming of the Precambrian Earth came from the frictional conversion to heat of the kinetic energy of the large tides during the early transfer of part of the Earth's diurnal spin angular momentum to the Moon. Thereby the Moon acquired a larger orbital angular momentum at a greater separation from the Earth, and the day and the month lengthened while the number of days in the years was reduced. The microlaminations of stromatolites approximately 2 billion years old have been used by some workers to estimate a 25 day month and a 800–900 day year in the mid-Precambrian, but others find a 32 day month and a 448 day year at that epoch. It is generally accepted that the Precambrian stromatolites are poorer clocks than the Phanerozoic fossil corals and bivalves whose growth rings indicate a 400 day year and 22 hour day in the Devonian period 450 million years ago.

Precise data on the lengthening of the day are recent, coming from historical times (Rosenberg and Runcorn 1975). The ever-ingenious Edmund Halley found in 1695 that the times and places he calculated for ancient eclipses were displaced from those recorded by increments that suggested a lengthening of the day over the intervening period. The 129 BC solar eclipse observed by Hipparchus at Rhodes appears displaced by 80° of longitude or 5.3 h of the Earth's diurnal rotation in the current chronology. A discrepancy of 8.8 h is estimated for an earlier solar eclipse in 1223 BC, recorded on a clay tablet found at Ugarit (in modern Syria): the value is consistent with the deceleration of the Earth's diurnal rotation, or the corresponding acceleration in the Moon's mean longitude of some -23 s/(century)^2 determined from lunar and planetary observations since *circa* AD 1700 or from recent lunar laser-ranging measurements (de Jong and van Soldt 1989).

10.3 The Earth's early biosphere and atmosphere

The earliest known terrestrial organic carbon has the form of a bituminous polymer, containing much graphite, found in the 3.8 billion year old Isua rocks. The Isua sediments were subject to temperatures of about 550°C and pressures of 5 kbar (~5000 atmospheres), which produced some metamorphic transformations. While the more robust minerals remain intact, the chemical and isotopic composition of the more sensitive materials became mixed. In all unmetamorphosed sediments less than 3.5 billion years in age the organic carbon, represented entirely by polymeric kerogen in the older rocks, is depleted in ^{13}C relative to the inorganic carbon of limestones and other carbonates generally by a mean difference, $\triangle\delta(^{13}C)$, of $-27\%_{00}$. But the corresponding difference for the Isua sediments lies at only $-11\%_{00}$, owing to the partial mixing of the oranic and the inorganic carbon at the high temperature and pressure prevailing during the metamorphosis.

The difference refers to the isotopic fractionation δ-values for carbonate and organic carbon, $\triangle\delta = [\delta^{13}C_{carb} - \delta^{13}C_{org}]$, the δ-values being defined for the heavier isotope from the ratio $R = {}^{13}C/{}^{12}C$ for the sample and a standard by $\delta^{13}C = [R_{sample}/R_{standard} - 1] \times 1000$, in parts per thousand ($\%_{00}$), as for other elements. For carbon the standard is Peedee Belemnite (PDB), a dolomite, $CaMg(CO_3)_2$, so that $\delta^{13}C \sim 0$ for most sedimentary carbonates. The carbonates of the 3.8 billion year old Isua sediments are a little lighter, averaging $\delta^{13}C = -2\%_{00}$, just as the graphitic kerogen is heavier than the later sedimentary organic material from the metamorphic mixing.

The near-uniform depletion of the heavy carbon isotope at the $\delta^{13}C = -27\%_{00}$ mean level in the organic material of sediments less than 3.5 billion years in age is generally arrributed to the fixation of atmospheric carbon dioxide by photosynthetic organisms, although all chemical and physical processes involving carbon compounds produce some isotope fractionation. The kerogen and soluble organic substances of the C1 and C2 carbonaceous meteorites are even more depleted in the heavy carbon isotope relative to the inorganic meteoritic carbonate ($\triangle\delta^{13}C = -60$ to $-80\%_{00}$), and the rate of bolide infall to the Earth was still high at the time of the Isua sedimentation, decreasing by a factor of about one-tenth from 4.0 to 3.5 billion years ago. Further, the abiotic mechanisms producing organic substances on the parent bodies of the carbonaceous meteorites would have been effective on the surface of the Earth during the period of the initial reducing atmosphere of gaseous hydrides, producing substantial carbon isotope fractionation into light organic substances and heavy carbon dioxide. The entrainment of the abiotically synthesized organic material, terrestrial or meteoritic, into the Isua sedimentation represents a possible mechanism, alternative to biochemical fractionation, to account for the observed residual fractionation in the organic carbon of the sediments, averaging $\triangle\delta^{13}C \sim -11\%_{00}$ over the range -2.1 to $-22.9\%_{00}$.

The fixation of atmospheric carbon dioxide by the main photosynthetic process, involving the Calvin cycle, gives rise to organic substances depleted in the heavy carbon isotope at the average level of $\delta^{13}C = -27\%_{00}$, corresponding to the mean value ($-27 \pm 7\%_{00}$) for the organic material of the post 3.5 billion years old bp sediments. Relative to the PDB standard, primitive carbon in the form of diamond crystals coming up from the mantle of the Earth is depleted in the heavy isotope by $\delta^{13}C = -5\%_0$, reflecting the enrichment of the crustal inorganic carbonate in ^{13}C. Taken together, the $\delta^{13}C$ values for carbonate and primitive and organic carbon indicate that some 20 per cent of the total steady-state reservoir of mobile carbon in the atmosphere (CO_2), hydrosphere (HCO_3^-), and biosphere has had the form of organic material for the past 3.5 billion years (Schidlowski 1988).

The organisms that generated the earliest stromatolites (3.5 billion years old) and left organic carbon as kerogen in the fossil remains are taken to be the ancestors of photosynthetic bacteria which do not produce oxygen. Present-day representatives of such photosynthetic bacteria, notably the green and the purple sulphur bacteria, live in anaerobic environments and use hydrogen sulphide, rather than water, as the hydrogen source for a photoreduction of carbon dioxide to the carbohydrate level [CHOH]. The vast domes of sulphur found in Louisiana and Texas may have been the products of these primitive lifeforms.

The sulphur photobacteria, and their analogues using hydrogen or organic substrates as reductants, possess a single photosynthetic unit, PS I (Photosystem I), whereas the oxygen producers developed a coupled additional unit, PS II (Photosystem II), which splits water into oxygen and reducing equivalents. According to the general principle that the biochemically more primitive organisms were the first to be evolved, the sulphur photobacteria with PS I alone were prior in time to the oxygen-generating photobacteria, like the blue–green cyanobacteria, with the two coupled photosystems PS I and PS II. Bioenergetically more primitive organisms were even earlier on the same principle. Modern representatives are the anaerobic fermenting bacteria that lack not only chlorophyll but all other intermediates containing the porphyrin nucleus, such as the cytochromes, and metabolize carbohydrates by redox dismutation with no overall oxidation.

In modern stromatolites chains of cyanobacteria cells grow as mucilaginous filaments perpendicular to the upper surface of the structure, usually dome shaped, trapping suspended sediments. The sediment particles stick to the gum coating of the filaments, burying the dead forerunners of the growing cells at the top of the cell chain. Sulphides in the trapped sediment and the organic remains in the subsurface layers provide substrates for the sulphur photobacteria and their non-sulphur analogues.

Stromatolites today make a negligible contribution to carbon dioxide fixation and oxygen production. The overwhelming mass of modern cyanobacteria are free floaters, found mainly as picoplankton (0.2–2 μm in diameter) and nanoplankton (2–20 μm in size). The smaller picoplankton approximate to the least possible size estimated for a photosynthetic organism from the minimum amount of genetic material and number of catalysts required to maintain life. The picoplankton population averages 10^5 cells/cm^3 in the seas, or 10^6 cells/cm^3 in bodies of fresh water. Measurements with radioactive $^{14}CO_2$ show that the picoplankton are responsible for up to 80 per cent or more of the production of oxygen and organic substances in tropical waters, decreasing progressively at higher latitudes to 10–25 per cent in the arctic (Fogg 1986). The picoplankton predominate at about 80 m below the water surface, a depth that would have protected them from the photolytic short-wavelength solar radiation in the ultraviolet region before the formation of an ozone layer.

In addition to the fossil stromatolites, some 40 classes of microfossils have been reported from 28 different sediments of the late Archaean period, ending 2.5 billion years ago. These specimens are less well defined than the more numerous and more fully characterized microfossils of the following early Proterozoic period. Representatives of some 58 genera of simple spheroidal forms have been identified, together with several segmented filamentous types and tubular microstructures that once enclosed multicellular strands. The spheroidal fossil cells are generally less than 15 μm in size, commonly about 5 μm in diameter, and resemble the blue–green cyanobacteria of the picoplankton and nanoplankton class (Schopf 1983).

Studies of the sensitivity of the modern representatives of primitive organisms that flourished about 2.5 billion years ago to environmental variables afford an indication of the probable composition of the atmosphere during the Archaean era. The organisms converting atmospheric nitrogen to ammonia are mostly anaerobic and intolerant of all but the lowest oxygen levels. The aerobic organisms able to fix atmospheric nitrogen, like some filamentous cyanobacteria, confine the nitrogen fixation to specialist cells devoid of the oxygen-producing PS II, namely the heterocysts, which have thickened walls to limit the ingress of oxygen produced by neighbouring cells in the strand. The inhibition of both nitrogen fixation and photosynthesis in cyanobacteria with heterocysts is minimal in solutions under an atmosphere containing 10 per cent oxygen, i.e., about one-half the present atmospheric level but a level probable some 2 billion years ago. Single-cell cyanobacteria fixing nitrogen are confined to deep waters, where both the oxygen content and the light intensity are low, and the oxygen slowly formed by water photolysis rapidly diffuses away.

The enzyme through which atmospheric nitrogen becomes fixed, nitrogenase, is found in modern representatives of the primitive anaerobic fermenters and was evolved early to provide ammonia for subsequent biosynthetic reactions. The biosynthesis of nitrogenase is inhibited even more severely by the ammonium ion than by oxygen in solution, suggesting that the partial pressure of ammonia in the atmosphere was reduced to a very low level at an early stage. In unprotected organisms, the biosynthesis of nitrogenase is inhibited by solutions equilibrated with an atmosphere containing oxygen only at $10^{-3}-10^{-1}$ of the present atmospheric level (PAL), depending upon the particular species. Thus, at the time nitrogen fixation evolved, the atmospheric oxygen could not have much exceeded 10^{-3} PAL, although even this level allows the beginnings of an ozone screen; with oxygen at 10^{-1} PAL, the steady-state ozone layer developed photochemically by the solar radiation becomes largely biologically protective.

The response of photobacteria to partial pressure changes of carbon dioxide limits the likely range of carbon dioxide levels in the Archaean atmosphere. The enzyme catalysing the assimilation of carbon dioxide in the Calvin photosynthetic cycle, ribulose-1,5-bisphosphate (RuBP) carboxylase, has an activity not greatly affected by a small increase in the partial pressure of carbon dioxide. However, the biosynthesis of RuBP carboxylase in purple sulphur bacteria is inhibited at carbon dioxide concentrations 100 times larger than the present atmospheric level. Thus model Archaean atmospheres invoking carbon dioxide pressures of 1000 PAL appear to be unrealistic, and the carbon dioxide levels probably lay in the range from 1 to 100 PAL.

In addition to light and carbon dioxide, the sulphur photobacteria, possessing only Photosystem I, require aqueous sulphide with a concentration greater than 10^{-5} molar. The cell yields increase with larger sulphide concentrations up to 3–5 millimolar (mM), then level off and decline with the onset of sulphide toxicity at 7 mM. Some strains of the photosynthetic cyanobacteria, with both Photosystem I and II, are able to fix carbon dioxide photochemically, using only PS I with sulphide over a comparable but less extensive concentration range. The oxygen-producing PS II is inhibited by very low sulphide levels, beginning around 10^{-5} molar. The capacity of cyanobacteria to switch between water and hydrogen sulphide as electron donors for the photoreduction of carbon dioxide suggests that they developed from the sulphur photobacteria. Through the evolution of PS II, the cyanobacteria were able to move out from localized sulphide-rich areas, such as the hydrothermal vent systems, into the open waters (Towe 1985).

The limits to the range of oxygen and carbon dioxide partial pressure in the Archaean atmosphere suggested by the viability ranges of modern representatives of primitive organisms are broadly supported by the changes found in the oxidation state and acid–base balance of the minerals from the period. The dating of the minerals and their rate of deposition places the rise of a significant partial pressure of oxygen in the atmosphere to between 2.2 and 1.8 billion years ago. The main indicators of the appearance of appreciable atmospheric oxygen are the banded iron formations (BIF) which contain both ferrous and ferric iron, often near to the 1:1 ratio of magnetite, Fe_3O_4, with more Fe(II) in the earlier deposits and less in the later. The BIF come to an end around 1.8 billion years ago and are followed by the red beds of ironstone, containing haematite, Fe_2O_3.

The BIF contain about 30 per cent iron and are widely worked as sources of iron ore. The bands, several metres thick, are made up of layers of silica as chert and iron as magnetite. The chert bands are microlaminated into iron-rich and iron-poor layers approximately 1 mm thick, considered to represent seasonal variations. The earlier BIF, 3.5–2.5 billion years old, contain additionally some ferrous iron as the carbonate $FeCO_3$, silicate Fe_2SiO_4, and the sulphides FeS (pyrrhotite) and FeS_2 (pyrite), but these Fe(II) minerals decrease to insignificant levels in the later BIF deposits laid down under more oxidizing conditions.

The BIF served as a sink for the oxygen produced by the photolysis of water vapour in the atmosphere and, later, by the photobacteria which added the water-splitting PS II to the carbon-dioxide-fixing PS I. The photo-oxidation of soluble ferrous species in the oceans, such as $[FeOH]^+$, was another source of ferric iron and of hydrogen. In terms of mass, the earlier BIF from 3.5 to 2.5 billion years ago are small and cover a few square kilometres up to 10 m thick, but the later formations from 2.5 to 1.8 billion years bp extend over ranges 10^2–10^3 km long and 10^2–10^3 m thick. The photochemical sources of oxygen and ferric iron adequately account for the volume of the early BIF, but the bulk of the later BIF and the red beds of ferric iron following them require the additional oxygen input from organic photosynthesis.

Two further geochemical systems taken in conjunction, the palaeosols and the uraninites, define the partial pressures of both oxygen and carbon dioxide over broad limits around 2.5 to 2.0 billion years ago. The palaeosols, or fossil soils, were formed by the attack of water containing carbon dioxide and oxygen on rocks made up of ferrous silicates and other minerals. In modern soils the Fe(II) released is soon oxidized to Fe(III), which precipitates out as $Fe(OH)_3$, so that the iron accumulates near to the soil surface. In ancient soils the iron is concentrated near to the bottom of the soil column since, with low oxygen levels but a plentiful carbon dioxide supply, the soluble Fe(II) percolates downwards with the ground water. The analysis of the Fe(II) and Fe(III) content of successive layers in a dated fossil soil column, and of other substances consuming oxygen or carbon dioxide, provides the ratio of the partial pressures of the two gases at the time, but not the absolute value of either.

Grains of the mineral uraninite UO_2, the insoluble oxide of uranium (IV), were weathered from the Archaean rocks, transported by the rivers, and deposited in the estuaries to form some of the main uranium ore lodes. Most uraninite deposits are older than about 2 billion years, when the atmospheric partial pressure of oxygen began to become appreciable, owing to the ready oxidation of U(IV) to U(VI). In the presence of water and carbon dioxide the insoluble UO_2 becomes oxidized to the soluble carbonato-complex, $[UO_2(CO_3)_2]^{2-}$, at a rate proportional to the product of the partial pressures of oxygen and carbon dioxide but independent of the acid–base balance of the water near to neutral conditions (pH 6–8).

From the ratio of the partial pressures of oxygen and carbon dioxide given by the analysis of the fossil soils, and the product of those pressures determined by uraninite survival measurements, it is estimated that about 2.5 billion years ago atmospheric oxygen was up to a concentration between 2×10^{-2} and 3×10^{-3} of the present atmospheric level while the carbon dioxide concentration might have decreased to the present level already or was larger, at most, by approximately 600 times (Holland 1984).

Some 1.8 billion years ago the soluble U(VI) carbonato-complex leached from the entire watershed of a West African river was localized and redeposited as uraninite UO_2 in the anaerobic reducing ooze of the river delta. The process formed a rich uranium ore, now mined at Oklo in the Gabon Republic of West Africa. Parts of the ore lode were rich enough to form half a dozen natural uranium fission reactors about 1.8 billion years ago, producing 10–100 kW of power for hundreds of thousands of years and some 6×10^3 kg of fission products. The fraction of ^{235}U was then higher (~3 per cent) than it is today (0.72 per cent), owing to the shorter half-life of the isotope (0.71 billion years) than that of ^{238}U (4.51 billion years). The ^{235}U isotope plays the major role in the nuclear chain reaction of a fission reactor, and its depletion down to the 0.44 per cent level in parts of the Oklo ore led to the discovery of the natural reactors in 1972. The discovery was confirmed by the detection in the ore of typical uranium fission products, such as the lanthanides, with isotopic distributions characteristic of a fissiogenic origin.

The possible or even probable existence of Precambrian natural uranium fission reactors had been predicted during the 1950s, but American nuclear scientists led by Enrico Fermi (1901–54), who had collaborated in the construction of the first man-made nuclear reactor beneath the Stagg Field athletic stadium at the University of Chicago in 1942, dismissed the conjecture (Cowan 1976):

Some of the world's best physicists had constructed the Stagg Field reactor with careful attention to mechanical detail, to the purity of the materials and to the geometry of the assembly. Could nature have acheived the same results so casually?

11

The energetics of living systems

11.1 Spontaneous generation and biochemical autonomy

The development of the microscope during the seventeenth century revealed a new world of micro-organisms. Antonie van Leeuwenhoek (1632–1723), a haberdasher and civic official of Delft, was the most eminent observer of the period, fashioning more than 400 biconvex lenses from which he constructed some 247 microscopes. In drops of water from the sea, rivers, ponds, and rain-water tubs Leeuwenhoek observed motile 'little creatures', and likewise in beer, saliva, dental plaque, and faeces. His observations were described and illustrated in a series of more than 200 letters to the Royal Society of London from 1673, with size estimates for the organisms ranging down to diameters one-thousandth that of a grain of sand. The spheroid, rod, and spiral forms of his minute 'beasties' correspond to bacterial types and the larger organisms to algae, protozoa, and fungi (yeasts). From their presence in rain-water, van Leeuwenhoek supposed that microbes are omnipresent in the atmosphere, as well as in the waters, soils, and the macroscopic plants and animals on the surface of the Earth (Dobell 1932).

The discovery of micro-organisms appeared to demonstrate and amplify the ancient belief that small living creatures spontaneously generate from organic material, or matter at large, for the boundary between the inorganic and the organic was still indistinct. The view persisted that mineral ores grew underground within the 'living rock' and developed from the base metal to the noble metal type, although the latter notion encountered increasing scepticism. Georg Stahl (1660–1734), the main author of the phlogiston theory of combustion, observed that the tin ore of the Cornish mines must be an 'addled egg', since it was mined as tin ore by the Romans and, he supposed, by the Phoenicians before them, yet it had made no discernible progress towards native gold or even silver ore. But Stahl's theory that a pneuma-like principle, phlogiston, escaped from bodies during combustion, calcination, respiration, or fermentation was itself the near-terminal expression of a traditional vitalist thought-style, embodied in the general body–spirit model which had dominated the theory and guided the practice of proto-chemical science since the time of the Alexandrian Greek alchemists.

The model interpreted the results of pyrotechnical analysis as the separation of a spirit from its body, as in the calcination of a metal to produce the calx or 'dead body' after the escape of the pneuma moiety. Distillation afforded the essential active potencies of substances in condensed form, like the 'spirit of vitriol' (sulphuric acid from the dry distillation of hydrated ferrous sulphate) and the other alchemical spirits and essences. The iatrochemist J. B. van Helmont (1579–1644) of Brussels appreciated that some of the essences and spirits released were non-condensable, a class for which he coined the term 'gas', from the Greek for 'chaos'. These 'wild and untameable spirits' were taken to be universally active and energetic: van Helmont supposed that gases mediated the chemical and vital processes of combustion, respiration, and fermentation, and spontaneously endeavoured to escape from any vessel designed to imprison them.

The ensuing study of gas reactions, the pneumatic chemistry of the eighteenth century, eliminated the remaining hold of the general body–spirit model upon the physical scienes and introduced the beginnings of a chemical reductionism into biology. For the Unitarian minister and ardent phlogistonist Joseph Priestley (1733–1804), the vital gas he discovered in 1774 by training his burning glass on to the red calx of mercury (mercuric oxide) was 'dephlogisticated air', the component of the atmosphere serving as an absorbent limbo for the phlogistic spirit ascending from spent combustibles and expired airs. With a revolution in chemistry already planned, Antoine Lavoisier (1743–94) took the new vital gas to be the universal constituent of the products of combustion and calcination, required to account for the weight increase observed in these processes. The accompanying luminous and thermal effects registered the disengagement of the weightless elements of light and of caloric, the imponderable matter of fire. Lavoisier held that the products of combustion were invariably acidic, and so termed the new vital gas 'oxygen', from the Greek for 'acid-generator'. Muriatic acid (hydrogen chloride) and muriate oxide (chlorine) presented problems early on, and Humphry Davy (1778–1829) gave them their current names on finding no oxy-derivatives in the products of their reactions with the alkali metals.

Lavoisier and his school, having eliminated vitalist interpretations from combustion and calcination, carried forward the chemical revolution into the other fields of classical proto-chemistry, respiration, and fermentation. From 1783, with the mathematician and astronomer Pierre Simon Laplace (1749–1827), Lavoisier used the ice calorimeter and carbon dioxide absorbents to show that the respiration of a guinea pig produced nearly the same ratio of heat to carbon dioxide as the combustion of charcoal:

Respiration is thus a combustion, very slow it is true, but perfectly similar to that of carbon; it occurs in the interior of the lungs, without disengagement of visible light since the matter of fire which becomes free is at once absorbed by the humidity of these organs.

From his studies of fermentation, Lavoisier introduced the Arabic term, 'alkohol', to replace the traditional name 'spirit of wine', since the latter was identical to the 'spirit of cider' from the Calvados region, or the 'spirit' of any fermented sugar. Measurements of the amounts of carbon dioxide and alcohol produced by the fermentation of a given weight of sugar led him to the view that the sugar split into two parts, one being oxidized at the expense of the other to give carbon dioxide, leaving the reduced portion as alcohol, with the consequence that 'if it were possible to reunite alkohol and carbonic acid together, we ought to form sugar'. The role of the yeast for Lavoisier was what was later to be termed 'catalytic'. On trying out the effects of iron filings as a possible fermenter of sugar solutions, Lavoisier was surprised to find only a little 'bark', from the explosion of the hydrogen evolved, and no change in the sugar.

Many examples of catalysis were known by 1835, when Berzelius (1779–1848) took over the term to cover generally cases where an agent active in a reaction is not consumed. Earlier the term 'catalysis' had been used with the original Greek meaning of 'down-loosening' for decompositions, while effects such as the promotion of gas reactions by metal surfaces were termed 'contact activity'. Berzelius now subsumed under the common head of 'catalytic force' these heterogeneous contact activities and homogeneous effects like the continuous production of ether and water when alcohol is run into heated sulphuric acid, or the hydrolysis of starch to glucose by either mineral acids or soluble enzyme extracts.

The fermentation of sugar by yeast was considered to be a case of heterogeneous catalysis, as appeared to be established by the experiments of Eilhard Mitscherlich (1794–1863), who had studied with Berzelius. In 1841 Mitscherlich reported that yeast in a tube closed at the lower end by filter paper gave rise to fermentation only within the tube when immersed in a sugar solution. Unlike the sugar molecules in the solution, the yeast particles were unable to pass through the filter paper, and it was concluded that the fermentation of sugar to alcohol and carbon dioxide took place only on the yeast surface, wholly analogous to the decomposition of hydrogen peroxide to oxygen and water on the surface of a platinum sponge.

Inspired by the work of Lavoisier, a school of chemical reductionists developed around Berzelius, including Mitscherlich and another former student, Friedrich Wöhler (1800–82), who produced urea 'without the aid of a kidney' by the isomerization of ammonium cyanate in 1828, together with the principal pontiff of early organic chemistry, Justus Liebig (1803–73). All opposed the view that yeast is a living organism which grows by fermenting sugar, proposed by Theodor Schwann (1810–82), Charles Cagniard-Latour (1777–1859), and others in the late 1830s from observations of the budding of yeast cells in sugar solutions. The reductionists held the yeast to be a decomposing organic precipitate which destabilized the otherwise-stable sugar and catalysed the internal rearrangement of the sugar to alcohol and carbon dioxide. Animalcules might well generate spontaneously from the decomposing yeast precipitate, as from any other decaying organic matter, but that was independent of the chemically determined course of the sugar fermentation. While Wöhler became disenchanted with the organic field and turned back to his minerals and the analysis of recovered meteorites, Liebig remained a biochemical reductionist to the end of his days, attacking the microbiological chemistry of Louis Pasteur (1822–1895) as late as 1870. In Paris Pasteur had a major opponent in Marcelin Berthelot (1827–1907), who sought 'to banish life from all explanations relative to oranic chemistry' by showing from 1855 onwards that a range of simple organic molecules could be synthesized directly from carbon and inorganic substances.

While teaching at Lille (1854–7) and carrying out imaginative research on the possible conversion of natural products into mirror-image structures by the continuous rotation of the plants producing them, Pasteur became concerned with the malfunctioning of a local beet-sugar fermentation. Instead of ethyl alcohol, lactic acid appeared to be a principal product, along with some amyl alcohol and butyric acid, while hydrogen appeared in the gases evolved. The microscope revealed a rich microflora in the fermenting liquor. In addition to the relatively large yeast globules, Pasteur observed smaller spheroids, rods, and comma-shaped organisms, some with flagella. A separation of the different microbe types and their individual cultivation in sugar solutions showed that each type gave different and characteristic products. Thus the course of a fermentation was governed by the biochemistry of the particular type of micro-organism, and not by the intrinsic affinities of the atomic groups in the sugar molecule, as the reductionists supposed. 'Fermentation is life without oxygen', Pasteur concluded.

The comma-shaped organisms produced butyric acid and hydrogen from sugar solutions, and they soon died on exposure to the oxygen of the atmosphere, being classified thereby as obligate anaerobes. The rods and small spheroids, or cocci, producing lactic acid from sugar, were tolerant of atmospheric oxygen but generally more competent without it, falling into the amphiaerobe or facultative aerobe class. Only the yeasts appeared to produce alcohol from sugar solutions, and only under anaerobic conditions. In the air the yeasts respired and reproduced efficiently without producing alcohol, oxidizing the sugar completely to water and carbon dioxide. The dry weight of yeast produced aerobically from a given amount of sugar turned out to be some 10 times greater than the corresponding anaerobic yield by fermentation (the 'Pasteur effect').

The spoiling of alcoholic fermentations and their products, or unprocessed organic materials such as milk, could be avoided by the heat sterilization of the substrates or products ('Pasteurization'), and by the use of pure yeast cultures and anaerobic conditions in fermentations. Recipes in cookery books of the eighteenth century describe the boiling of sealed bottles of

foodstuffs for their preservation but, before the time of Pasteur, the procedure was often ascribed to a Paris confectioner, Nicolas Appert (1750–
1841), who published an account of the method in 1810. In that year
Gay-Lussac (1778–1850) concluded, from an analysis of the residual gases
in Appert's preserves, that the success of the method was due to the absence
of oxygen in the gas remaining. However, Schwann showed in 1837 that, if
first passed through a heated tube, air or other oxygen-containing gases left
the preserves unaltered.

Pasteur found that heat-sterilized sugar solutions remained unchanged in
contact with air filtered through gun-cotton or asbestos; moreover, a gun-
cotton filter, dissolved in an alcohol–ether mixture, left a residue containing microbes trapped from the air which were visible under the microscope.
The relative populations of airborne microbes in different localities were
estimated by opening and then resealing sets of heat-sterilized sugar solutions, e.g. at various heights during the ascent of a mountain, where the
percentage of a set showing subsequent microbial growth decreased with
increasing elevation.

In 1861 Pasteur published on account of his early microbiological work,
prefaced by a history of previous studies from the time of Leeuwenhoek
and of the conflict of views on the spontaneous generation of living organisms. The Paris Academy of Sciences in 1862 awarded Pasteur the Alhum-
bert Prize for the best work on the subject of spontaneous generation, a
competition occasioned by the appearance in 1859 of a work on heterogenesis by Felix Pouchet (1800–72) at the Rouen museum of natural history.
Pouchet upheld the ancient vitalist tradition, maintaining that the seeds of
life are never extinguished and grow heterogeneously, without genetic determination, wherever organic materials are available. Pasteur's conclusion
that microbes are present throughout the atmosphere, termed the 'doctrine
of panspermia' by Pouchet, would if true, the latter supposed, result in a
perpetual pea-soup fog, impeding vision, respiration, and even motion.

Pouchet's supporter, Henry Bastian (1837–1915), a pathologist at University College London, subsequently clarified the issues discussed by
contrasting the tenets of heterogenesis with other views of generation.
Homogenesis describes the normal reproduction process of macroscopic
organisms where like breeds like. For micro-organisms there are two
additional alternatives: first, the random and undirected process of spontaneous heterogenesis from once-living and specifically organic matter due
to the persistence of a vital organizing force within it; and, second, archaebiosis, the primordial abiotic generation of organisms from non-living and
ultimately inorganic materials such as carbon dioxide, ammonia, water,
etc. Up to the end of the nineteenth century, the discussion of spontaneous
generation generally centred upon heterogenesis versus homogenesis, and
archaebiosis was rarely discussed: as Pouchet put it, 'it is then not mineral
molecules but organic particles which are called into life'.

The spontaneous generation question became particularly controversial in France during the 1860s due, in part, to the appearance in 1862 of a translation of Charles Darwin's *Origin of species* (1859) with an anti-clerical preface. As Pasteur indicated in a Sorbonne lecture of 1864, the theory of the evolution of the organic species by natural selection required an original act of spontaneous generation to begin the process. The two theories are necessarily connected and the doctrine of spontaneous generation, Pasteur noted, 'has followed the developmental pattern of all false ideas'. The question continued to haunt Pasteur over the following two decades. 'Spontaneous generation?' he wrote in 1878, 'I have been looking for it for twenty years but I have not found it, although I do not think that it is impossible.' If found, Pasteur felt, the secret would lie in the chiral force of nature he had sought since his 1848 optical resolution of racemic sodium ammonium tartrate into its enantiomers. 'Life is dominated by these dissymmetric actions of whose enveloping and cosmic existence we have some indication', remarked Pasteur in an 1883 lecture to the Paris Chemical Society, 'What can one say of the development of plant and animal species if it becomes feasible to replace cellulose, albumin, and their analogues in the living cell by their optical enantiomers?'

In Britain Charles Darwin (1809–82) and his supporters, such as Thomas Henry Huxley (1825–95) and John Tyndall (1820–93), decoupled the question of the origin of life and the evolution of the organic species from that of present-day spontaneous generation. Any protein molecule abiotically produced today, Darwin observed in 1871, 'would be instantly devoured, or absorbed, which would not have been the case before living creatures formed'. At the Royal Institution, London, Tyndall followed up the observations of Pouchet and Bastian, that infusions of hay and other organic materials, even after prolonged boiling in sealed vessels, produced a rich population of microbes after a few days. In 1877 Tyndall showed that a second boiling, after a period to allow the heat-resistant spores to germinate, afforded a sterile infusion. Followed by a third boiling for greater certainty, the process became known as 'Tyndallization'.

In 1898 Francis Japp (1848–1925), an organic chemist at the University of Aberdeen, addressed the British Association for the Advancement of Science on the subject of 'Stereochemistry and Vitalism', taking up a demarcation criterion between the chemistry of the laboratory and the chemistry of life laid down by Pasteur in his 1860 Sorbonne lectures on molecular dissymmetry. Optically active organic molecules with a handed structure non-superposable upon the corresponding mirror-image form, the enantiomer, are biosynthesized by living organisms. In contrast the chemist in the laboratory could synthesize only optically inactive mixtures of the two oppositely handed enantiomers (racemic mixtures) without the aid of natural organic products as differentiating reagents. This observation, Japp suggested, indicated that the organic world is infused by a *vis vitalis*, 'a force of precisely the same character as that which enables the intelligent operator, by the exercise of his Will, to select one crystallised enantiomorph and reject its asymmetric opposite'.

In an extension of such chemical animism, the asymmetric carbon atom appeared to provide the primary expression of the unique vital force, so that all optically active molecules were expected to be organic, or at least to contain carbon. The first chiral metal coordination compounds synthesized by Alfred Werner (1866–1919) at Zürich indeed contained organic diamines. Stimulated by the vitalist challenge Werner, in a *tour de force*, synthesized and resolved into its optical enantiomers in 1914 a wholly inorganic tris-chelate coordination compound containing no carbon—the dodecammine hexa-μ-hydroxotetracobalt (III) complex ion, $[Co\{(OH)_2Co(NH_3)_4\}_3]^{6+}$ (Fig. 11.1).

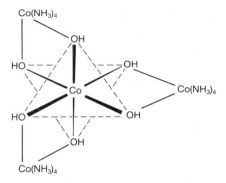

Fig. 11.1 The chiral polynuclear coordination compound containing no carbon atoms (the dodecammine-μ-hydroxotetracobalt(III) ion), synthesized and optically resolved into its enantiomers by Alfred Werner in 1914.

11.2 Fermentation and glycolysis

During the nineteenth century some two dozen 'soluble ferments' were isolated, generally as precipitates deposited on the addition of ethyl alcohol or acetone to an aqueous extract of plant or animal material. The soluble ferments were characterized by their specific catalytic activity, usually the hydrolysis of a particular carbohydrate, protein, or lipid. Often the reaction was promoted just as well, but non-specifically, by an inorganic acid or alkali. These proteinaceous catalysts were distinguished as 'unorganized ferments' from the intact cells of microbial organisms, such as yeast, which were termed 'organized ferments', although the two categories were sometimes confused in controversy, as in the case of the dispute between Liebig and Pasteur on the nature of fermentations. In order to clarify the relation between the two types, Willy Kühne (1837–1900) at Heidelberg introduced the term enzyme (Greek, 'in yeast') in 1878 to cover the soluble unorganized ferments which, he proposed, are present in and produced by the organized ferments, whether microbial or more complex. All plants and animals as well as microbes, Kühne pointed out, must be designated 'organized ferments' according to the older nomenclature.

The most remarkable feature of the enzymes known by the 1890s was their substrate specificity, since the hydrolytic and oxidative types of reaction they mediated were commonplace in the chemical laboratory. The development of stereochemistry from Kekulé's flatland molecular structure theory for aromatic substances (1865) to the generalized three-dimensional 'chemistry in space' of Le Bel and van't Hoff (1874) brought an increasing appreciation of the delicacy and discrimination of enzymatic substrate specificity, especially through the work of Emil Fischer.

The predictions by van't Hoff of the number and types of stereoisomers resulting from a chain of bonded asymmetric carbon atoms $R-[CXY]_n-R'$, were initially tested and subsequently used as a guide by Fischer in his studies of the sugar series. Fischer found in the 1890s, as van't Hoff had foreseen in 1874, that there are 2^n stereoisomers if $R-$ and $-R'$ are not identical atoms or groups. All of the 2^n stereoisomers are optically active, rotating the plane of polarized light, and they are made up of 2^{n-1} chemically distinct diastereoisomers. Each diastereomer consists of a pair of enantiomeric molecules, distinguished by their oppositely signed optical rotations, which are equal in magnitude, but apparently not by their reactions with achiral reagents. The two enantiomers have structures which differ only in that each one has the non-superposable mirror-image form of the other, the molecules lacking any reflection–rotation axis in their symmetry elements. If $R-$ and $-R'$ are identical groups and the number of chiral centres (asymmetric carbon atoms) in the chain is even, $R-[CXY]_n-R$, there are 2^{n-1} optically active stereoisomers, grouped into pairs of enantiomers, and $2^{(n-2)/2}$ optically inactive *meso*-isomers with an internally compensated chirality, one-half of the molecule having the non-superposable mirror-image structure of the other. The expectations for the analogous case of an odd number of chiral centres in the chain, a total of 2^{n-1} stereoisomers of which $2^{(n-1)/2}$ are inactive *meso*-forms, were supplied and verified experimentally by Fischer himself (1892).

With his extensive range of natural and synthetic sugar derivatives, Fischer found the action of microbes and enzyme preparations to be highly selective, not only between the diastereoisomeric substrates of a given series, but also between their two mirror-image forms, the L- and the D-enantiomers of his configurational convention. The action of 12 different pure strains of yeasts on a set of 14 sugars showed that only four monosaccharides are fermentable, three from the aldohexose series (D-glucose, D-galactose, and D-mannose) and one ketohexose (D-fructose). The common stereo-

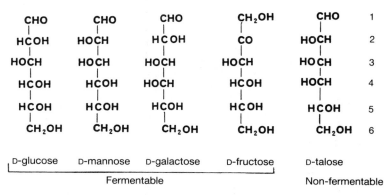

Fig. 11.2 The hexose monosaccharides found by Emil Fischer (1894) to be ferment-able with pure yeast strains (D-glucose, D-mannose, D-galactose, and D-fructose) or non-fermentable (D-talose).

chemical configuration at carbon-5 of these hexose sugars, specifying Fischer's D-series, was a necessary but not sufficient condition for fermentability. D-talose proved to be a non-fermentable aldohexose, yet all of the chiral carbon atoms individually, those at the 5-, 4-, 3-, and 2-position, have the same stereochemical configuration as the corresponding carbon atom of one or more of the fermentable hexose sugars. Only the sequence of carbon atom configurations in D-talose is distinctive, with an ordering different from that of two of the fermentable aldohexoses in only a single chiral position (Fig. 11.2).

By means of the enzyme preparations emulsin, first isolated from sweet and bitter almonds by Liebig and Wöhler in 1837, and 'invertin' (containing maltase and invertase), obtained from yeast by Berthelot in 1860, Fischer found an even more specific substrate discrimination in the hydrolysis of the two types of D-glucoside found in natural products or synthesized in the laboratory (Fig. 11.3). The treatment of glucose with methyl alcohol and acid gives two isomers, α- and β-methyl glucoside, through the introduction of an additional chiral centre by semi-acetal formation. Emulsin hydrolyses only a β-D-glucoside while maltase is active solely with an α-D-glucoside substrate. Both enzyme preparations are inactive to the corresponding mirror-image L-glucoside substrates and to other semi-acetal sugar derivatives.

These and analogous observations led Fischer to propose that an enzyme and its specific substrate have complementary molecular structures, allowing close contact and intimate interaction. 'To use a picture', he wrote in 1894, 'I would say that the enzyme and the glucoside must fit each other like a lock and key, in order to effect a chemical reaction on each other.' In this respect the enzymes did not differ from standard laboratory materials.

β-methyl-D-glucoside (R=Me)
β-D-glucose (R=H)

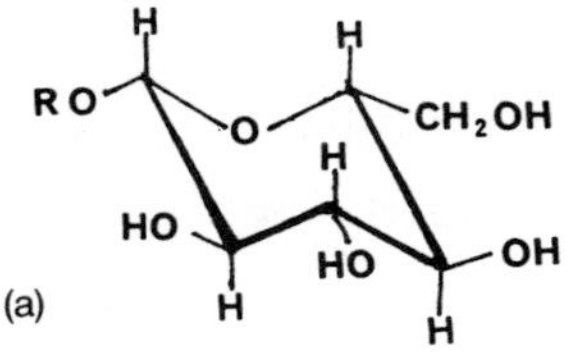

α-methyl-D-glucoside (R=Me)
α-D-glucose (R=H)

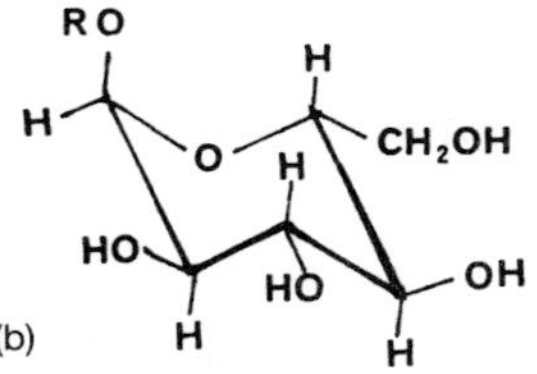

Fig. 11.3 The structure of β- and α-D-glucoside isomers.

Fischer discovered a substantial selectivity in the ascent of the sugar series using achiral reagents. From an aldopentose, arabinose, the addition of a further chiral centre afforded two aldohexose diastereoisomers, but in grossly unequal yields. Mannose was the major product and the proportion of the other, glucose, was so small that the latter had remained undetected in the earlier work of Kiliani, whose reaction Fischer employed: HCN + R–CHO → R–CH(OH)–CN → R–CH(OH)–COOH → R–CH(OH)–CHO. While the Kiliani reaction was only stereoselective, in contrast to the respective stereospecificity of emulsin and maltase for β- and α-D-glucosides, the similarities led Fischer to conclude 'that the difference frequently assumed in the past to exist between the chemical activity of living cells and of chemical reagents, in regard to molecular asymmetry, is nonexistent'.

The discovery of cell-free fermentation in 1897 by Eduard Buchner (1860–1917), an organic chemist at Tübingen, extended Fischer's chemical perspective by opening up the study of molecular intermediates on the pathway between the sugar substrate and the fermentation products. With a filter press, Buchner obtained a cell-free juice from the paste formed by brewer's yeast ground up with quartz sand and kieselguhr. On adding a concentrated sucrose solution, to serve as a preservative for the labile press juice, Buchner noted the formation of carbon dioxide after less than an hour at icebox temperatures, the gas evolution continuing for many days. After 3 days he isolated some 1.5 cm^3 of ethyl alcohol from 50 cm^3 of the cell-free juice expressed from 100 g of yeast. Buchner attributed the cell-free fermentation to a single enzyme, which he termed 'zymase'.

Brewing technologists were generally sceptical of Buchner's results, invoking the authority of Pasteur for the orthodox view that only 'organized ferments', intact microbial cells, could bring about the fermentation of sugars. The disciples of Pasteur on the other hand hailed Buchner's discovery as a scientific breakthrough. 'For a long time enzymes were believed to carry out only hydrolysis', observed Emile Duclaux (1840–1904), director of the Pasteur Institute in Paris, 'Buchner's alcoholic enzyme ... is the first to cause changes in carbon chains and the rearrangement of groups.'

The 'zymase' of Buchner was found to be a complex series of enzymes by Arthur Harden (1865–1940) at the Lister Institute for Preventative Medicine in London. Harden, with William John Young (1878–1942), discovered in 1906 that inorganic phosphate promoted the alcoholic fermentation of glucose by yeast juice in proportion to the amount of added phosphate, and that a dialysable heat-stable 'co-ferment' in the juice is essential for fermentation. In 1908 Harden and Young isolated from the fermentation mixture an organic phosphate which they identified as a hexose diphosphate, shown to be fructose-1,6-diphosphate 20 years later by Phoebus Levene (1868–1940) with Albert Raymond at the Rockefeller Institute. With Robert Robison (1883–1941), Harden found in 1914 a hexose monophosphate in fermentation mixtures, and Robison concluded in 1922 that the product was an isomeric mixture of fructose and glucose monophosphate.

Otto Meyerhof (1884–1951) at Berlin obtained in 1926 a cell-free press juice from muscle tissue which converted glycogen through glucose to lactic acid. The glycolysis of glucose to lactate was promoted by the addition of inorganic phosphate, like the alcoholic fermentation of glucose by yeast press juice. His colleague Gustav Embden (1874–1933) at Frankfurt/Main isolated a hexose monophosphate mixture from the muscle extract similar to the mixture of fructose and glucose monophosphate obtained by Robison from yeast juice, and the addition of the Harden–Young hexose diphosphate to muscle press juice was found to give a substantial increase in the production of lactic acid. By 1930 Embden and Meyerhof had come to the view that the glycolytic conversion of glucose to lactic acid in muscle and the breakdown of glucose to ethyl alcohol and carbon dioxide by yeast fermentation have marked similarities, both processes being mediated by the hexose phosphates and a common heat-stable 'co-ferment'. The conversion of glucose to lactic acid by the lactic acid bacteria, discovered by Pasteur, reinforced the analogy between glycolysis and fermentation.

The characterization of the Harden–Young ester as fructose-1,6-diphosphate in 1928 suggested that cleavage to the triose esters, glyceraldehyde-3-phosphate and dihydroxyacetone phosphate, might be the following stage in the anaerobic metabolism of glucose. Stimulated by the proposal, Hermann Fischer (1888–1960) at the Chemical Institute of Berlin University (like his father, Emil, before him) synthesized racemic glyceraldehyde-3-phosphate with Erich Baer in 1932 as a test substrate for fermentation. Yeast metabolized only one-half of the racemic mixture, and the reactive enantiomer was subsequently shown to be the 3-phosphate ester of D-(+)-glyceraldehyde, the parent triose of Emil Fischer's D-series of sugars, dominant in the economy of the organic world.

During the 1930s the sequence of intermediates in the Embden–Meyerhof pathway of the anaerobic breakdown of glucose became clear, each stage being catalysed by a soluble enzyme in muscle extract and in yeast press juice. After the esterification of D-glucose to D-fructose-1,6-diphosphate, via the 6-phosphate ester of first glucose and then fructose, enzymatic cleavage by aldolase results in the two monoesters D-glyceraldehyde-3-phosphate and dihydroxyacetone phosphate which are interconverted by triose phosphate isomerase. The oxidation of D-glyceraldehyde-3-phosphate to phosphoglyceric acid, followed by the transformation of the latter to pyruvic acid, is coupled to and balanced by the reduction of pyruvate to lactate in glycolysis, or the reduction of acetaldehyde from the decarboxylation of pyruvate to ethanol in yeast fermentation (Fig. 11.4).

The 'co-ferment' which Harden and Young had isolated in 1906 by the dialysis of boiled and filtered yeast press juice was shown during the 1930s to be a mixture of substances essential for oxidative metabolism as well as anaerobic glycolysis and fermentation. Otto Warburg (1883–1970) and Walter Christian (1907–55) at Berlin-Dahlem obtained a 'co-ferment' required for the oxidative metabolism of glucose from red blood cells in 1934 and what they termed the 'fermentation co-ferment' from the same source 2 years later. Both substances were shown to be composed of a nicotinamide, an adenine, and two pentose components, with two phosphate groups in the latter and three in the former. Known initially as coenzymes I and II, they came to be termed diphosphopyridine and triphosphopyridine nucleotides (DPN and TPN) and, after their laboratory synthesis in 1957, nicotinamide adenine dinucleotide (NAD^+) and nicotinamide adenine dinucleotide phosphate ($NADP^+$) (Fig. 11.5).

Warburg found that the nicotinamide component of the cofactors was the active moiety in the enzymatic redox reactions requiring NAD^+ or $NADP^+$. Such reactions were readily monitored spectrophotometrically from the new absorption band near 340 nanometres (nm, 10^{-9} metres) appearing in the electronic spectrum of nicotinamide when the pyridine ring is transformed to the dihydro form in the reduced cofactors NADH or NADPH. By this spectroscopic method Warburg showed in 1936 that the reduction of acetaldehyde to ethanol during fermentation requires the

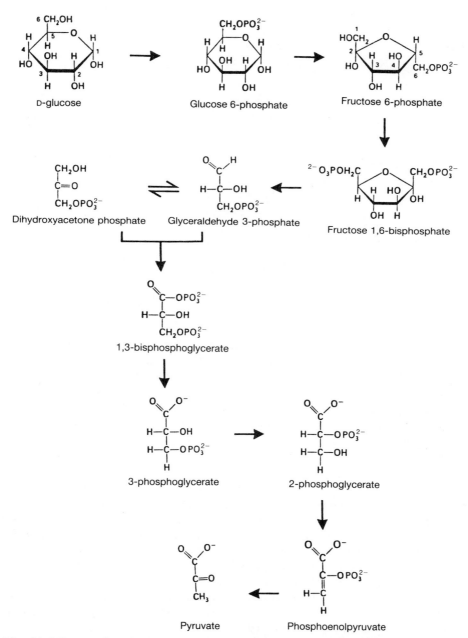

Fig. 11.4 Intermediates on the Embden–Meyerhof pathway for the glycolysis of glucose to pyruvate. See Fig. 11.8 for the recycling of the redox carrier (NAD+/NADH, Fig. 11.5) and the phosphorylation balance (ADP/ATP, Fig. 11.7).

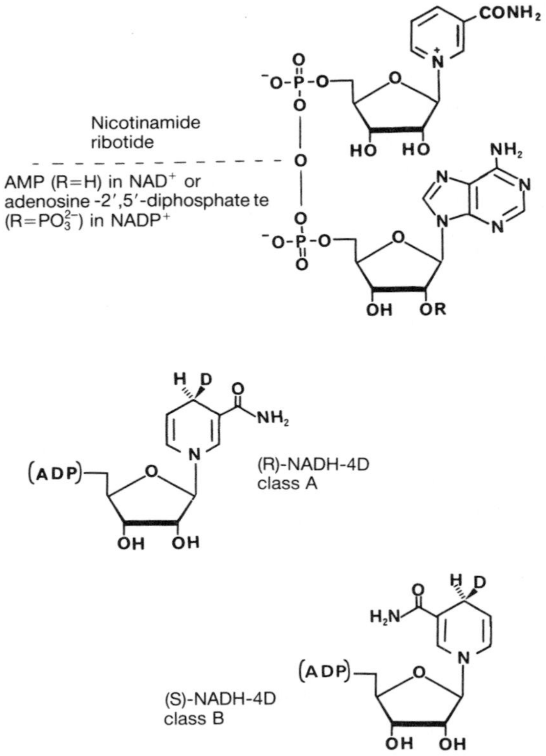

Fig. 11.5 Structures of nicotinamide adenine dinucleotide (NAD⁺) and the corresponding phosphate (NADP⁺), made up of a nicotinamide ribotide and an adenine ribotide moiety; and of the reduction product of NAD⁺ from a deuterated substrate catalysed by a dehydrogenase of class A (R-NADH-4D) or of class B (S-NADH-4D) (ADP = adenosine-5'-diphosphate).

concomitant oxidation of NADH to NAD⁺. Two years later Warburg found that the oxidation and acyl phosphorylation of D-(+)-glyceraldehyde-3-phosphate in fermentation and glycolysis to 1,3-diphospho-D-glyceric acid involves the parallel reduction of NAD⁺ to NADH.

Warburg's associates, E. Negelein and H. Wulff, crystallized in 1937 the enzyme catalyst of the acetaldehyde reduction in fermentation, yeast alcohol dehydrogenase. The enzyme catalysing the oxidative stage, glyceraldehyde phosphate dehydrogenase, was similarly isolated in pure crystalline form from yeast press juice in 1939.

The detailed steric course of these enzymatic reactions was later established, from the early 1950s, by deuterium substitution for hydrogen in the substrate or the cofactor, or by the replacement of the aqueous solvent by

D_2O. The catalysis by yeast alcohol dehydrogenase of the reduction of acetaldehyde by NADH in heavy water gives a deuterium-free alcohol product, indicating that the hydrogen transferred to the substrate does not come from the solvent. The same enzyme–cofactor system with the deuterated substrate CH_3CDO gives specifically the chiral 1-deutero-ethanol, (S)-(−)-CH_3CHDOH. The enantiomeric product, (R)-(+)-CH_3CHDOH, results from the enzymatic reaction of the normal acetaldehyde substrate with the reduced cofactor deuterated in the 4-position of the pyridine ring in the nicotinamide moiety, NADH-4D.

The 4-carbon atom of the pyridine ring in the reduced coenzyme NADH (or NADPH) is prochiral and the two bonded hydrogen atoms (H_R and H_S) are inequivalent, giving two stereoisomers on monodeuteration, (R)-NADH-4D and (S)-NADH-4D. These stereoisomers distinguish two types of enzyme reaction dependent upon NAD^+ (or $NADP^+$) in the 100 or more dehydrogenases subsequently characterized (Fig. 11.5). In type A the $4D_R$ atom is transferred to the substrate, or from the substrate to the 4-pyridine position in the reverse reaction (e.g. the reduction of acetaldehyde catalysed by the alcohol dehydrogenases, or the reverse oxidation of 1-dideutero-ethanol, CH_3CD_2OH, to CH_3CDO). In the type B enzyme reaction the $4D_S$ atom is specifically transferred to the substrate, or to the 4-pyridine position of NAD^+ from the substrate in the reverse reaction (e.g. the oxidation and acyl phosphorylation of D-glyceraldehyde-3-phosphate to 1,3-diphospho-D-glyceric acid catalysed by glyceraldehyde phosphate dehydrogenase).

The class A dehydrogenases catalyse the reduction of the more reactive carbonyl compounds, while the reduction of their less reactive analogues is mediated by the class B enzymes. The crystal structures of the dehydrogenase–cofactor complexes determined by X-ray diffraction indicate that the NAD^+ is bound to the class A enzymes with the nicotinamide ring in the *anti* conformation about the glycosidic bond to the adjacent ribose, whereas the class B enzymes have the nicotinamide ring bound in the *syn* conformation. The correlation of the relative substrate reactivities of the class A and B dehydrogenases with the corresponding enzyme–cofactor crystal structures suggests that NADH becomes a stronger reducing agent in the *syn* conformation (Retey and Robinson 1982; Fersht 1985).

The X-ray diffraction crystal structures of these and analogous enzyme–substrate or substrate–analogue complexes support the stereochemical 'key and lock' hypothesis of Emil Fischer (1894) for the mechanism of enzyme catalysis in the later revised and extended forms of J. B. S. Haldane (1892–1964) and Linus Pauling. Pauling (1948a) developed the concept introduced by Haldane (1930) that the steric strain on the substrate due to binding by the enzyme promotes the chemical transition to the product. As Haldane put it: 'Using Fischer's lock and key simile, the key does not fit the lock perfectly but exercises a certain strain on it.' The strain concept was

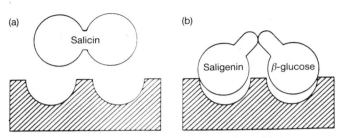

Fig. 11.6 The concept of steric strain in enzyme action (Haldane 1930), illustrated by the reversible emulsin-catalysed hydrolysis of the β-D-glucoside, salicin (R = o-HOPh–CH$_2$– in Fig. 11.3(a)). The binding site on the enzymes: (a) stretches the substrate towards the fission configuration for separation into the products, saligenin (o-hydroxybenzyl alcohol) and β-D-glucose; and (b) compresses the products towards the configuration required for reunion back to the substrate.

made more precise by Pauling who proposed that the complementarity of the enzyme's stereochemical fit is more complete for the transition state intermediate of the reversible reaction than for either the initial substrate or the ultimate product. Thus, on binding to the enzyme, the substrate becomes sterically deformed along the direction of the reaction coordinate towards the transition state common to both the forward and the reverse reaction catalysed by the enzyme (Fig. 11.6).

Comparisons of the salient structural features of a free enzyme with those of the corresponding substrate–enzyme complex led subsequently to the appreciation that the steric adjustments of enzyme to substrate are mutually induced. In the absence of the substrate, the enzyme does not have a structure complementary to the transition state for the substrate reaction. The polypeptide chains of the enzyme are labile, and it is only when a relatively rigid substrate molecule is bound that the catalytic groups of the enzyme are lined up in their optimal orientations to form a catalytic mould with the complementary morphology of the substrate's transition state for reaction.

The role in fermentation, glycolysis, and respiration of a second major constituent of the dialysable heat-stable 'co-ferment' isolated from yeast juice by Harden and Young in 1906 was characterized during the 1930s. 'Phosphagen', an organic phosphate detected in muscle preparations, appeared to be a source of muscle energy since the phosphagen disappeared during muscular contraction and then was restored rapidly at rest through oxygen-consuming reactions. The view was supported by the calorimetric measurements of the heat of hydrolysis of creatine phosphate, from vertebrate muscle, and arginine phosphate, from invertebrates, by Karl Lohmann and Otto Meyerhof, who found these reactions to be strongly

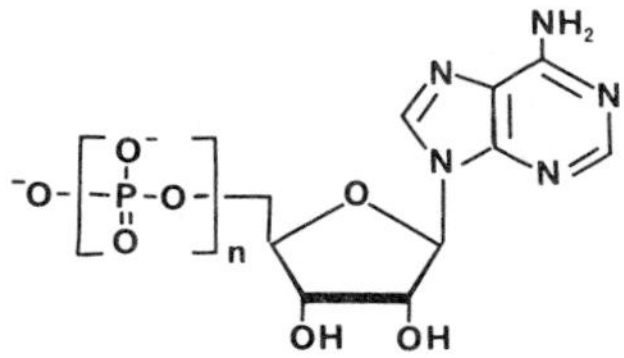

Fig. 11.7 Structure of adenosine-5'-monophosphate, -diphosphate, and -triphosphate: AMP ($n = 1$); ADP ($n = 2$); ATP ($n = 3$).

exothermic. In 1929 Lohmann traced back the origin of the inorganic pyrophosphate $H_4P_2O_7$ found in both muscle and yeast to another labile organic phosphate, containing adenine, and 6 years later he characterized the substance structurally as adenosine-5'-triphosphate (ATP). By that time Lohmann had isolated adenosine-5'-disphosphate (ADP), which was later shown to undergo a general enzymatic disproportionation to ATP and adenosine-5'-monophosphate (AMP) (Fig. 11.7).

The hydrolysis of ATP was found calorimetrically to liberate approximately twice as much heat as the hydrolytic cleavage of creatine phosphate, accounting for the phosphorylation of creatine by ATP on the assumption that heat-content changes reflect free-energy changes. Already in 1931 Meyerhof and Lohmann considered ATP to be an essential 'coenzyme' for glycolysis, serving as the primary phosphorylating agent of the intermediates on the pathway between glucose and lactic acid. The ATP hydrolyses during the process to inorganic phosphate and AMP, from which ATP is regenerated by harnessing the chemical energy derived from the cleavage of glucose to lactic acid through the organic phosphate intermediates. Meyerhof's group in 1938 showed that ATP production from ADP is coupled to the reduction of NAD^+ to NADH in the dehydrogenation of D-glyceraldehyde-3-phosphate to 3-phospho-D-glyceric acid in the anaerobic degradation of glucose. The enzymatic isomerization of the 3- to the 2-phospho-D-glyceric acid had been found by Lohmann in 1935 to be followed by a dehydration to phosphoenolpyruvic acid (PEP) which phosphorylates ADP to ATP with the liberation of pyruvic acid (Fig. 11.4). The ATP component of the Harden–Young co-ferment plays the same role in yeast fermentation up to the pyruvic acid stage. Here the anaerobic redox balance is restored by the coupling of the oxidation of NADH back to NAD^+ with the reduction of acetaldehyde, from the decarboxylation of pyruvate, to ethanol in fermentation, or with the direct reduction of pyruvate to lactate in glycolysis (Fig. 11.8). By the early 1940s ATP had come to be perceived as a general biochemical carrier of free energy and NAD^+/NADH as a general transfer couple for reducing equivalents, at least in anaerobic metabolic turnover.

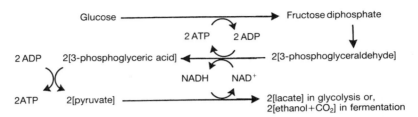

Fig. 11.8 The net production of two molecules of ATP in the glycolysis or the fermentation of a glucose molecule, with the recycling of NADH/NAD⁺ and of ADP/ATP.

22.3 Intracellular respiration

Studies of the oxidation of carbohydrate in respiration encountered problems absent from the investigation of the anaerobic degradation of glucose. The mechanisms of fermentation and glycolysis, mediated by soluble enzymes isolated from cell-free yeast or muscle extracts, could be studied by the classical kinetic and analytical methods, using homogeneous solutions. But the oxidative process of respiration, dependent upon membrane-bound enzymes, requred the use of intact cells or, after their isolation in the late 1940s, intact mitochondria, the membrane-enclosed organelles serving as the powerhouse of the cell.

Pasteur, using intact microbial cells, brought to light the substantially larger bioenergetic efficiency of respiration over anaerobic metabolism. The 'Pasteur effect' refers to his discovery that the dry-weight ratio of glucose consumed to yeast cells produced is approximately ten-fold larger by the fermentation pathway than through the oxidative respiration route. The ratio roughly corresponds in present-day molecular terms to the optimum ATP yield quotient via the two routes: 38 molecules of ATP per molecule of glucose oxidized to carbon dioxide in respiration, compared with 2 molecules of ATP per molecule of glucose anaerobically glycolized into lactate, or fermented into ethanol and carbon dioxide, by the Embden–Meyerhof pathway.

Studies of respiratory processes followed two main directions during the period of 1920–40: first, the spectrochemical investigation of the coloured catalysts mediating electron transfer to molecular oxygen; and second, the biochemical characterization of the intermediates linking the respiratory product, carbon dioxide, back to the penultimate stage of glycolysis, pyruvate.

The well-established vector of the breath of life, haemoglobin, had attracted the attention of the pioneer spectrochemists during the 1860s. George Gabriel Stokes (1819–1903), Lucasian professor of mathematics at

Cambridge, mapped out in 1864 the spectroscopic differences between the pigment of scarlet oxygenated arterial blood and purple deoxygenated venous blood. Stokes found that scarlet oxyhaemoglobin is characterized by two absorption bands in the visible wavelength region while the purple haemoglobin has only one, and he showed that a mild reducing agent (ferrous ammonium tartrate, 'Stokes reagent') transformed the spectrum of the former to that of the latter, whereas oxygen had the converse effect.

The studies of Stokes, termed 'chromatology' by Henry Sorby (1826–1908), a gifted amateur inventor–scientist of Sheffield, were extended by Ray Lankester (1847–1929), a zoologist at London and Oxford, and Charles MacMunn (1852–1911), a Birmingham physician and amateur chromatologist. In 1867 Sorby invented the microspectroscope by combining the microscope and spectrometer, and with this new instrument Lankester looked for haemoglobin-like pigments in a variety of organisms. In 1872 Lankester found 'muscle haemoglobin', later given its present name of myoglobin, in molluscs with no haemoglobin in their blood and he characterized spectroscopically a green respiratory pigment found in some marine worms. From 1884 MacMunn identified a respiratory pigment, termed myohaematin or histohaematin, widely distributed in most orders of the animal kingdom with four absorption bands in its visible spectrum, as opposed to the single band of haemoglobin or the two of oxyhaemoglobin.

The respiratory pigment of MacMunn was generally regarded as a decomposition product of haemoglobin until 1925 when David Keilin (1887–1963) at Cambridge showed with the microspectroscope that the pigment was present in bacteria and yeasts as well as insects and higher animals, and that it was a mixture of three substances. These substances Keilin termed cytochrome a, b, and c, and he showed that the four-banded spectrum of the MacMunn pigment arose from the superposition of the two-banded spectra of the individual cytochromes over the visible wavelength region, each having a common band at approximately 520 nm and a unique characteristic band at 605, 565, and 550 nm, respectively.

Over the 40 year period between the studies of MacMunn and those of Keilin the structure of the blood pigment haem, readily separable from the protein moiety, globin, with dilute acid, had been worked out by the now-classical methods of degradation and synthesis. Drastic methods were required for the degradation of the rather stable haem molecule, digestion with hydrogen iodide affording no less than eight different substituted pyrroles. However, on the basis of the degradative evidence then available William Küster (1863–1929) was able to propose in 1913 the correct cyclic tetrapyrrole structure enclosing iron, specifically Fe(II), by square–planar coordination. Hans Fischer (1881–1945) at Munich established the structure of haem by the synthesis of a wide range of porphyrins from 1926 on, including the dihydroporphyrins of the green leaf pigments, the chlorophylls,

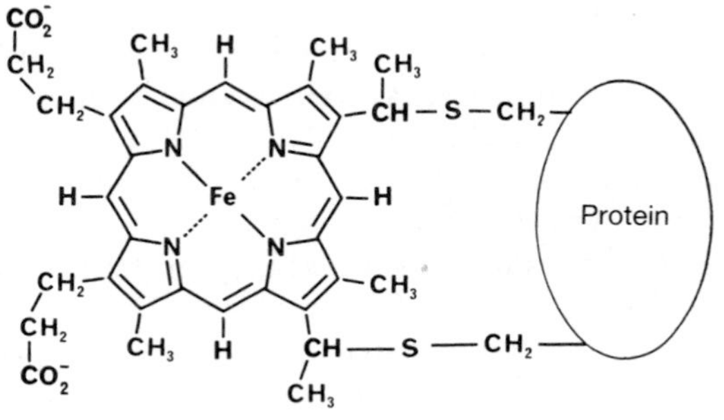

Fig. 11.9 The structure of cytochrome c: the two vinyl substituents of the porphyrin ring in haem are transformed into thioether links to the protein in cytochrome c by reaction with the HS-groups of two cysteine residues in the polypeptide.

and the tetrahydro-derivatives, the bacteriochlorophylls of the photosynthetic bacteria.

The same haem moiety was found to be bound to protein in cytochrome c, the sole cytochrome that could be extracted intact into aqueous solution (Fig. 11.9). The other cytochromes remained membrane bound, but their reactions were readily monitored by the microspectroscope. Keilin showed that the coordinated iron atom of the cytochromes changes from ferrous Fe(II) to ferric Fe(III) form on oxidation and back to Fe(II) on reduction at different redox potentials in the three members of the series. The values of the potentials indicate that electrons are transported along the cytochrome chain in the order $b \rightarrow c \rightarrow a$, and then to molecular oxygen, which becomes reduced to water. None of the cytochromes alone were found to react directly with oxygen, but in 1939 Keilin with E. F. Hartee showed that the enzyme cytochrome oxidase, containing cytochrome a and a copper protein, is auto-oxidized with molecular oxygen and terminates the cytochrome electron-transfer chain.

The graded redox steps of the cytochrome chain led to the view, developed from about 1940, that the energy from the oxidation of substrates in respiration is liberated in small steps from an input of reduced nicotinamide adenine dinucleotide (NADH) or the corresponding phosphate (NADPH) produced by the dehydrogenation of substrates. The first step involves a flavoprotein redox catalyst with a coloured prosthetic group, either the 'old yellow ferment' flavin mononucleotide (FMN), i.e. riboflavin D-ribityl phosphate, isolated from yeast in 1932 by Warburg and Christian, or the 'new yellow ferment' flavin adenine dinucleotide (FAD), characterized by Warburg and others in 1938 (Fig. 11.10). The amount of free energy released at each step of the respiratory chain was taken to be proportional to the difference between the redox potentials of the two systems bridged by the step. The redox potentials, adjusted to the biochemical

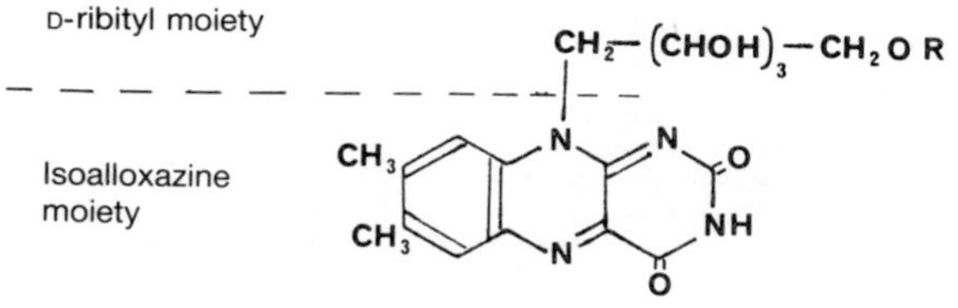

D-ribityl moiety

Isoalloxazine moiety

$CH_2-(CHOH)_3-CH_2OR$

Flavin mononucleotide (FMN), $R=PO_3^{2-}$

Flavin adenine dinucleotide (FAD), R=ADP

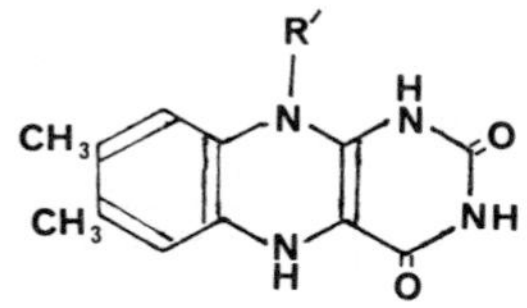

FMNH$_2$ and FADH$_2$, R' = side chain of FMN and FAD, respectively.

Fig. 11.10 Structure of riboflavin, R = H, flavin mononucleotide (FMN), R = PO$_3^{2-}$, flavin adenine dinucleotide (FAD), R = ADP, and the dihydro-derivatives, FMNH$_2$ and FADH$_2$.

standard (pH 7) from the physico-chemical standard of unit hydrogen ion activity (pH ~ 0), were estimated to have the values, in volts at 25°C: NAD$^+$ (or NADP$^+$) -0.3, FAD -0.1, cytochrome b $+0.1$, cytochrome c $+0.2$, cytochrome a $+0.3$, falling into the range between the hydrogen and the oxygen electrode potentials of -0.42 and $+0.82$.

Developments from 1950 led to a fuller characterization of the redox intermediates in the respiratory chain, and the division of the chain into three sets of enzyme systems, localized at specific membrane-bound sites and linked by mobile (soluble) electron carriers (Fig. 11.11). The first site, made up of the NADH dehydrogenase and the succinate dehydrogenase systems, is linked to the second site, the cytochrome reductase system, by coenzyme Q, ubiquinone, so named from its ubiquity in biochemical redox processes (Fig. 11.12). The soluble carrier, cytochrome c, transports electrons from the second to the third site, the cytochrome oxidase complex, which reacts directly with molecular oxygen.

Two electrons are transferred from NADH to flavin mononucleotide (FMN), which becomes reduced to FMNH$_2$ with the regeneration of NAD$^+$, in the NADH-dehydrogenase system. Iron–sulphur cluster proteins in the system mediate the subsequent transfer of electrons from FMNH$_2$ to the mobile intersystem carrier, ubiquinone (Q), which becomes reduced

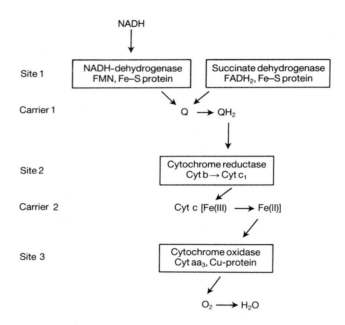

NADH

| Site 1 | NADH-dehydrogenase
FMN, Fe–S protein | Succinate dehydrogenase
FADH$_2$, Fe–S protein |

Carrier 1   Q $\longrightarrow$ QH$_2$

Site 2   Cytochrome reductase
Cyt b → Cyt c$_1$

Carrier 2   Cyt c [Fe(III) $\longrightarrow$ Fe(II)]

Site 3   Cytochrome oxidase
Cyt aa$_3$, Cu-protein

O$_2$ $\longrightarrow$ H$_2$O

Fig. 11.11 Electron transfer from reducing equivalents (NADH, FADH$_2$) through the three membrane-bound sites of the mitochondrial respiratory chain, linked by mobile redox carriers, to the terminal electron acceptor, molecular oxygen. The boxed enzyme systems and bound carriers remain associated when the membrane is partly dissociated.

first to the semiquinone (QH.) and then to ubiquinol (QH$_2$). Alternatively, reduced flavin adenine dinucleotide (FADH$_2$) from the succinate dehydrogenase system transfers electrons directly to the soluble ubiquinone carrier, with the regeneration of FAD.

At the second site, an iron–sulphur cluster protein of the cytochrome reductase complex serves as a one-electron acceptor from ubiquinol and a donor to membrane-bound cytochrome c_1, which transmits the electron to the next mobile intersystem carrier, the soluble cytochrome c. Cytochrome b in the cytochrome reductase complex catalyses the dismutation of the product of the initial one-electron transfer, the semiquinone (QH.), to ubiquinone (Q) and ubiquinol (QH$_2$).

The final reduction of molecular oxygen to water is completed at the third site, the cytochrome oxidase complex, which accepts four electrons by step-wise one-electron donation from the mobile cytochrome c in reduced (ferrous) form. The third complex contains two forms of cytochrome iron (cyt a and a_3) and two types of protein-bound copper. The concerted effect of the four types of coordinated metal ion results in the accumulation

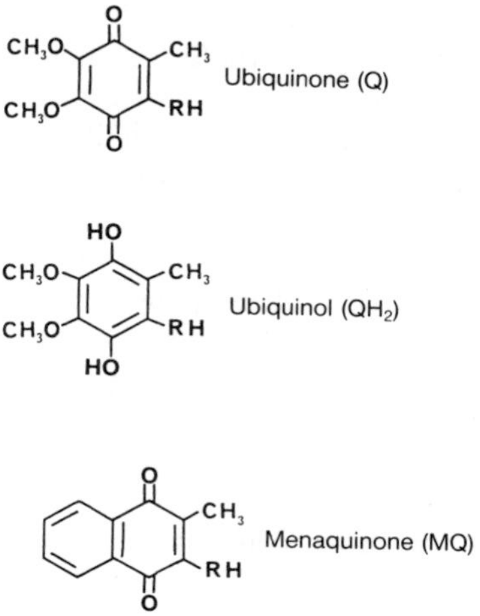

Fig. 11.12 The structures of quinone redox carriers: ubiquinone (Q) and its reduction product, ubiquinol (QH$_2$); and menaquininone (MQ). The number of isoprenoid units in the side chain (R = [$-CH_2-CH=C(CH_3)-CH_2-]_n$) varies in different organisms from $n = 6$ to $n = 10$.

of specifically *four* electrons by the cytochrome oxidase complex in the reduction cycle of one molecule of oxygen to two molecules of water. The accumulation plays a protective role. The firm retention of partly reduced intermediates in the reduction cycle precludes the release of the highly reactive superoxide ion (O$_2^-$) and other toxic oxygen radical species.

The primary source of the NADH fed into the respiratory chain, the tricarboxylic acid cycle, was characterized from the late 1930s principally by Hans Krebs (1900–1981), a biochemist from Germany who, like other Jewish refugees but unlike his one-time mentor, Otto Warburg, was spared the distinction of honorary Aryan status (Krebs 1981*a, b*).

The method of analysing the products from whole cells provided with potential substrates, introduced by Pasteur using microbes, was extended to the specialized cells of animal organs by Warburg, who employed thin tissue slices a few cells thick. The use of slices from tissues with a high respiration rate, such as pigeon breast muscle, enabled Albert Szent-Györgi (1883–1986) at Szeged to show (1934–7) that succinate, fumarate, malate, and oxaloacetate promote the oxidation of carbohydrate. These dicarboxylic acids, Szent-Györgi proposed, are components of a hydrogen-transport chain from carbohydrate to molecular oxygen, oxaloacetate being reduced to malate, which is dehydrated to fumarate, which in turn is reduced to succinate.

Krebs with W. A. Johnson at Sheffield in 1937 showed that citrate and analogous tricarboxylic acids similarly promote the respiration rate of

Chemical evolution

pigeon breast tissue slices, and they formulated the first version of the citric acid respiratory cycle, which includes the dicarboxylic acid chain of Szent-Györgi. From the observed enhancements of respiration rate, Krebs and Johnson suggested that citrate is dehydrated to *cis*-aconitate, which rehydrates to isocitrate. Dehydrogenation of the latter produces oxalosuccinate which, after two successive decarboxylations, first to α-ketoglutarate and then to succinate, leads to Szent-Györgi's chain: succinate → fumarate → malate → oxaloacetate. The salient proposal made by Krebs and Johnson was the completion of the cycle by the regeneration of citrate from oxaloacetate by the incorporation of a two-carbon residue from pyruvate, provided by the glycolysis of carbohydrate. The two carbon atoms served to compensate for the two molecules of carbon dioxide evolved in the decarboxylation stages (Fig. 11.13).

The two-carbon fragment was taken to be acetate, derived from the decarboxylation of pyruvate, but how it became incorporated into the citric acid cycle remained unclear until 1945 when Fritz Lipmann (1899–1986) discovered the cofactor that promotes the transfer of acetyl groups, coenzyme A (A for acetylation). Lipmann had investigated the phosphate ester intermediates in glycolysis with Meyerhof's group at Heidelberg, moving to Copenhagen in 1932 and thence, in 1939, to the USA, ultimately to the Rockefeller University. Lipmann found in 1939 that the immediate product of the oxidation of pyruvate by some bacteria is acetyl phosphate, but animal tissue experiments showed that this phosphate was not a general form of 'active acetate'. The search for activated acetate resulted in the isolation from pigeon liver slices of a heat-stable dialysable cofactor for acetylation, coenzyme A (CoA–SH), which forms an active thioester with fatty acids generally ($CoA-S-COC_nH_{2n+1}$).

By the early 1950s it was established that the pyruvic oxidase system of animal tissues generates acetyl-coenzyme A ($CoA-S-COCH_3$) and NADH by the reaction between pyruvate, NAD^+, and the coenzyme ($CoA-SH$) with the evolution of carbon dioxide: thence the $CoA-S-COCH_3$ formed transfers acetate to the tricarboxyllic acid cycle by transforming oxaloacetate to citrate. Parallel studies of the chemical degradation of coenzyme A showed that adenosine-3'-phosphate-5'-pyrophosphate, pantothenic acid, and β-mercaptoethylamine are constituent components. The structure deduced from the degradation studies was subsequently confirmed by the synthesis of the coenzyme A molecule from its component units (Fig. 11.14).

The oxidative metabolism of fats and proteins, as well as sugars, was found to proceed through the tricarboxylic acid cycle with acetyl-coenzyme A as the transfer mediator in each case. Each of the four oxidative steps within the cycle, namely the oxidation of isocitrate, 2-oxoglutarate, succinate, and malate, generate a molecule of reduced nicotinamide adenine dinucleotide or its phosphate NAD(P)H, or the equivalent reduced flavin

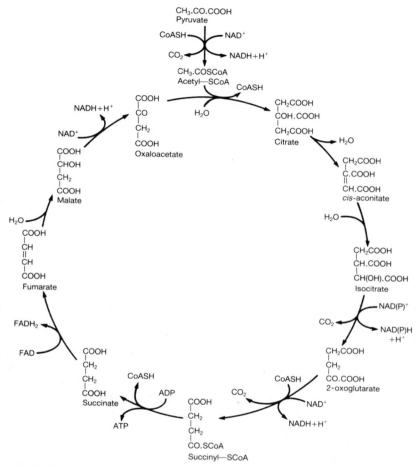

Fig. 11.13 The oxidative tricarboxylic acid cycle (TCA): five reducing equivalents [NADH, NADPH, or FADH$_2$] are generated by the oxidation of each pyruvate molecule to carbon dioxide, four by the reactions of the cycle and the fifth by the decarboxylation of pyruvate and the formation of acetyl coenzyme A (CH$_3$CO–SCoA) which enters the cycle in the transformation of oxaloacetate to citrate. Each reducing equivalent, on donating two electrons to the mitochondrial respiratory chain with the regeneration of the oxidized form [NAD$^+$, NADP$^+$, FAD], gives rise to the production of three molecules of ATP.

adenine dinucleotide FADH$_2$ in the case of the succinate to fumarate dehydrogenation. A further molecule of NADH results from the external conversion of pyruvate with CoA–SH and NAD$^+$ to acetyl-coenzyme A. The reducing equivalents from the complete oxidation of one molecule of pyruvate to carbon dioxide, fed into the membrane-bound electron-transfer

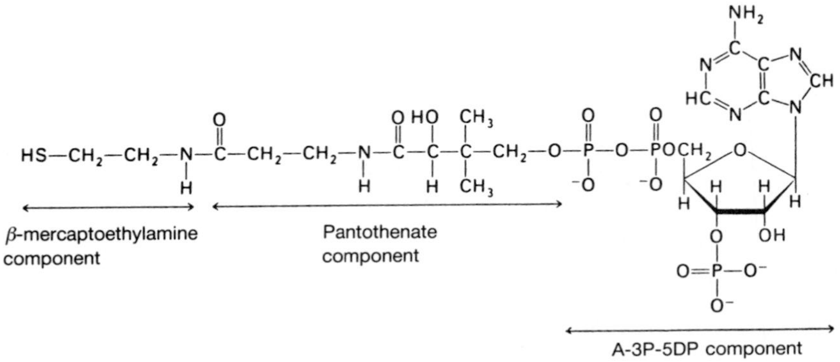

Fig. 11.14 The structure of coenzyme A, made up from the three components: adenosine-3'-phosphate-5'-diphosphate (A-3P-5DP); pantothenic acid; and β-mercatoethylamine.

cytochrome complexes of the respiratory chain, result at an optimum in the formation of 15 molecules of ATP.

The optimum is achieved in the mitochondrial organelles of eukaryote (Greek, 'true nucleus') cells, in which the genetic material is membrane enclosed, like the energy-generating mitochondria or the photosynthesizing chloroplast organelles. The three sets of enzyme systems in the respiratory chain, while general to the inner membrane of mitochondria, are not always complete within the inner cell membrane enclosing the cytoplasm of pro-karyote aerobic bacteria, where the ATP yield may be lower. Eukaryotic microbes, like yeast, which use the Embden–Meyerhof glycolytic pathway anaerobically and the tricarboxylic acid cycle in addition under aerobic conditions, obtain only two molecules of ATP by the fermentative dismuta-tion of one molecule of glucose to two of ethanol and two of carbon dioxide. In contrast their aerobic metabolism generates through the tricar-boxylic acid cycle an additional 30 molecules of ATP from the two pyru-vate molecules produced at the penultimate stage of glycolysis and another six ATP molecules from the two NADH molecules otherwise used to reduce acetaldehyde to ethanol in fermentation.

Shortly after finding refuge at Cornell University in the USA, Lipmann wrote a seminal review of the biochemical role of organic phosphates, proposing in 1941 that ATP and analogous molecules containing the 'high-energy phosphate bond' served as general bioenergetic intermediates in both anaerobic and oxidative metabolism. The review introduced addi-tionally the alternative term 'group transfer potential', which expressed more precisely the free energy potentially derivable by a coupled reaction from the hydrolysis of an acyl phosphate or phosphate ester. The generally accepted chemical meaning of the term 'bond energy', the energy required to dissociate a bond to specified fragments, appeared to be at variance with Lipmann's concept of the 'high-energy phosphate bond', resulting in an extended controversy with the contenders often debating at cross-purposes. Particular criticism came from some proponents of classical thermodynamics, accustomed to a reversible equilibrium view of closed chemical systems, rather than the kinetic flow view of open biochemical assemblies driven by an irreversible dissipation of free energy.

From a biochemical viewpoint it is fortunate for aerobic organisms that some expectations of equilibrium thermodynamics remain kinetically unfulfilled. Lewis and Randall in their classical textbook *Thermodynamics* (1923) pointed out that the atmosphere and the oceans of the Earth in principle ought to equilibrate thermodynamically to dilute nitric acid with the loss of almost all of the free molecular oxygen: they added the reflection, 'It is to be hoped that nature will not discover a catalyst for this reaction.' The best catalysts nature has discovered so far, the nitrogen-fixing and nitrifying bacteria, have a limited performance. The former require no less than 16 molecules of ATP for each molecule of nitrogen reduced to ammonia, with the necessary and inefficient generation of at least one molecule of hydrogen as a byproduct, while the latter have a very low growth yield, returning only one molecule of ATP for each molecule of ammonia oxidized to nitrite.

The prevalence of monophosphate esters as intermediates in biochemical metabolism, and the role of polyphosphate esters (particularly ATP) in bioenergetic transfer processes, appear to be more dependent upon kinetic and specific chemical properties than those of a general thermodynamic nature. The phosphate esters are soluble in water, unlike the esters of the higher fatty acids, and they are relatively stable in solution, so that enzyme-catalysed reactions do not suffer serious competition from uncatalysed side reactions. In water the polyphosphates have only a moderate reactivity, intermediate between those of the adjacent oxyanion periodic neighbours, the inert polysilicates, and the unstable polysulphates. Both inorganic and organic phosphates undergo ionic dissociation in water with pK_a values in the region of neutrality, serving as buffers for fluctuations in hydrogen ion concentration around pH 7 *in vivo* and *in vitro*.

The particular values of the free energies for hydrolysis of the polyphosphates provide no guide to the factors underlying the specific choice of ATP for bioenergetic transfers during the course of biochemical evolution. The standard free energies of hydrolysis, $-\triangle G^{\circ\prime}$, where the prime refers to physiological conditions (pH 7) rather than unit hydrogen ion activity (pH~0), have values in the limited range of 33–37 kJ mol^{-1} at 20°C for the hydration of pyrophosphate to phosphate, or of ADP to AMP and phosphate, and of ATP by either of two routes: to AMP and pyrophosphate, or to ADP and phosphate. These are all classified as 'high-energy compounds', as opposed to AMP or the monophosphates of glucose or fructose with $-\triangle G^{\circ\prime}$ values for hydrolysis in the 'low-energy' range of 10–20 kJ mol^{-1}. The distinctive role of ATP hydrolysis *in vivo* derives from its quasi-irreversibility through the rapid removal of a hydrolysis product, whereby the coupled reactions of biosynthesis and metabolic turnover are driven unidirectionally. Two widely distributed enzymes promote product removal in ATP hydrolysis. One catalyses a rapid hydrolysis of pyrophosphate, produced with AMP in one of the hydrolytic routes of ATP, while another, in the case of the second route, promotes a fast phosphate transfer between two molecules of ADP to form one molecule of ATP and one of AMP.

11.4 Chemiosmosis

By the early 1940s the main reactions of the oxidative tricarboxylic acid cycle were established, at least in general outline. Reducing equivalents NAD(P)H and FADH$_2$, generated from carbohydrate and other nutrients, are fed into the cytochrome-mediated redox processes of the respiratory chain where their oxidation leads to the phosphorylation of ADP to ATP. Another quarter of a century was to elapse, however, before a testable general mechanism for the conversion of the reducing equivalents to ATP became available. The analogy of anaerobic glycolysis and fermentation in homogeneous solution, catalysed by soluble enzymes, suggested the mediation of high-energy compounds in the coupling of ATP production to the oxidation of the reducing equivalents, but no such energy-rich intermediates could be convincingly demonstrated in the ensuing extensive searches.

The membrane-bound character of the major enzyme systems involved in the tricarboxylic acid cycle and in the respiratory chain directed attention to the structure and the function of the cytoplasmic membrane bounding the cell, enclosing inorganic salts and organic substances at concentrations substantially different from those prevailing in the external medium. One theory deriving from the reorientation, the view that free energy may accumulate in high-energy conformations of the proteins in the phospholipid bilayer of the plasma membrane, proved to have only an ancillary significance, but another, based upon the transport properties of the membrane, turned out to be fruitful at a fundamental level.

Seminal physico-chemical studies on the structure and permeability of the plasma membrane were carried out from 1935 by James Danielli (1911–84) and Hugh Davson at University College London and subsequently by Danielli elsewhere. Danielli's first graduate student at Cambridge in the mid-1940s, Peter Mitchell, developed from 1961 a largely physico-chemical mechanism of ATP generation, linking the membrane-bound respiratory chain of redox electron carriers to the semi-permeability of the mitochondrial or cytoplasmic membrane and the electrical potential developed across the membrane due to the different concentration of ionic species at each surface. Mitchell proposed that the electron flow through the respiratory chain from the reducing equivalents to molecular oxygen gives rise to an efflux of hydrogen ions from the outer side of the plasma membrane. The proton extrusion sets up a pH and potential difference at the cell boundary between the internal cytoplasm and the external medium of the periplasm between the plasma membrane and the cell wall in a bacterium, or between the inner and outer membrane of the mitochondrion organelle in a eukaryote cell. The plasma membrane is largely impermeable to hydrogen ions or hydroxyl ions, except at specific channels through the membrane which lead to an enzyme system (ATP-synthase) catalysing reversible ATP production from ADP and inorganic phosphate (P_i) at the inner surface terminus of the pore conducting the protons. Here the returning influx of hydrogen ions down the pH gradient drives the formation of ATP during oxidative phosphorylation (Fig. 11.15).

In contrast to the isotropy of the homogeneous solution reactions of anaerobic metabolism, the redox reactions of oxidative metabolism are anisotropic. The latter reactions are virtually two-dimensional, confined to electron transfer by the mobile carriers between the two surfaces of the membrane through the prosthetic groups of active proteins embedded in the phospholipid bilayer, fed from one side and drained from the other by the perpendicular transmembrane flow of protons and other species. Lipmann's concept of a scalar group transfer potential in anaerobic metabolism, represented by the free energy of hydrolysis of phosphate intermediates, is complemented for oxidative metabolism by Mitchell's vectorial proton-motive potential ($\triangle p$) orientated perpendicular to the membrane surface and measured by the free energy of the hydrogen ion concentration gradient ($\triangle$pH) and potential difference ($\triangle\psi$) across the plasma membrane:

$$\triangle p = \triangle\psi - Z\triangle\text{pH}$$

where Z is the conversion factor, $2.303RT/F$, with the value of approximately 60 mV at 25°C when the potentials are expressed in millivolts. Conventionally $\triangle$pH refers to the difference, pH(outer) − pH(inner).

The chemiosmotic hypothesis provided a range of testable specific expectations stimulating much subsequent experimental research, which generally has supported the theory. Additionally the hypothesis clarified long-known puzzling effects, notably the action of 'uncouplers' such as 2,4-dinitrophenol and a variety of structurally unrelated molecules which stimulate the respiratory oxidation of substrates yet inhibit the synthesis of ATP. The common features of uncoupling molecules are physico-chemical. All of the uncouplers are weak acids and lipid soluble in both dissociated and undissociated forms. Both the acid and its conjugate base freely circulate across the plasma membrane transporting hydrogen ions and

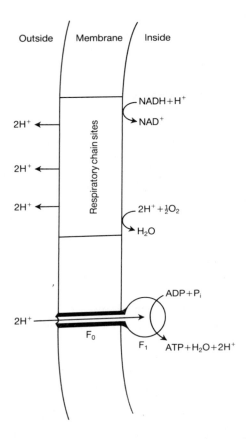

Fig. 11.15 The chemiosmotic production of ATP by oxidative phosphorylation in the respiratory chain of the mitochondrial inner membrane (or the cytoplasmic membrane of prokaryotes). Each electron pair from a reducing equivalent produced by the TCA cycle (Fig. 11.13), on transfer through the three membrane-bound sites of the respiratory chain to the terminal acceptor (molecular oxygen, Fig. 11.11), produces the extrusion of six hydrogen ions. The return of two hydrogen ions through the proton-conducting channel (F_0) generates a molecular of ATP from ADP and inorganic phosphate, catalysed by the ATP-synthase enzyme system (F_1) at the inner terminus of the channel.

thereby eliminate the electrochemical gradient providing the proton-motive potential. In consequence ATP synthesis is inhibited, but the respiratory electron-transfer chain is liberated from the mass-action restraint of the electrochemical gradient normally generated, and the chain is set free for the runaway oxidation of substrates.

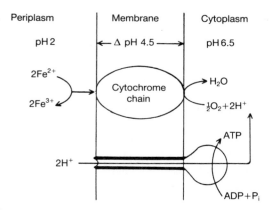

Fig. 11.16 The chemiosmotic production of ATP by *Thiobacillus ferro-oxidans*. In an acidic environment (pH 2), external ferrous ions donate electrons to the membrane-bound cyctochrome chain. The electrons, transferred to internal oxygen, produce water with the consumption of internal hydrogen ions, setting up a pH gradient across the membrane ($\triangle$pH -4.5). External hydrogen ions under the proton-motive potential migrate through the proton-conducting channel in the membrane to the ATP-synthase system at the channel's inner terminus, where the production of ATP from ADP and P_i is catalysed by the enzyme. The redox potentials of the couples [Fe^{3+}/Fe^{2+}] and [$\frac{1}{2}O_2/H_2O$] are close ($E_0' = +0.78$ and $+0.82$ V, respectively).

Measurements of the cytoplasmic pH of aerobic bacteria, or of intact mitochondrial organelles suspended in near-neutral media, indicate that the internal hydrogen ion concentration has a value lower by approximately an order of magnitude than that of the suspension solution. The $\triangle$pH value is substantially more negative (-4.5) for organisms flourishing in an acidic environment (pH 2), such as *Thermoplasma acidophilum*, a bacterium bounded solely by its plasma membrane, lacking a cell wall, and first isolated from a steamy, smouldering coal-waste tip, or *Thiobacillus ferro-oxidans*, which generates ATP chemiosmotically by oxidizing ferrous to ferric iron (Fig. 11.16). Variations in $\triangle$pH from one organism to another are offset to a degree by compensating changes in the membrane potential, giving the total proton-motive potential ($\triangle p$) for respiratory membranes values in the range from approximately 100 to 250 mV.

The vectorial character of the proton-motive potential is demonstrated by experiments employing closed vesicles constructed from the respiratory membranes of mitochondria or aerobic bacteria turned inside out. These inside-out vesicles actively pump in hydrogen ions during oxidative metabolism and the internal pH becomes more acidic than that of the suspension medium. The proton-conducting channels now lead to the ATP synthase enzyme complex at the outer surface and, with a potassium ion gradient providing a membrane potential, the pH gradient drives ATP formation from ADP and P_i in the suspension medium.

The photosynthetic membrane (thylakoid) of the intact chloroplast organelle, like the inside-out vesicles constructed from respiratory membranes, has the ATP synthase enzyme complex located on the outer surface, linked by a proton-conducting channel to an interior that becomes more acidic when the chloroplast is illuminated. Just as pulses of dissolved oxygen give rise to transitory increases in the hydrogen ion concentration of suspensions containing respiring mitochondria, so flashes of light give rise to a corresponding loss of $[(H_3O)^+]$ in the solution containing the chloroplast thylakoids illuminated. Typically the internal hydrogen ion concentration of the thylakoids in a chloroplast is some three orders of magnitude larger than that externally, and the overall proton-motive potential, while appoximately of the same magnitude, has a sign opposite to that of the mitochondrial respiratory membrane.

Early support (1966) for the chemiosmotic theory was provided by an 'acid bath' experiment, in which chloroplasts were equilibrated in the dark at pH 4, so that the thylakoid interior attained the acid pH otherwise generated by photons on illumination. A rapid change in the hydrogen ion concentration of the suspension solution, containing ADP and P_i, to pH 8 produced a burst of ATP synthesis in the solution at the external surfaces of the membranes, owing to the migration of hydrated protons down the pH gradient from the thylakoid interior to the environmental medium (Fig. 11.17).

Experiments with stirred suspensions of intact mitochondria, or aerobic bacteria with a similar respiratory chain, subjected to pulsed increments of

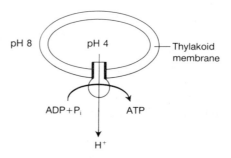

Fig. 11.17 The chemiosmotic 'acid bath' experiment. The thylakoid units of chloroplasts equilibrated in an acid bath in the dark attained an internal hydrogen ion concentration of pH ~4. A pH gradient ($\triangle$pH +4) across the thylakoid membrane, produced by a rapid adjustment to pH 8 of the suspension solution containing ADP and P_i, gave rise to a pulse of ATP synthesis, owing to the migration of hydrogen ions through the proton-conducting channel in the membrane to the ATP-synthase system at the external terminus of the channel.

dissolved oxygen demonstrate that some six protons are extruded from the outer surface of the inner or plasma membrane for each molecule of NADH oxidized to NAD^+ or, what is equivalent, for each oxygen atom reduced to water. The particular ratio suggests that each of the three electron-transfer systems of the respiratory chain spanning the plasma membrane translocate an average of two protons to the outer medium for the transmission of a pair of electrons through the system's site (Figs. 11.11 and 11.15).

Organisms that lack cytochrome c and the normal cytochrome reductase site, such as the work-horse of molecular biology, *Escherichia coli*, are limited to only two energy-coupling sites and in oxidative metabolism extrude only four protons for each oxygen atom reduced to water. In mitochondria each molecule of NADH produced by the degradation of pyruvate to carbon dioxide in the tricarboxylic acid cycle generates, on oxidation to NAD^+, three molecules of ATP. The stoichiometry indicates that one molecule of ATP is synthesized from ADP and P_i for an influx of two hydrogen ions through the proton-conducting channel to the ATP synthatase unit located on the inner surface of the respiratory mitochondrial membrane (Fig. 11.15).

In his Nobel Prize lecture, Peter Mitchell (1979) pointed out that the early opponents of his chemiosmotic hypothesis (1961) had, through their own searching studies, 'not only come to accept it, but have actively promoted it to the status of a theory'. The general acceptance of the chemiosmotic theory within a decade of its formulation effectively falsified, Mitchell indicated, the pessimistic view of Max Planck (1948), that 'a new scientific truth does not triumph by convincing its opponents and making them see the light, but rather because its opponents eventually die, and a new generation grows up that is familiar with it'. The general adoption of the theory of continental drift founded on plate tectonics during the 1960s, as a result of the global palaeomagnetic surveys of the previous decade, followed a similar course in a wholly unrelated field; few geophysicists, apart from Harold Jeffreys, exemplify Planck's contention (Hallam 1973, 1983).

Planck had long held a somewhat gloomy view of the advance of science. In 1913 he counselled the young Einstein, then in hot pursuit of the general theory of relativity: 'As an older friend I must advise you against it, for in the first place you will not succeed; and even if you succeed, no one will believe you.' (Pais 1982.) Planck's sombre perception of the controversies surrounding relativity and quantum theory was developed into a general theory of the growth of science by Thomas Kuhn (1962), with exemplifications taken mainly from the earlier phases of modern science. Kuhn held that science develops through a series of 'revolutions' (Planck's 'deep reconstructions') in which rival 'paradigms' (opposing 'world-views' for Planck) become 'incommensurable' ('incompatible' according to Planck) and one of the contending theories becomes dominant as its opponents die out.

Kuhn's theory derived additionally from his earlier study (1957) of the astronomical revolution of Copernicus (1473–1543), in isolation from the salient movements of the period (the great geographical discoveries, the Protestant Reformation, the rise of the absolute monarchies) at a time when secure scientific traditions and institutions were not yet established and before the foundation of the first enduring scientific societies (the Royal Society of London in 1660 and the Paris Academy of Sciences in 1666).

The natural sciences during the sixteenth and the seventeenth centuries were embroiled in the ideological conflicts of the period (Mason 1953), but with the progressive growth of an autonomous scientific tradition they achieved the virtually global consensus widely perceived today, with well-supported new theories generally accepted within a decade. Some sociologists take Kuhn's theory to imply that their subject has a status equivalent to that of a physical science, both the social and the natural sciences consisting apparently of a series of ideological enclaves, dominated by the factionalism of 'incommensurable paradigms' (Barnes 1983).

11.5 Oxygenic photosynthesis

During the early phase of the industrial revolution, the Unitarian theologian and pneumatic chemist Joseph Priestley (1733–1804) was already concerned by the degradation of our atmosphere, the providential sustainer of the breath of life. His abiding interest in the factors governing the 'goodness of the air' led him to not only the discovery of the atmospheric component supporting vitality, oxygen, but also the finding that air spoiled through such processes as combustion, fermentation, and respiration is restored by green plants converting fixed air (carbon dioxide) to dephlogisticated air (oxygen). By the time that Joseph Pelletier (1788–1842) and his student J. B. Caventou (1795–1877) at the Paris École de Pharmacie had isolated the green pigment of plants, naming it chlorophyll (1817), it was established that water, carbon dioxide, and sunlight are the primary nutrients of plants, which produce oxygen and 'organic matter'.

The primary organic product was shown during the 1860s to be the glucose polymer, starch, by the botanist Julius Sachs (1832–97) at Würzburg, who found that plants in the dark respire normally, taking up oxygen to produce carbon dioxide at the expense of their starch reserves. At the same time Stokes and his successor chromatologists investigated the spectrum of chlorophyll, showing the pigment to absorb light specifically in the red and the blue regions of the visible spectrum, largely transmitting green light. In 1881 the physiologist Theodor Engelmann (1843–1909) at Utrecht showed that red and blue wavelengths are photosynthetically active, and the green wavelength relatively inactive, by irradiating the green chloroplast band of the alga *Spirogyra* under the microscope with the solar spectrum dispersed by means of a prism. The relative amounts of oxygen produced by the several spectral colours were estimated by counting the population of oxygen-requiring motile bacteria, which Engelmann had discovered, clustering round successive regions along the chloroplast band of an irradiated *Spirogyra*.

In addition Engelmann discovered that some coloured bacteria assimilate carbon dioxide when illuminated but do not evolve oxygen. The two main groups of non-oxygenic photosynthetic bacteria, one green and the other purple, contain the photocatalytic pigment bacteriochlorophyll based upon the tetrahydroporphin ring system, in contrast to the dihydroporphin photopigment, chlorophyll, of oxygenic organisms, whether prokaryote like the cyanobacteria or eukaryote with chloroplast organelles, such as the algae or higher plants. The photosynthetic bacteria obtain reducing equivalents from organic materials or inorganic substances, such as hydrogen or hydrogen sulphide, for the photoreduction of carbon dioxide to carbohydrate and for the photogeneration of ATP to drive metabolic turnover and biosynthesis.

These bacteria are termed photoautotrophs and form a subsection of the general class of autotrophs (self-supporting on carbon dioxide and inorganic substances), identified by the Russian microbiologist Sergey Vinogradsky (1856–1953). By the use of limited and selective growth media, Vinogradsky isolated in 1889 examples of another subsection, the chemoautotrophs, which obtain free energy and reducing equivalents for the conversion of carbon dioxide to carbohydrate by the chemical oxidation of sulphur to sulphate, or of ferrous to ferric iron. The heterotrophs, like the nitrifying bacteria discovered by Vinogradsky, require carbohydrate or other reduced organic materials both as an energy source and a feedstock for biosynthesis, although some, such as the methane-producing bacteria, can assimilate carbon dioxide by dark thermal reactions in addition.

Until the 1930s it was supposed that in photosynthesis the oxygen evolved came from the fission of the carbon dioxide molecule, despite the carbon dioxide assimilation without oxygen production by the green and purple photosynthetic bacteria. The Delft microbiologist Cornelis van Niel (1897–1985), working in California, suggested in 1931 that anoxic and oxygenic photosynthesis have a unity expressed by the photochemical transfer of hydrogen from a donor to carbon dioxide in each of the two cases:

$$CO_2 + 2H_2A + light \rightarrow [CHOH] + 2A + H_2O$$

where [CHOH] refers to generalized carbohydrate, and 2A may be oxygen if the hydrogen donor is water, or sulphur where the donor is hydrogen sulphide, or absent if hydrogen is taken up directly. Chlorophyll-catalysed photosynthesis thus involved two distinct and separable processes: carbon dioxide assimilation and oxygen production. Van Niel pointed out that the chemoautotrophs and some heterotrophs, as well as the photoautotrophs, fix carbon dioxide without producing oxygen.

Light-saturation and flash-irradiation studies supported the two-step view of oxygenic photosynthesis. During the 1880s it was found that oxygen evolution in photosynthesis is proportional to the light intensity only at low radiation levels, becoming constant at high levels, independent of intensity. For a given partial pressure of carbon dioxide, the saturation rate of oxygen evolution at a constant high-intensity illumination increases if the temperature of the suspension containing the photosynthesizing cells is raised. The temperature sensitivity indicates that a dark thermal enzyme reaction, requiring an activation energy, follows the initial photoreaction and that the dark reaction becomes the limiting step in the overall rate of photosynthesis at high light intensities.

The two primary reactions of photosynthesis were quantitatively characterized by Robert Emerson (1903–59) in California and, from 1947, in Illinois, using flash-illumination methods. In 1932 Emerson, with William Arnold, irradiated suspensions of the green alga *Chlorella* with 10^{-5} s flashes from a discharge tube and measured the oxygen evolved as a function of the radiation intensity, the duration of the dark interval between successive flashes and the temperature of the cell suspension. Under conditions of light saturation, the yield of evolved oxygen was found to increase with the duration of the dark interval up to a limit of 6×10^{-2} s and thereafter remained constant. From the limit Emerson and Arnold estimated the half-life of the dark enzyme reaction to be about 2×10^{-2} s at 25°C. A single intense flash, producing saturation, gave one molecule of oxygen for each group of about 25 000 molecules of chlorophyll present in the cell suspension. From 1937 Emerson measured the quantum yield of photosynthesis, showing that eight photons are required for each molecule of carbon dioxide assimilated and for each molecule of oxygen evolved. The quantum yield and and the chlorophyll/oxygen ratio obtained earlier indicated that the photosynthetic unit of green plants contains some 300 chlorophyll molecules.

In 1957 Emerson obtained evidence that two photosystems, absorbing light optimally at different wavelengths, are active in oxygenic photosynthesis. The quantum yield of the oxygen liberated from suspensions of the green alga *Chlorella* was found to fall on irradiation at long wavelengths beyond the 660 nm absorption band maximum of chlorophyll *a*, particularly beyond about 685 nm where the chlorophyll still has an appreciable light absorption. Emerson discovered that the original quantum efficiency is restored, even if the *Chlorella* suspension is irradiated at 700 nm, where chlorophyll *a* shows much less absorption of light, when the suspension is subject to supplementary illumination at a shorter wavelength (650 nm).

The total quantum yield for the two-wavelength irradiation turned out to be larger than the sum of the individual contributions, suggesting a cooperative interaction between two photosystems.

Biochemical evidence for two photoreactions in oxygenic photosynthesis had been discovered in 1937 by the plant physiologist Robert Hill at Cambridge. Hill separated by differential centrifugation the photosynthesizing particles (chloroplasts) from the respiratory particles (mitochondria) in a coarsely filtered macerate of green leaves. The chloroplasts isolated were not wholly intact and failed to take up carbon dioxide when illuminated in suspension, but oxygen was liberated by the photolysis of water if a ferricyanide salt or other electron acceptor (oxidant) was added to the suspension. One molecule of oxygen was evolved for every four electrons accepted by the oxidant in the photoreduction. Hill's reaction indicated that photochemical oxygen evolution is distinct and separable from the reduction of carbon dioxide in photosynthesis. The inference was confirmed, after Emerson's discovery of the quantum yield enhancement by two-wavelength irradiation, by the observation that the photochemical production of oxygen in the Hill reaction is driven by light of 650 nm or shorter wavelength, but not by far-red light in the 700 nm wavelength range.

In 1960 Hill, with Fay Bendall, proposed that oxygenic photosynthesis is dependent upon two photosystems which overlap on the redox potential scale and connect in series through dark reactions mediated by cytochromes (Fig. 11.18). The Z-scheme, as the Hill and Bendall model was termed, unified numerous previous observations on oxygenic photosynthesis, and subsequent studies amply supported and elaborated the scheme. In the Z-scheme Photosystem I (PS I) governs the photoreduction of carbon dioxide to carbohydrate. PS I is made up largely of chlorophyll a and a little chlorophyll b with a reaction centre composed of a specialized photopigment (P) absorbing light optimally at 700 nm (P_{700}). Photosystem II (PS II), liberating oxygen by the phototransfer of electrons from water, has a similar composition with a larger proportion of chlorophyll b and a reaction centre pigment with maximum light absorption at 680 nm (P_{680}). The ground electronic state of PS II lies close to the mid-redox potential (+0.82 V) of the water–oxygen couple. Light absorption activates the PS II reaction centre P_{680} to a photoexcited state at a negative redox potential, appropriate for electron donation through cytochrome carriers to PS I. The electrons restored to PS II are taken up from the reaction converting water to hydrogen ions and molecular oxygen. The ground electronic state of PS I has a redox potential of some +0.4 V, and its reaction centre P_{700} undergoes a transition on light absorption to a photoexcited state near to the mid-redox potential of the carbon dioxide–glucose couple (−0.43 V) or the H_3O^+/H_2(gas) couple at pH 7 (−0.42 V). The photoexcited reaction centre P_{700}^* thence indirectly donates electrons to carbon dioxide, leading ultimately to the formation of carbohydrate and water.

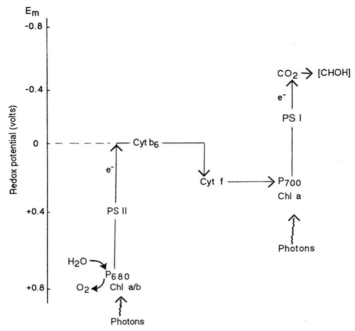

Fig. 11.18 The two light reactions of oxygenic photosynthesis, involving two photosystems (PS I and PS II) in the Z-scheme of Hill and Bendall (1960) (*Nature*, **186**, 136).

Subsequent studies showed that, in the absence of carbon dioxide, illuminated chloroplasts generate ATP and NADPH and that such pre-illuminated chloroplasts convert carbon dioxide to carbohydrate in the dark. The carbon dioxide reductant, NADPH, forms photochemically by electron transfer to $NADP^+$ from the photoexcited reaction centre P^*_{700} of PS I through soluble mobile ferredoxin, analogous iron–sulphur proteins that are membrane bound, and other electron-transfer species detected and characterized spectroscopically. In the absence of $NADP^+$, cyclic electron flow from the photoexcited reaction centre P^*_{700} of PS I through the electron-transfer chain linking PS II to PS I regenerates the ground electronic state P_{700} and translocates hydrogen ions across the thylakoid membrane. The transmembrane proton-motive potential generated, with $\triangle pH +3$, drives ATP production chemiosmotically (Figs. 11.19 and 11.20).

Like the inner membrane of the mitochondrion, the thylakoid membrane of the chloroplast is spanned by three membrane-bound bioenergetic systems linked by mobile electron carriers, together with ATP-generating units consisting of a proton-conducting channel terminated by an ATP synthetase enzyme system. As well as the two photosystems, thylakoid membranes contain a bound cytochrome complex mediating electron transfer from PS II to PS I. The thylakoid membrane, some 6 nm thick, embeds a complex made up of cytochrome b, cytochrome f (f, from fronds, a c-type cytochrome), and an iron–sulphur protein. The iron–sulphur protein serves as a one-electron acceptor from the mobile electron carrier, soluble plasto-quinone (Fig. 11.21), from PS II. The photoexcited reaction centre P^*_{680} of PS II donates an electron to pheophytin, a chlorophyll molecule lacking coordinated magnesium, and thence to the soluble plastoquinone through two membrane-bound forms of plastoquinone, Q_A and Q_B. The soluble blue copper protein, plastocyanin, transports electrons from cytochrome f in the bound cytochrome complex to PS I. In cyclic electron flow, when $NADP^+$ is in short supply, cytochrome b in the bound complex accepts an electron from the photoexcited state P^*_{700} of PS I through the mobile carrier, ferredoxin, which otherwise converts $NADP^+$, when the latter is in ample supply, to NADPH non-cyclically.

Only the non-cyclic photoreaction path requires a supply of electrons to PS I from the water-splitting reaction through the photoexcitation of PS II and the cytochrome electron-transfer link. The 'water-oxidizing' protein of PS II contains four manganese ions which, by a concerted change in oxidation state from [3Mn(IV),Mn(V)] to [3Mn(III),Mn(IV)], take up four electrons from two water molecules to form an oxygen molecule and four hydrogen ions. The four-electron transfer is inferred from the observation that, in flash-illumination experiments, the largest pulse yield of molecular oxygen per flash is obtained after every fourth successive flash. The con-certed accumulation of specifically *four* electrons by the water-oxidizing protein, in the formation of molecular oxygen from water, protects the photosystem and the organism as a whole from the highly reactive super-oxide ion (O_2^-) and from other toxic oxygen radical species. Step-wise changes in the oxidation state of the individual metal ions in the manganese cluster give a series of one-electron transfers from the water-splitting protein to the reaction centre P_{680} of PS II. The one-electron transfers are mediated by an unusual plastaquinone cation radical with a midpoint redox potential of more than $+1.0\,V$ (Prince 1986).

Both PS I and PS II contain a light-harvesting complex of a few hundred chlorophyll molecules and other antenna pigments organized around the reaction centres P_{700} and P_{680}, respectively, so that the energy of light incident over a wide area and with a range of photon energies is funnelled towards the reaction centre. Adjacent pigments in a light-harvesting com-plex have absorption maxima of successively longer wavelength inwards so that the photoexcitation energy of light absorbed anywhere in the complex over the 400–700 nm range migrates downhill to the reaction centre.

In the chlorophylls the fine-tuning of the optimum absorption wave-length arises from a change of the 3-aldehyde substituent in chlorophyll b to a 3-methyl group in chlorophyll a, shifting the absorption maximum

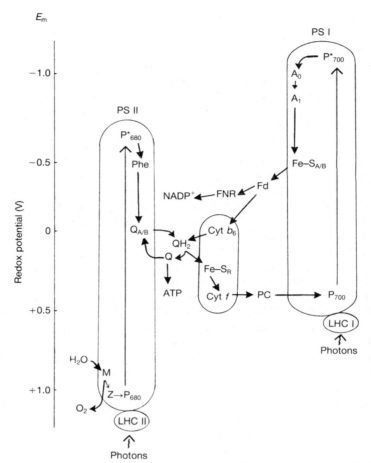

Fig. 11.19 The electron transfer processes in chloroplast photosynthesis. The reaction centres of PS I and PS II (P_{700} and P_{680}), which are dimers of chlorophyll a', trap the photoexcitation energy gathered by the respective light-harvesting complexes (LHC I and LHC II). The photoexcitation of P_{680} to P_{680}^* results in ionization to P_{680}^+ and an electron, which transfers to pheophytin (Phe, chlorophyll lacking coordinated magnesium), and thence to two forms of membrane-bound plastoquinone (Q_A and Q_B). An electron is restored to P_{680}^+ from the water-splitting enzyme (M), containing a [4Mn-4O] cluster, through the plastoquinone radical cation intermediate (Z).

The photoelectron of PS II transfers from the bound quinones ($Q_{A/B}$) to the soluble, mobile plastoquinone (Q), which is reduced to plastoquinol (QH_2) with the uptake of hydrogen ions. The membrane-bound cytochrome b,f complex contains an iron–sulphur protein (Fe–S_R) which accepts electrons from QH_2 (thence converted back to Q) and donates electrons to the cytochrome f of the complex. The blue copper-containing protein, plastocyanin (PC), transports an electron from cytochrome f to the cation P_{700}^+ formed by the ionization of the photoexcited reaction centre of PS I, P_{700}^*, to regenerate P_{700}.

The electron liberated by the photoionization of P_{700}^* is captured by a mono-meric chlorophyll a molecule (A_0) and passed on through the secondary acceptor A_1 (probably phylloquinone) to a set of iron–sulphur centres (Fe–$S_{A/B}$) bound to PS I. The photogenerated electron transfers from PS I to another iron–sulphur protein, the mobile carrier ferredoxin (Fd). If NADP$^+$ is in ample supply, it is reduced to NADPH with electrons carried by Fd, catalysed by ferredoxin-NADP$^+$ reductase (FNR). If NADP$^+$ is in short supply, the electrons carried by Fd are donated to the cytochrome b,f complex. From the complex the electrons are transferred to the mobile Q/QH$_2$ pool where a proton-motive potential across the thylakoid membrane is generated, resulting in the production of ATP. From the Q/QH$_2$ pool, the electrons originating from the ionization of P_{700}^* but not used in the reduction of NADP$^+$ return to the cytochrome b,f complex and thence, through PC, to PS I where the charge of P_{700}^+ is compensated cyclically without an electron input from PS II. The boxed systems, PS I, PS II, and the cytochrome b,f complex are enclosed in the thylakoid membrane, to which they are bound (Fig. 11.20).

---

measured in organic solvents from 643 to 660 nm (Fig. 11.22). The environmental medium influences the position of optimum absorption, and the protein matrix of a light-harvesting complex shifts the chlorophyll absorption to a range of longer wavelengths up to 700 nm. The reaction centre itself, P_{700} of PS I or P_{680} of PS II, consists of two specialized chlorophyll a molecules (chlorophyll a') which differ only in the absolute stereochemical configuration at the 10-position. The major pigment of chloroplasts, chlorophyll a, has the R-configuration at the 10-position whereas the minor constituent, chlorophyll a' ($a/a' \sim 300$), has the epi-meric S-configuration (Fig. 11.22). The minor structural modification pro-duces a small reduction in the energy of the chlorophyll excitation, so that a dimer of chlorophyll a' becomes the ultimate trap of the photoexcitation energy gathered by the antenna pigments in the light-harvesting complex (Watanabe *et al.* 1985).

The dark enzyme reactions converting carbon dioxide to carbohydrate from the NADPH and ATP produced photochemically take place in the yellow stroma surrounding the thylakoid membranes in a chloroplast (Fig. 11.20). The chemical course of the dark reactions was worked out from 1946 by Melvin Calvin and his coworkers at Berkeley, California, using the newly available β-radioactive isotope ^{14}C (half-life approximately 5700 years). Suspensions of unicellular green algae, such as *Chlorella*, were exposed to ^{14}CO$_2$, injected as a bicarbonate solution into the suspension, over a range of conditions of concentration, illumination, and exposure time. The reactions of samples drawn from the suspension were rapidly quenched with hot methanol and the radioactive substances formed were detected and identified by means of two-dimensional paper chromatography. A brief exposure of the photosynthesizing algae to ^{14}CO$_2$ followed by immediate sampling gave the initial stable product of carbon dioxide

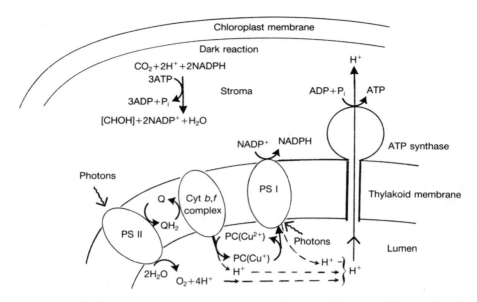

Fig. 11.20 The location of the three photoelectron transfer systems (PS I, PS II, and the cytochrome b,f complex) and the proton-conducting channel terminated by the ATP-synthase enzyme system in the membrane of the inner thylakoid vesicles of the chloroplast.

On irradiation, the transport of electrons through the three carrier systems increases the concentration of hydrogen ions inside the thylakoid space (the lumen) and depletes that concentration outside, in the stroma between the thylakoid and the chloroplast membrane. Water-splitting produces an internal hydrogen ion for each electron donated to PS II. The donation of each pair of electrons to the mobile carrier plastoquinone (Q), directly from PS II or indirectly from PS I in cyclic electron transfer when $NADP^+$ is in short supply, consumes two external hydrogen ions from the stroma in the reduction to plastoquinol (QH_2). Two hydrogen ions are released into the internal lumen when Q is regenerated from QH_2 by the direct transfer of electrons to the cytochrome b,f complex and thence to PS I through plastocyanin (PC). Each pair of electrons photoexcited in PS I consume an external hydrogen ion in the reduction of $NADP^+$ to NADPH.

ATP is generated chemiosmotically from ADP and inorganic phosphate in the stroma by the extrusion of hydrogen ions from the lumen through the proton-conducting channel in the thylakoid membrane to the ATP-synthase enzyme system at the outer surface. The dark reaction in photosynthesis, the reduction of carbon dioxide to carbohydrate [CHOH], take place in the stroma with the consumption of two molecules of NADPH and three of ATP for each carbon dioxide molecule assimilated.

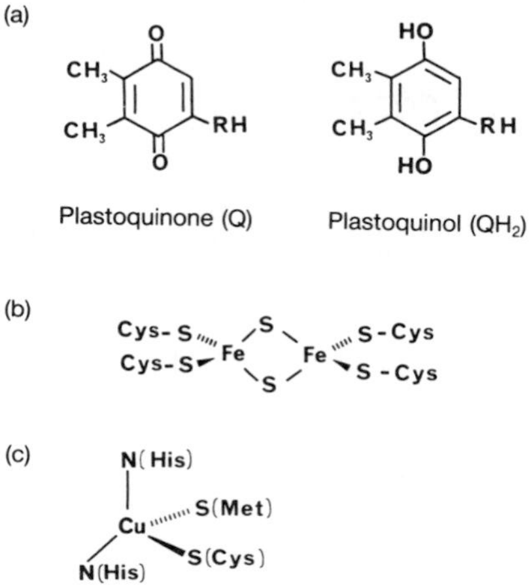

(a)

Plastoquinone (Q)   Plastoquinol (QH$_2$)

(b)

(c)

Fig. 11.21 Structures of electron-transporting molecules and protein-bound co-ordination complexes active in oxygenic photosynthesis. (a) Plastoquinone and plastoquinol: in different organisms the number of isoprenoid units in the side chain R (R=[–CH$_2$–CH=C(CH$_3$)–CH$_2$–]$_n$) varies for n between 6 and 10. (b) Ferredoxin contains an active [2Fe–2S] redox cluster coordinated to cysteine residues of the protein: the active centres of other iron–sulphur proteins are the cubane [4Fe–4S] cluster or the tetrahedrally coordinated [Fe(S–R)$_4$] unit. (c) The copper ion transporting electrons in plastocyanin has a distorted tetrahedral co-ordination to the imidazole rings of two histidine residues and to the sulphur atom of a cysteine residue and of a methionine residue in the protein.

fixation as the principal radioactive product, shown to be 3-phospho-glyceric acid. A more extended exposure to $^{14}CO_2$ indicated that radio-active triosephosphate and higher sugar phosphates are formed quite rapidly, but radioactivity in polysaccharides, fatty acids, and amino acids appears only after long periods of photosynthesis.

The dark reactions of the photosynthetic carbon cycle were found by Calvin's group to divide into three stages between the input of carbon dioxide with photogenerated ATP and NADPH and the output of 3-carbon intermediates for further biosynthesis (Fig. 11.23). The first stage consists of the carboxylation of the 5-carbon sugar, ribulose-1,5-bisphosphate (RuBP), with carbon dioxide, catalysed by RuBP carboxylase to form two molecules of 3-phosphoglyceric acid (PGA) through the fission of a 6-carbon intermediate. Each PGA molecule formed is reduced in the second stage with NADPH and ATP to 3-phosphoglyceraldehyde, which equili-brates with dihydroxyacetone phosphate to give the mixed triose phos-phate product. Much of the triose phosphate enters the third stage where RuBP is regenerated through a series of sugar phosphates containing three to seven carbon atoms, the remainder entering biosynthetic pathways.

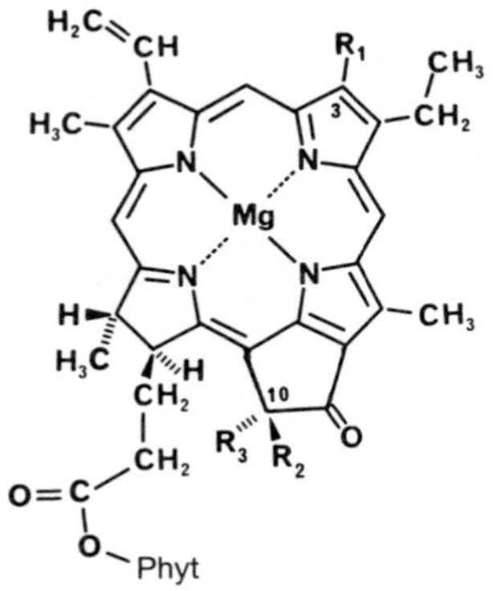

Phyt-OH (phytol)

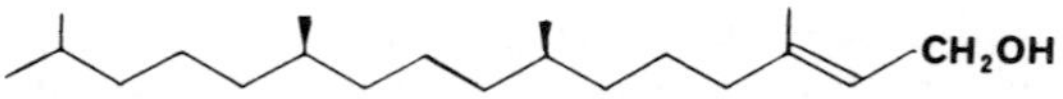

	R_1	R_2	R_3
Chlorophyll *a*	$—CH_3$	—H	$—CO_2CH_3$
Chlorophyll *a′*	$—CH_3$	$—CO_2CH_3$	—H
Chlorophyll *b*	—CHO	—H	$—CO_2CH_3$

Fig. 11.22 Structures of the chlorophyll photopigments active in oxygenic photosynthesis: chlorophyll *a′* serves as the ultimate trap (the reaction centre) of the excitation energy transferred to the photosystems from chlorophyll *a* and chlorophyll *b* in the light-harvesting complexes.

Overall, three molecules of ATP and two of NADPH are required to reduce carbon dioxide to the level of carbohydrate [CHOH]. Most of the free energy formally needed for the reduction of carbon dioxide to [CHOH] in water ($\triangle G^{\circ\prime}$ + 480 kJ mol^{-1}) derives from the coupled oxidation of the two NADPH molecules to $NADP^+$ ($\triangle G^{\circ\prime}$ − 440 kJ mol^{-1}) while the linked hydrolysis of the three ATP molecules makes up the residual free-energy requirement and provides a small overall excess.

The photochemical stages of carbon dioxide fixation in photosynthesis are less efficient than the subsequent dark enzyme reactions. In principle the combined energy of three photons of visible radiation in the red region (~680 nm wavelength) should suffice, with about a 10 per cent excess, to drive the reduction of a carbon dioxide molecule to [CHOH] in water, but it is found experimentally that at least eight of such photons are needed to produce the NADPH and ATP required for the reduction. That is, the absorption of a minimum of eight quanta of red light are required to fix one molecule of carbon dioxide or, equivalently, to liberate one molecule of oxygen, limiting the efficiency to some 30 per cent. The overall limita-

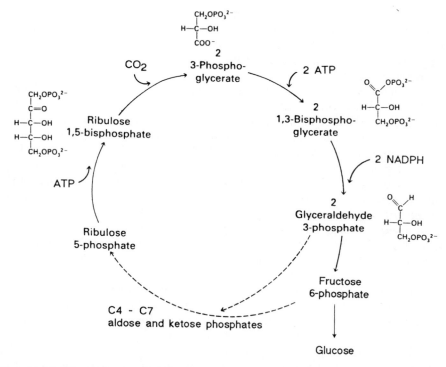

Fig. 11.23 The Calvin cycle of dark reactions converting carbon dioxide into carbohydrate: the production of one molecule of glucose from six carbon dioxide molecules requires 12 reducing equivalents (NADPH) and the coupled free energy of hydrolysis of 18 molecules of ATP.

tion of photosynthesis to an average efficiency of less than 1 per cent derives principally, however, from the restricted ability of the whole plant to capture all of the incident sunlight over the photoactive wavelength region (Hall and Rao 1987).

11.6 Non-oxygenic photosynthesis

Three types of coloured bacteria, distinguished by their photopigments, harness the energy of sunlight for metabolism and biosynthesis without evolving oxygen. The green and the purple photosynthetic bacteria use bacteriochlorophyll, a dihydro-analogue of the chlorophyll employed in oxygenic photosynthesis, to transform photons trapped at the reaction centre into electrons for transfer to a chain of redox carriers. Electron transport back to the photopigment through the chain translocates hydrogen ions, producing ATP chemiosmotically, and reduces NAD^+ to NADH directly or indirectly by taking electrons from a terminal donor with a less-positive redox potential than water (sulphide, thiosulphate, or an organic substrate). The third type, the halobacteria, developed a wholly different type of photopigment, a polyene bound to protein (bacteriorhodopsin), which is remarkably similar to the visual purple pigment of the vertebrate eye.

The halobacteria belong to the kingdom of archaebacteria, which is distinct from the other prokaryote kingdom of eubacteria ('true bacteria') that includes the green and the purple photosynthetic bacteria. The cytoplasmic membrane of the archaebacteria is made up of distinctive lipids composed of the diether formed by D-glycerol-1-phosphate with a branched-chain alcohol, principally phytanol, the dihydro-analogue of phytol, which forms the hydrocarbon side chain of the chlorophylls (Fig. 11.22). In contrast the lipids forming the cytoplasmic membrane of eubacteria, and the membranes of organisms belonging to the third kingdom, the eukaryotes, are composed of diesters formed by L-glycerol-1-phosphate with straight-chain fatty acids. In addition the archaebacteria differ from the eubacteria in cell wall composition, nucleic acid organization, and a number of detailed biochemical respects.

Halobacteria normally grow in aerobic environments of high salinity (3–5 M in NaCl), such as salt lakes, marshes, and marine lagoons. Under conditions of low oxgen partial pressure and high light intensity, these organisms develop purple patches of bacteriorhodopsin on the cytoplasmic membrane, covering 50 per cent or more of the membrane surface. The energy of light absorbed by the pigment results in the extrusion of hydrogen ions from the cytoplasm interior, so augmenting the proton-motive potential across the membrane normally provided by the oxidation of organic substrates through the respiratory chain. The ATP required for the metabolism and growth of the bacterium is generated by the influx of hydrogen ions back into the cytoplasm through proton-conducting channels terminated on the inner side of the membrane by an ATP synthase complex. In addition the hydrogen ion inflow drives the motor of the flagellum that impels the organism through its aqueous environment and maintains the ionic and osmotic balance of the cytoplasm, e.g. by accumulating potassium ions to replace the exported sodium ions.

Bacteriorhodopsin consists of a single polypeptide some 248 amino acid residues long with the chromophore retinal (vitamin A aldehyde) bonded as a protonated Schiff base to the ϵ-amino group of a lysine residue at position 41 from the NH_2-terminus of the molecule (Fig. 11.24). The polypeptide has the form of seven α-helix segments, each approximately 30 residues long and each spanning the lipid bilayer of the membrane, linked by non-helical residues at the inner and the outer membrane surface. The helical segments enclose the retinal chromophore within the membrane closer to the surface bounding the cytoplasm than to the outer surface.

The protonated retinal imine chromophore has the all-*trans* structure in the electronic ground state. The absorption of light in the yellow region (λ_{max} 568 nm) transforms the chromophore, through excited intermediates detected by flash spectroscopy, to the neutral isomeric 13-*cis* structure with the loss of a hydrogen ion which is extruded from the outer surface of the plasma membrane in the bacterium. The neutral 13-*cis* structure (λ_{max} 412 nm) relaxes thermally with the uptake of a hydrogen ion from the cytoplasm to reform the protonated all-*trans* retinal imine chromophore, again through intermediates detected and characterized spectroscopically (Stoeckenius 1980; Fodor *et al.* 1988).

The ionic dissociation constant (pK_a value) of many organic acids changes substantially on photoexcitation. The protonated all-*trans* retinal imine chromophore becomes a stronger acid with a $\triangle pK_a \sim 5$ pH units on photoexcitation, corresponding to a chemiosmotic potential of approximately 300 mV. In bacteriorhodopsin the proton is lost from the cationic 13-*cis* chromophore immediately formed on photoexcitation at a location in the protein (opsin) different from that of the hydrogen ion taken up by the neutral 13-*cis* form, owing to a change in the protein conformation from '*trans*-type' to '*cis*-type', consequent on the change in the shape of the chromophore cavity due to the photoisomerization (Fig. 11.24).

In a classical experiment supporting the chemiosmotic theory, purified bacteriorhodopsin was incorporated into an artificial closed lipid vesicle, together with units consisting of the proton-conducting channel and ATP synthase complex isolated from mitochondria. On illumination, the vesicles produced ATP from ADP and inorganic phosphate, driven by the flow of hydrogen ions through the synthase complex under the proton-motive potential generated across the membrane by the bacteriorhodopsin on photoexcitation.

The plasma membrane of halobacteria contains three other retinal-protein complexes analogous to bacteriorhodopsin. One, halorhodopsin, serves as a halide pump, translocating principally chloride ions. The other two are sensory rhodopsins activating the flagella, one for motion towards regions of high yellow–green light intensity (optimum wavelength approximately 570 nm) and the other for the avoidance of lethal ultraviolet radiation (Oesterhelt and Tittor 1989).

The non-oxygenic green and purple photosynthetic eubacteria have a membrane-bound photosystem with a more complex organization than that of the halophilic archaebacteria. The bacteriorhodopsin of the halobacteria is a single molecule in which the retinal imine chromophore serves both as a light-harvesting pigment and a photoreaction centre. The carotenoids of the halobacteria protect the organism from ultraviolet radiation but they do not transfer their photoexcitation energy to the bacteriorhodopsin. In contrast the carotenoids of the green and the purple photosynthetic bacteria are part of a light-harvesting assembly, passing on their photoexcitation energy to other antenna pigments which in turn transfer

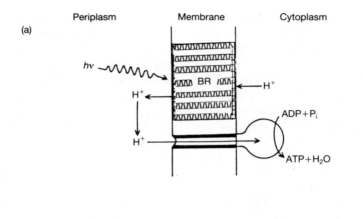

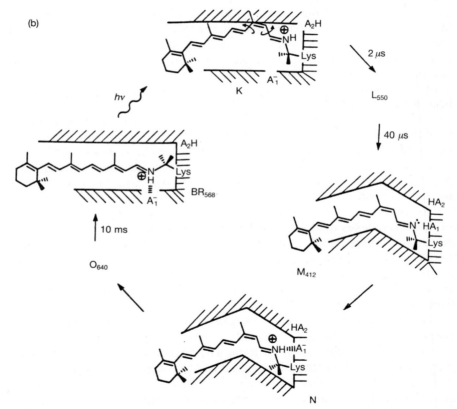

Fig. 11.24 The photogeneration of ATP by the bacteriorhodopsin hydrogen ion pump of halobacteria.

(a) Each photon absorbed by the retinal imine chromophore of bacteriorhodopsin results in the extrusion of a hydrogen ion into the periplasm and the uptake of a hydrogen ion from the cytoplasm of the halobacterium: the return of hydrogen ions to the cytoplasm through a proton-conducting channel to the ATP synthase enzyme system at the inner terminus produces an ATP molecule from ADP and P_i for each pair of protons passing through the system.

(b) The retinal imine chromophore in the electronic ground state, with the all-*trans* structure, is protonated and enclosed in the stable '*trans*-type' protein pocket (BR_{568}). Irradiation at 568 nm photoisomerizes the chromophore to the 13-*cis* structure (K) in which the hydrogen bond from the N—H$^+$ cationic centre to the carboxylate side chain (A_1^-) of the protein in the '*trans*-type' conformation is broken. Hydrogen bonding with another carboxylic acid side chain (A_2H) in the intermediate L_{550} leads to the dissociation of the —COOH group, with the relay of the H$^+$ released to the periplasm, and the transfer of the chromophoric N—H$^+$ hydrogen ion to reform the carboxylic acid side chain (HA_2). The relaxation of the protein conformation to '*cis*-type' accommodates the neutral 13-*cis* chromophore (M_{412}) and brings the lone pair of electrons on the imine nitrogen atom close to the carboxylic acid side chain (HA_1) of the modified (*cis*-type) protein pocket. The hydrogen ion reforming the acid (HA_1) from its conjugate base (A_1^-), left free in the excited state (K), is relayed from the cytoplasm. Proton transfer reforms the 13-*cis* imine cation, hydrogen bonded to the carboxylate ion (N—H$^+$...A_1^-) in the intermediate N, where the protein retains the '*cis*-type' conformation. The thermal isomerization of the retinal imine chromophore from 13-*cis* to all-*trans*, and the relaxation of the protein conformation from '*cis*-type' to '*trans*-type', regenerates the ground state BR_{568} from N through the intermediate O_{640}, where the all-*trans* chromophore becomes a free cation in the '*trans*-type' protein conformation, owing to the temporary fission of the hydrogen bonding (N—H$^+$...A_1^-) consequent on the chromophore and protein structural changes (after Fodor *et al.* 1988).

---

the radiation energy to the reaction centre of the bacterial photosystem. The non-oxygenic photosynthetic bacteria possess a single photosystem and show no supra-additive enhancement of quantum yield on two-wavelength irradiation, unlike the oxygenic cyanobacteria which have two photosystems and display the same Emerson enhancement effect as the eukaryote chloroplast.

In addition to the carotenoids, the light-harvesting complex of the photosynthetic bacteria contains many molecules of antenna bacteriochlorophyll (BChl) per reaction centre, some 40 in the case of the purples and up to 1500 in the case of the greens. The colour of these, as of other photosynthetic organisms, is due mainly to selective light absorption by the antenna molecules, owing to the small fraction of reaction centres in the total pigment. The reaction centres of both the green and the purple photosynthetic bacteria are made up of a tetrahydroporphyrin, BChl *a* or *b*, and so too are the antenna pigments of the purple species; but most of the antenna pigment of the green species consists of a dihydroporphyrin (i.e. a chlorin), BChl *c*, *d*, or *e*, like the chlorophyll of oxygenic photosynthetic organisms, the blue–green cyanobacteria and green plants (Fig. 11.25).

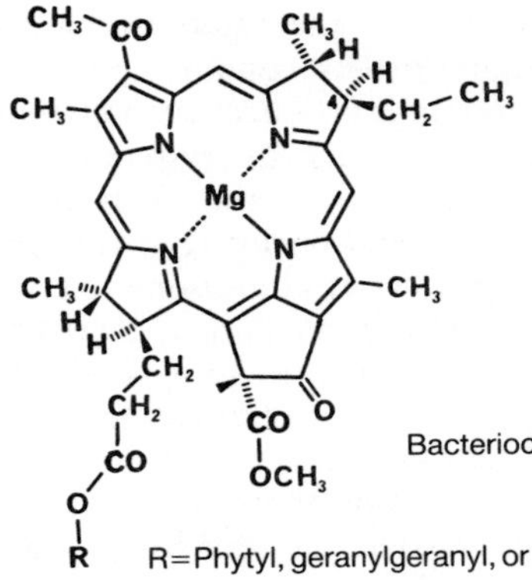

Bacteriochlorophyll *a*

R=Phytyl, geranylgeranyl, or farnesyl

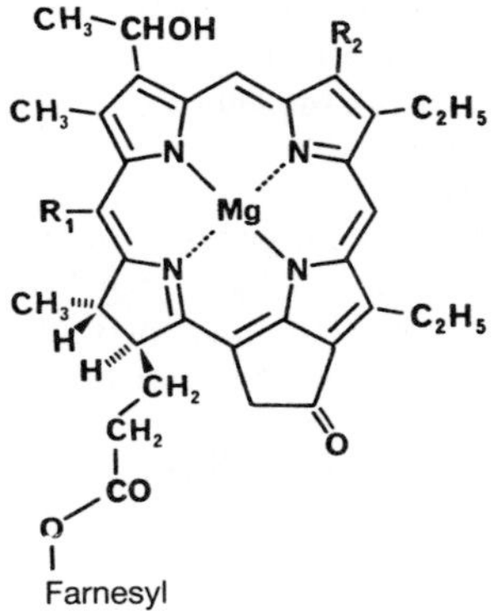

Farnesyl

Farnesol

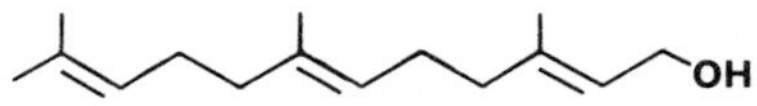

	R₁	R₂
Bacteriochlorophyll *c*	—CH₃	—CH₃
Bacteriochlorophyll *d*	—H	—CH₃
Bacteriochlorophyll *e*	—CH₃	—CHO

Fig. 11.25 Structures of the bacteriochlorophylls. The hydrogen and ethyl sub-stituents at the 4-position in bacteriochlorophyll *a* are replaced by a methyl vinyl group (=CH—CH₃) in bacteriochlorophyll *b*. These molecules, which are tetra-hydroporphyrins, form the ultimate light-trap, the reaction centre, of photo-synthetic bacteria and the antenna pigments additionally of the purple species. Bacteriochlorophyll *c*, *d*, and *e*, which are dihydroporphyrins (chlorins) like the chlorophylls (Fig. 11.22), form the antenna pigments of the green species, giving them their characteristic colour.

The bulk of the light-harvesting complex of green photosynthetic bacteria consists of a membrane-enclosed vesicle in the cytoplasm, about 40 nm in diameter and 140 nm long containing BChl *c, d,* or *e,* the chlorosome, which is connected to the reaction centre of BChl *a* in the cytoplasmic membrane by a baseplate composed of protein and antenna BChl *a.* The baseplate complex of one of the green photosynthetic species, making up some 5 per cent of the total antenna assembly, has been crystallized and characterized chemically, spectroscopically, and by an X-ray diffraction structural study (Matthews and Fenna 1980). The soluble baseplate antenna complex consists of a trimer containing three identical subunits, related by a three-fold rotational symmetry that ensures the isotropic absorption of light incident parallel to the trigonal axis normal to the symmetry plane of the plasma membrane surface. Each subunit is made up of a polypeptide with a molecular weight (MW) of about 50 000 and seven molecules of BChl *a.* The baseplate trimer, with maximum light absorption at 809 nm, is not itself photochemically active but relays its own photoexcitation energy, and that from the bulk chlorosome antenna BChl *c* (λ_{max} 750 nm) to the reaction centre BChl *a* (λ_{max} 840 nm).

The reaction centres in the photosynthetic membrane of several species of purple eubacteria have been isolated by selective detergent extraction and crystallized. In 1984 Deisenhofer and coworkers analysed by X-ray diffraction crystallography the structure of the reaction centre of *Rhodopseudomonas viridis* and subsequently the structure of the corresponding centre of another purple species, *Rhodobacter sphaeroides*, was similarly solved (Feher *et al.* 1989). The crystallized reaction centres remain photoactive and the structural evidence, in conjunction with flash spectroscopic studies, provide both the temporal and the spatial sequence of electron-transfer events in the centre following photoexcitation (Fig. 11.26).

The membrane-enclosed region of the reaction centre in the purple bacteria studied has an approximate two-fold rotational symmetry. The two-fold axis passes through an atom of ferrous iron close to the cytoplasmic surface of the membrane and the dimer or 'special pair' of BChl molecules, located near to the outer (periplasmic) surface. The [BChl]$_2$ special pair in *Rps. viridis*, earlier termed photopigment P$_{960}$ from its light absorption λ_{max} 960 nm, serves as the ultimate trap of the photoexcitation energy relayed from the antenna pigments. Additionally the two-fold axis relates two monomeric BChl molecules, each located some 1.1 nm from the nearer molecule of the [BChl]$_2$ dimer, with two bacteriopheophytin (BPhe) molecules, each approximately 1.1 nm from its neighbouring BChl monomer. BChl and BPhe differ only in that the former contains a coordinated magnesium ion which is replaced by two hydrogen ions in the latter.

Approximate two-fold symmetry is maintained by two quinone molecules, each close to a BPhe molecule, and by two polypeptides, the L subunit (light, MW 24 000; 273 residues) and M subunit (medium, MW 28 000;

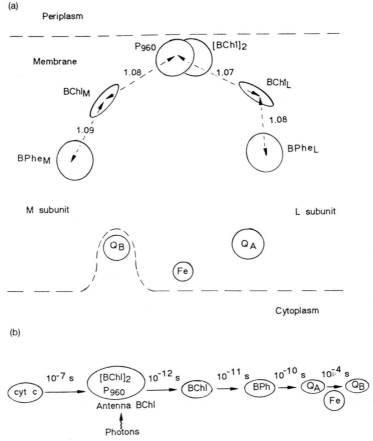

Fig. 11.26 (a) The arrangement of the electron-transfer species in the reaction centre (RC) of purple photosynthetic bacteria (distances in nm), based on the structural analysis of the *Rps. viridis* RC and that of the *Rb. sphaeroides* RC for the Q_B location (Q_B is lost during the crystallization of the former RC). The cytochrome c unit lies on the periplasm surface of the membrane with the nearest cyt c haem approximately 2 nm from the centre of the special pair $[BChl]_2$. An approximate two-fold rotational axis normal to the membrane surfaces, passing through the special pair and the Fe(II) ion, relates the species of the L subunit to those of the M subunit. An electron liberated by the photoionization of $[BChl]_2$ transfers through the species of the L subunit to the membrane-bound quinone Q_A, and thence through the Fe(II) to the mobile quinone Q_B, which is released into the cytoplasm on two-electron reduction to the quinol H_2Q_B. (b) The spectroscopically determined time sequence (in seconds) of electron transfer through the species in the reaction centre of purple photosynthetic bacteria following the photoionization of the special pair $[BChl]_2$ and the restoration of an electron to the $[BChl]_2^+$ radical cation produced from cytochrome c. The half-lives of the radical anion transfer species become progressively longer with an increasing separation from the special pair (after Knapp 1988; Feher *et al.* 1989).

323 residues), each with five transmembrane α-helix segments of 20–30 peptide residues connected by sequences, some helical and others non-helical, at the cytoplasm and the periplasm surfaces. One of the quinone molecules, Q_A menaquinone in *Rps. viridis* and ubiquinone in *Rb. spaeroides*, is bound while the other, Q_B ubiquinone in both species, is mobile, the two quinones being approximately equidistant (~ 0.7 nm) from the ferrous iron atom.

Overall the membrane-enclosed region of the *Rps. viridis* reaction centre has the form of an ellipsoidal cylinder some 4×7 nm in cross-section and approximately 5 nm thick, spanning the lipid bilayer of the plasma membrane. On the elliptical surface of the cylinder lie two further polypeptides, each with a globular form. One, the H subunit (heavy, MW 33 000), is not essential for photoactivity, which is modified only in detail by the removal of the subunit. The other globular polypeptide (MW 38 000) contains four cyctochrome *c* molecules which serve to restore an electron to the special pair—the $[BChl]_2$ dimer—following the photoionization of the latter to $[BChl]_2^+$. The nearest cyt *c* haem lies some 2 nm from the centre of the $[BChl]_2$ special pair. The cytochrome-containing globular polypeptide is absent from the reaction centre assembly in other species of purple photosynthetic bacteria where its role is taken over by mobile cytochrome *c*.

The two-fold symmetry to which the structure of the reaction centre approximates is wholly lacking in the electron-transfer sequence subsequent to the photoionization of the $[BChl]_2$ dimer, following the supply of a quantum of photoexcitation energy (λ_{max} 960 nm) from the antenna pigments of the light-harvesting complex. Only one of the two possible electron-transfer chains to a quinone is employed, the chain associated with the light polypeptide subunit leading to the bound quinone Q_A through $BChl_L$ and $BPhe_L$. Successive intermediates relaying an electron to Q_A have progressively longer half-lives as the negative charge attains an increasingly larger separation from the cation radical $[BChl]_2^+$ remaining from the photoionization (Fig. 11.26). Such an organization minimizes the probability of a prompt recombination of the cation and the electron liberated by the absorption of the photoexcitation energy supplied by the antenna pigments, and it ensures around 100 per cent efficiency in the separation of charge over a distance of approximately 3.5 nm between the cation $[BChl]_2^+$ and the semiquinone anion Q_A^-. The initial electron transfer to form the monomer anion $BChl^-$ (λ_{max} 830 nm) at a distance of approximately 1.1 nm occupies a picosecond (ps, 10^{-12} s), with subsequent transfers to form BPh^- (λ_{max} 800 nm) a further 1.1 nm away over approximately 7 ps and then the semiquinone anion Q_A^- over a further 150 ps.

The later electron transfers have longer time constants, involving more stable intermediates. An electron is donated to the dimer cation $[BChl]_2^+$ by the globular cytochrome *c* protein in about 0.3 microseconds (μs, 10^{-6} s), and the relatively stable bound semiquinone anion Q_A^- transfers an electron, probably through the sandwiched iron atom, in about 100 μs to the mobile quinone Q_B which remains enclosed in its polypeptide pocket between two transmembrane α-helix segments of the protein. A further photoexcitation of the $[BChl]_2$ dimer results in the transfer of a second electron to Q_B^- which, on the bonding of two hydrogen ions, is liberated as the quinol H_2Q_B.

Cyclic electron transfer is completed by a membrane-bound redox chain consisting of cytochromes and iron–sulphur proteins. Electrons are donated to the iron–sulphur proteins by the mobile quinol H_2Q_B which reverts to quinone form Q_B. With the net translocation across the membrane of two hydrogen ions per electron, the electrons are transported back to the globular cytochrome c polypeptide, which returns an electron to the special pair $[BChl]_2$ following its photoionization. The proton-motive potential produced by the translocation of the hydrogen ions generates ATP from ADP and inorganic phosphate. In the purple photosynthetic bacteria generally, the reducing equivalents of NADH required for the fixation of carbon dioxide through the dark enzyme reactions of the Calvin cycle are produced indirectly, by the consumption of ATP, with electrons taken from the inorganic or organic substrate donors (Fig. 11.27).

The membrane-bound elements of the photosystem in the green photosynthetic bacteria, although less well characterized, appear to be similar to those of the purple analogues. The mobile carrier accepting electrons produced by photoionization in the reaction centre of the green bacteria is the iron–sulphur protein, ferredoxin, with a more negative redox potential than the corresponding mobile carrier, ubiquinone, of the purple bacteria. The redox potential of the ferredoxin suffices for the direct reduction of NAD^+ to NADH without the consumption of ATP. The electron balance is restored non-cyclically by a supplementary electron donation from sulphide, thiosulphate, or from an organic substrate, to the membrane-bound cytochrome chain transporting electrons back cyclically to the photoionized pigment of the reaction centre (Fig. 11.28). The translocation of hydrogen ions across the membrane, consequent upon the flow of electrons through the cytochrome chain, gives rise to the proton-motive potential that generates ATP.

The dark reactions which assimilate carbon dioxide in the green photosynthetic bacteria are distinctive for, unlike their purple analogues or the oxygenic cyanobacteria and green plants, they do not employ the ribulose biphosphate pathway of the Calvin cycle. Instead the green photosynthetic bacteria assimilate carbon dioxide by what is essentially the Krebs citric acid cycle running in reverse, i.e. the reductive tricarboxylic acid cycle. The ATP and NAD(P)H photogenerated in the green photosynthetic bacteria are used to transform carbon dioxide to acetate in its activated form, acetyl-coenzyme A, for further biosynthesis. Two carbon dioxide molecules enter the reductive TCA cycle in the conversion of the cycle intermediate,

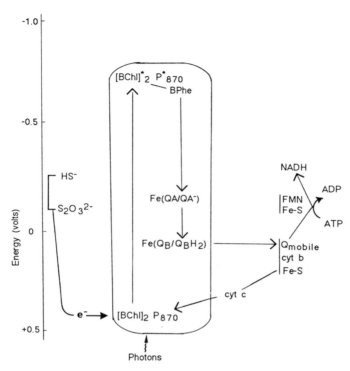

Fig. 11.27 The energetics of photoexcitation and electron transfer in purple photosynthetic bacteria. The redox potential of the mobile quinol H_2Q_B released from the membrane-bound reaction centre (boxed) is insufficient alone to reduce NAD^+ to NADH: the free energy of hydrolysis of ATP to ADP is also required, catalysed by flavin mononucleotide (FMN) and iron–sulphur (Fe–S) proteins. Electrons lost in the reduction are restored to the reaction centre from organic or sulphur-containing substrates. Alternatively, an electron is restored to the reaction centre cyclically, through cytochrome *c*, with the translocation of two hydrogen ions across the membrane into the chromatophore (an invaginated region of the cytoplasmic membrane with bound reaction centres and antenna pigments, enclosing a volume of periplasm). The flux of hydrogen ions back into the cytoplasm, across the membrane from the chromatophore interior through a proton-conducting channel terminated by an ATP-synthase system at the cytoplasmic surface, generates ATP chemiosmotically (2 to $3H^+$ per ATP molecule).

succinate, first to oxoglutarate and then to isocitrate. With coenzyme A, the citrate formed from isocitrate breaks down to give the acetyl-coenzyme A product and the cycle intermediate, oxaloacetate. The reductive TCA cycle is completed by the reduction of oxaloacetate, through malate and fumarate, to succinate (Fig. 11.29). Oxaloacetate for biosynthesis and the replenishment of the cycle is formed by the assimilation of two further

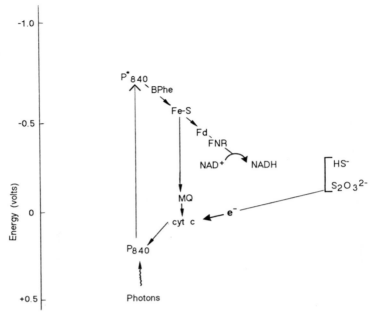

Fig. 11.28 The energetics of photoexcitation and electron transfer in green photosynthetic bacteria. The redox potential of the mobile electron carrier ferredoxin (Fd) suffices for the direct reduction of NAD$^+$ to NADH, catalysed by ferredoxin NAD$^+$ reductase (FNR). The electron, produced by the ionization of the photoexcited bacteriochlorophyll centre P$^*_{840}$, is transferred to ferredoxin by way of a bacteriopheophytin (BPhe) molecule and a low-potential iron–sulphur protein (Fe–S). The neutral photopigment P$_{840}$ is regenerated from the cation produced in the photoionization by an electron from an organic or a sulphur-containing substrate transferred through c-type cytochromes. Alternatively, an electron is restored to the cationic photopigment cyclically from the Fe–S protein to cytochrome c, through menaquinone (MQ), with the chemiosmotic generation of ATP.

carbon dioxide molecules: one gives pyruvate with acetyl-coenzyme A, and the other adds to phosphoenolpyruvate to give oxaloacetate (Evans *et al.* 1966).

11.7 The evolution of bioenergetic systems

It is generally considered that the earliest forms of life were the anaerobic heterotrophic bacteria, fermenting the prebiotically generated organic substances relatively inefficiently (~2ATP/glucose) in homogeneous solution catalysed by soluble enzymes. These organisms were followed by the anoxic and then the oxygenic photoautotrophic bacteria which assimilated carbon

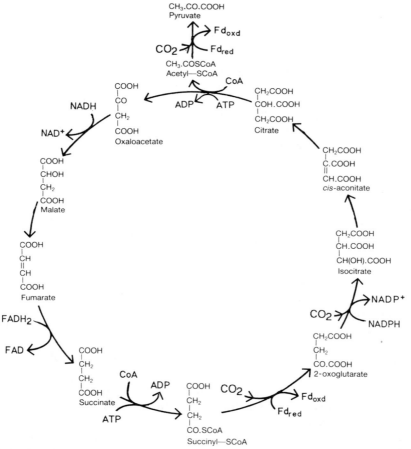

Fig. 11.29 The reductive tricarboxylic acid cycle used by green photosynthetic bacteria and some chemoautotrophic bacteria for the assimilation of carbon dioxide to produce acetate and thence pyruvate and more complex intermediates (e.g. oxaloacetate to replenish the cycle). The reductive TCA cycle consumes the reductants generated by the corresponding oxidative cycle [NADH, NADPH, FADH$_2$] (Fig. 11.13). In addition, reduced ferredoxin (Fd$_{red}$) is converted to the oxidized form (Fd$_{oxd}$) through the oxidation state change Fe(II) $\rightarrow$ Fe(III) in the reductive assimilations of carbon dioxide transforming succinate to 2-oxoglutarate, and acetate to pyruvate (Evans *et al.* 1966).

dioxide by means of membrane-bound catalytic systems, driven by energy captured from the solar radiation. The oxygen generated photosynthetically was adapted in the subsequent stage of prokaryote bioenergetic evolution to the highly efficient (up to 38ATP/glucose) oxidation of substrates by the aerobic heterotrophic bacteria, sustained by the oxidative tricarboxylic acid cycle which feeds reducing equivalents into the membrane-bound respiratory chain.

The succession of stages follows an order of increasing efficiency in energy conservation, driven by the pressure of natural selection. An evolutionary continuity, and a development by the processes of takeover and adaptation, is suggested by the similarities between the bioenergetic systems of the successive stages.

The analogies between the photosystems of the anoxic and the oxygenic photosynthetic bacteria are marked. Photosystem I of oxygenic photosynthetic organisms and the photosystem of the green anoxic photobacteria both directly reduce $NAD(P)^+$ to $NAD(P)H$ by photogenerated electrons carried by ferredoxin and other electron transporters of the same molecular and functional type in the two cases (Figs. 11.19 and 11.28). Corresponding and even closer analogies are found between the reaction centre structure of purple photosynthetic bacteria (Fig. 11.26) and that of Photosystem II, which generates molecular oxygen from water in cyanobacteria and the eukaryote chloroplast.

The structure of the reaction centre unit of Photosystem II shows a similar symmetry, with a two-fold rotational axis passing through an iron atom near to the outer (stroma) surface of the thylakoid membrane in a chloroplast, and a special pair of chlorophyll a' molecules $[Chl]_2$, the P_{680} photoexcitation trap, close to the inner (lumen) surface. Between the dimer $[Chl]_2$ and the iron atom lie two sets of electron-transporting intermediates, each consisting of a monomer chlorophyll a, followed by a pheophytin a and a plastoquinone Q.

Again the iron atom is sandwiched between two quinones Q_A and Q_B, which are enclosed, together with the dimer $[Chl]_2$ and other electron-transfer intermediates, by two polypeptides, each composed of 353 amino acid residues forming five transmembrane α-helix segments. The amino acid sequences of these two polypeptides in a PS II unit have a substantial homology with the corresponding sequences of the L and M subunits of the purple *Rps. viridis* reaction centre, particularly in the regions of the transmembrane α-helix segments (the fourth and the fifth) which bind the iron atom, the quinones Q_A and Q_B, and the chlorophyll special pair (Barber 1987). The homologies of amino acid sequence in the polypeptides, and the analogies between the constituents and their organization in the two types of reaction centre, indicate an evolutionary link between the photosystem of the purple photosynthetic bacteria and PS II of oxygen-evolving photosynthetic organisms.

The structural two-fold symmetry of the reaction centre in the purple *Rps. viridis*, and the use of only one of the two potential electron-transfer paths, suggest that the structure arose from the duplication of a primitive photocentre consisting of a single polypeptide enclosing a monomeric photopigment which, on photoionization, transferred an electron to a mobile quinone. A duplication to produce the L and M subunits facilitated the formation of the special pair $[BChl]_2$, giving a more efficient photo-excitation trap, and stabilized the charge separation through the additional electron-transfer step from bound Q_A to mobile Q_B.

There are analogies in addition between the membrane-bound electron-transfer chains of photosynthesis and respiration (Figs. 11.11 and 11.20). Both consist of cytochromes and iron–sulphur proteins and both are fed by mobile hydroquinones, resulting in the translocation of hydrogen ions across the membrane to generate the proton-motive potential required for ATP production. Since aerobic organisms appeared only after non-oxygenic photosynthetic bacteria had evolved into oxygen-evolving cyanobacteria, it is probable that the respiratory chain developed from the electron-transport chain of photosynthetic bacteria (Hall and Rao 1987).

Similarly it is probable that the major source of reducing equivalents fed into the respiratory chain of aerobic organisms, the oxidative tricarboxylic acid (TCA) cycle (Fig. 11.13), arose from a reversal of the dark enzyme reactions assimilating carbon dioxide in the ancestors of the green photosynthetic bacteria before the appearance of a significant partial pressure of oxygen in the atmosphere, i.e. the reductive TCA cycle (Fig. 11.29). The Calvin cycle (Fig. 11.23) employed by the purple photosynthetic bacteria and by the oxygenic photosynthetic organisms is biochemically unrelated, and it was an independent development. The reductive TCA cycle operates in chemoautotrophs as well as photoautotrophs, and it is found in remotely related organisms belonging to the two different bacterial kingdoms. The archaebacterium *Sulfolobus*, which lives in hot acidic springs (60–80°C, pH 2) by oxidizing sulphur to sulphuric acid, employs the reductive TCA cycle to assimilate carbon dioxide, like the green photosynthetic eubacteria. The wide divergence between the archaebacteria and the eubacteria suggests that the reductive TCA cycle may be an ancient pathway of carbon dioxide fixation, and that the Calvin cycle is a more recent development (Stanier *et al.* 1987).

Sections of the reductive TCA cycle were probably evolved in early anaerobic bacteria to enhance the efficiency of fermentation. Subsequently the full reductive TCA cycle, powered by solar radiation in place of the increasingly depleted prebiotic fossil fuels, became operational with the development of the green photosynthetic bacteria. In the normal course of glycolysis, glucose splits into two molecules of pyruvate with the production of two molecules of NADH, and redox balance is restored by the reduction of pyruvate to lactate (or ethanol and carbon dioxide equivalently) with the regeneration of NAD^+ (Figs. 11.4 and 11.8). One of the two molecules of pyruvate formed by the glycolysis of a glucose molecule in some present-day bacterial fermentations is often conserved for biosynthetic service from its original role—that of an electron (or hydride) sink in its reduction to lactate by the conversion of NADH back to NAD^+.

The conservation is achieved by the transformation of the other pyruvate molecule to oxaloacetate, through the assimilation of carbon dioxide. Both of the NADH molecules produced in the glycolysis of a glucose molecule are regenerated to NAD^+ through the successive reduction of oxaloacetate to malate and thence of fumarate (from the dehydration of malate) to succinate. The reductive sequence found in some modern fermenting bacteria

$$CO_2 + \text{pyruvate} \rightarrow \text{oxaloacetate} \rightarrow \text{malate} \rightarrow \text{fumarate} \rightarrow \text{succinate}$$

consists essentially of a reversal of the oxidative C_4 dicarboxylic acid chain discovered by Szent-Györgi, and subsequently incorporated by Krebs into the oxidative TCA cycle.

Further, some modern fermenting bacteria, while obtaining most of their ATP by glycolysis through soluble enzyme catalysis in homogeneous solution, generate additional ATP from a membrane-bound electron-transfer chain of iron–sulphur proteins, quinones, and cytochrome *b*. NADH donates electrons to the chain, thus regenerating NAD^+, and the output from the chain is accepted by fumarate, which is reduced thereby to succinate. The additional production, with the yield of one ATP molecule generated for each NADH molecule oxidized, originates chemiosmotically. Hydrogen ions are extruded through the cytoplasmic membrane during the transfer of electrons along the chain and produce a proton-motive potential. The subsequent return of the hydrogen ions through proton-selective channels, each terminated at the inner membrane surface by an ATP-synthase enzyme system, transforms ADP and inorganic phosphate to ATP (Gest 1987).

Vectorial chemiosmotic energy coupling is considered to have evolved through three main stages. The typical products of fermentation are acidic, and the maintenance of neutral pH conditions in the bacterial cell required mechanisms for the removal of excess hydrogen ions. At the earliest stage, selective protein channels across the lipid bilayer membrane were developed to allow the ingress of nutrients and the voiding of wastes, including the hydrogen ion, by passive diffusion governed by the gradients in concentration.

When the passive diffusion of the hydrogen ions produced became the limiting factor on the fermentation rate, an active proton pump was evolved by the localization of a soluble enzyme hydrolysing polyphosphates on the inner surface of the membrane at the terminus of a proton-conducting channel. The free energy of hydrolysis of the ATP (or other activated phosphate), coupled to the extrusion of protons through the specific channel, provided a more flexible and efficient mechanism than passive diffusion for the disposal of excess hydrogen ions generated by fermentation.

In the third stage, independent mechanisms for the translocation of hydrogen ions from the cell were evolved, setting up a hydrogen ion concentration larger outside than inside the cytoplasmic membrane of the

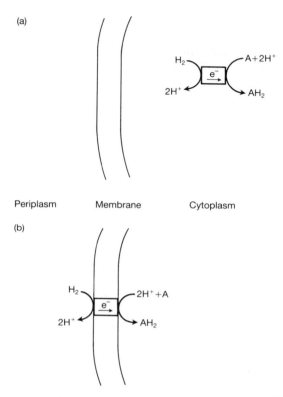

Fig. 11.30 The evolution of a proton-translocating hydrogenase. (a) The hypothetical ancestral soluble hydrogenase, catalysing the transfer of electrons from molecular hydrogen to a reducible substrate in the bacterial cytoplasm. (b) The membrane-bound hydrogenase of *E. coli*, catalysing the transfer of electrons from molecular hydrogen in the periplasm, with the production of two hydrogen ions, to a reducible substrate in the cytoplasm, with the consumption of two hydrogen ions. Each unit reaction effectively translocates two hydrogen ions from the cytoplasm to the periplasm (after Garland 1981).

cell. The inversion of the former pH gradient transformed the proton pump driven by ATP hydrolysis into a system synthesizing ATP under a chemiosmotic proton-motive potential. The return of hydrogen ions into the cell, through the proton-specific channel to the ATP-ase enzyme attached to the inner terminus, now promoted the synthesis of ATP from ADP and inorganic phosphate, the reverse of the hydrolytic reaction catalysed under the former conditions of excess internal acidity.

The independent mechanisms developed for the translocation of hydrogen ions from the cell involve membrane-bound systems of two main types, one driven by solar radiation and the other by the chemical free energy available from the redox degradation of nutrients. The bacteriorhodopsin system of the halobacteria provides the simplest example of the photon-driven case, where each photon absorbed results in the ejection of a hydrogen ion from the cell (Fig. 11.24).

The reactions catalysed by the hydrogenase enzyme embedded in the plasma membrane of *E. coli* illustrate the process of proton extrusion by the redox degradation of substrates. The hydrogenase, an iron–sulphur protein with a possibly soluble ancestral form, catalyses the transformation of hydrogen into two hydrogen ions at the outer surface of the plasma membrane and transfers the two electrons liberated to the inner surface where they convert the substrate (A) to its reduced form (AH_2), with the consumption of two internal hydrogen ions (Fig. 11.30). Effectively each set of unit reactions translocates two protons out of the *E. coli* cell overall (Garland 1981).

12
Organic replication and genealogy

12.1 Organic substances and organized matter

During the first half of the nineteenth century a distinction was drawn between 'organic substances', characterized by a constant elementary composition and a sharp, specific, melting point or boiling point, and 'organized matter' of indefinite composition, which normally could not be crystallized. Initially 'organized matter' covered the 'immediate or proximate principles' of living organisms, generally divided into the three categories of 'oleaginous', 'saccharinous', and 'albuminous', which are made up of the 'essential elements' of hydrogen, carbon, oxygen, and nitrogen and the 'incidental elements' of sulphur and phosphorus (Prout 1834). The oleaginous materials were soon transferred to the 'organic substances' division through the saponification studies in 1811–23 of Michel Chevreul (1786–1889) in Paris, who showed these materials to be the glycerol esters of the higher fatty acids. The application of analogous hydrolytic procedures to the saccharinous and albuminous materials showed little more than that simple sugars and amino acids, respectively, are extractable from them, and they remained in the limbo of 'organized matter' virtually to the end of the nineteenth century.

Berzelius, who had proposed the names of cysteine and glycine for two amino acids isolated from hydrolysates, suggested in 1838 to Geradus Mulder (1802–80) at Utrecht the term *protein* (Greek, 'primary substance') for the basic unit of all albuminous materials. Mulder followed the dualistic theory of Berzelius, according to which all compounds, whether organic or inorganic, were regarded as binary combinations of an electropositive and an electronegative unit. In mineral substances both units were taken to be the atoms or simple atomic groupings of the chemical elements, whereas the electropositive unit of an organic compound was a complex polyatomic radical which, in principle, could be isolated in free form, like the basic elements of inorganic compounds. Mulder supposed from his elementary analyses that all albuminous substances consist of the protein radical, with the empirical formula $C_{40}H_{62}N_{10}O_{12}$, combined in various ways with an atom or atoms of sulphur, or of phosphorus, or with both types of atom, to give an individual egg or serum albumin, fibrin, or gluten, and the like. Mulder's theory failed to stand up to the results of further elementary analyses and, after a brief period of popularity in the 1840s, all that remained of the theory was the term 'protein', to cover the general class of albuminous materials.

By the 1860s a number of organic chemists, headed by August Kekulé (1829–96), felt that 'organized matter' was no longer part of their domain of study, and that the investigation of the amylaceous and albuminous materials should be transferred to the physiologists of the medical schools and to the students of plant and animal natural products. The proper object of organic chemistry, it appeared, should be the investigation of the testable consequences of the new organic structure theory and the use of the theory for the synthesis of target molecules. The success of this research programme, with its fruitful application to the production of synthetic dyes and the extension of organic structure theory from the flatland aromatic models of Kekulé (1865) to the three-dimensional chemistry in space of Le Bel and van't Hoff (1874), provided the basis for a new perspective towards the end of the nineteenth century.

The development of organic chemistry by the 1890s allowed Emil Fischer, a one-time student of Kekulé, to plan, and in significant measure to achieve, not only the stereochemical characterization of the monomers, the D-sugars and the L-α-amino acids, of the main biopolymers then known, but also the synthesis of protein-like polypeptides. By 1907 Fischer had accomplished the synthesis of an octadecapeptide with a molecular weight of 1213, a value comparable with the lower current estimates for the molecular weights of natural proteins. Other estimates ranged up to more than an order of magnitude larger, but all such estimates were considered to rest on the insecure assumption that the natural proteins are homogeneous substances. Some of the synthetic polypeptides were hydrolysed by proteolytic enzymes, dependent, Fischer noted, 'partly on the nature of the amino acids, partly on their sequence, also on the chain length, and most especially on the configuration of the molecule'. That is, polypeptides composed of the L-α-amino acids, but not those made up of the D-enantiomers or racemic mixtures, are enzymatically cleaved.

Doubts as to the homogeneity of the natural proteins had been emphasized through the distinction drawn by Thomas Graham (1805–69), from his dialysis studies at University College London in the 1850s, between 'crystalloids' and 'colloids'. In 1861 Graham reported that crystalline compounds, such as cane sugar, rapidly diffuse in aqueous solution from a vessel closed at the lower end by a parchment paper membrane into a reservoir of water beneath, whereas 'vegetable and animal extractive matters', like starch, albumin, or gelatin, cross the membrane extremely slowly. The traditional distinction between 'organic substances' and 'organized matter' was now reinforced by the diffusion-rate demarcation criterion between 'crystalloid' and 'colloid' (from the Greek, *colla*, glue). Graham supposed that colloid particles are inhomogeneous aggregates of crystalloid molecules with a varying degree of hydration, dependent upon the salt concentration of the colloidal solution.

The dependence of the solubility of proteins upon the salt concentration was established during the 1850s, allowing the ready separation of proteins which are selectively salted in or salted out. The general view that proteins are colloidal inhomogeneous aggregates was not greatly changed by the isolation of crystalline proteins from salt-solution extracts of plant seeds during the 1880s. These were further examples of exceptions of the 'non-crystallizable' criterion for 'organized matter', since 'blood crystals' of haemoglobin from various animal species had been described since the 1840s. By the 1900s crystallinity was no longer a guide to homogeneity for, as Emil Fischer observed, mixed crystals and isomorphic replacements are commonplace in both organic and inorganic chemistry. Nor were molecular weight determinations any guide, for these record average values. The quantitative analysis of haemoglobin crystals gave a molecular weight of about 16 000, based on the assumption that each molecule contains a single iron atom. But, Fischer noted in 1916, 'it should always be remembered that the haematin [haem], from all that we know of its structure, can bind several globin units'.

Evidence for the homogeneity of natural proteins came with the development of the ultracentrifuge by Theodor Svedberg (1884–1971) from 1924 at the University of Uppsala. Avogadro's number and the mean weight of relatively large colloidal particles, visible under the microscope, had been determined from their diffusion rate and sedimentation equilibria in the Earth's gravitational field by Jean Perrin (1870–1942) at Paris in 1908, and subsequently by others, including Svedberg. The first ultracentrifuge, producing a field of up to 500 times the force of gravity, was developed for the study of small colloidal particles, detectable optically only by their bulk properties, such as light absorption or refraction. Initially Svedberg held the general view that proteins are inhomogeneous colloidal aggregates, like the polydisperse gold sols he had studied previously, and he was surprised to observe with haemoglobin in the ultracentrifuge a sharp sedimentation edge move along the centrifuge tube, characteristic of a homogeneous monodisperse system. After two days in the ultracentrifuge, sedimentation equilibrium was established with a haemoglobin concentration distribution corresponding to a uniform molecular weight of about 68 000 for the protein. Compared with earlier results based upon quantitative elementary analysis, the value showed the presence of four iron atoms in the haemoglobin molecule, suggesting a possible tetrameric subunit structure.

By 1933 the ultracentrifuge had been developed to give fields of $4 \times 10^5 g$ routinely, and with the instrument Svedberg showed that most natural proteins are homogeneous and that many dissociate into discrete subunits on changing the salt or hydrogen ion concentration. His measurements led him to suppose that most proteins are weakly bound aggregates of an integral number of uniform subunits; each basic unit was assigned a molecular weight of about 35 000, composed of some 288 amino acid residues (both values were halved in a later amendment). Subsequent studies, while confirming that the larger proteins are often composite, showed that the subunits are generally far from uniform in either molecular weight or chain length.

The weak intermolecular bonds between the subunits of a large protein appeared to have a character similar to the corresponding intramolecular bonds involved in protein denaturation. Changes in the salt or hydrogen ion concentration of a protein solution produced a marked viscosity change, followed by a precipitation of the protein, which proved to be reversible under mild conditions. The molecular weight of native egg albumin appeared to be unchanged by denaturation, and the theory of the viscosity of solutions suggested a transformation of the protein solute from a spherical shape to rod form on denaturation. In 1936 Linus Pauling and Alfred Mirsky proposed that the polypeptide chain of a native protein molecule is folded into a unique regular conformation, held together by hydrogen bonding between the amide N—H group and the oxygen atom of the C=O group of the peptide bonds. On denaturation the regular conformation of the protein molecule is lost, and the polypeptide chain assumes the form of an elongated random coil.

Over the next decade or so, Pauling and his coworkers at Pasadena determined the bond lengths and bond angles of amino acids, amides, and dipeptides and tripeptides by X-ray diffraction crystallography with the aim of building reliable structural models of the regular conformations of native protein molecules. In 1950 Pauling and Robert Corey (1897–1971) reported two regular protein conformations, both rod-like with the polypeptide backbone forming a cylindrical helix maintained by hydrogen bonds orientated along the direction of the helix axis between the N—H group and the C=O group of near-neighbour peptide residues. The structure free from steric strain, the α-helix conformation, contains 3.7 amino acid residues per turn of the polypeptide helix, while the other, the γ-helix, is made up of 5.1 residues per turn. The non-integral number of residues per turn was an entirely novel feature of the structures, since earlier helix models for protein molecules, inspired by the fibre X-ray diffraction patterns of silk and wool, had been based upon integral numbers under the constraint of the rotational axes of symmetry permitted crystallographically (one-, two-, three-, four-, and six-fold axes).

One of the previous advocates of an integral number (four) of peptide residues per turn, Max Perutz at Cambridge, who had begun a study of the haemoglobin crystal structure in 1938 with the pioneer of protein X-ray crystallography, J. D. Bernal (1901–71), found immediately an expected feature of the α-helix structure in the X-ray diffraction patterns of several fibrous proteins and a synthetic L-polypeptide. After Perutz had solved the problem of the phase relation between the diffracted waves in X-ray protein crystallography by the method of isomorphous heavy-atom replacement in 1953, the first direct evidence for the α-helix structure in a native globular protein came from a colleague in the Cavendish group at Cambridge, John Kendrew. With his research associates, Kendrew solved the three-dimensional structure of myoglobin (molecular weight approximately 17 500) at near-atomic resolution in 1960, showing some 75 per cent of the polypeptide chain to be helical. The α-helix segments in the myoglobin were found to be right-handed, as expected from the absolute configuration of the L-α-amino acids on steric grounds, and from the changes of optical activity observed in the transition from the random coil to the α-helix conformation of synthetic L-polypeptides in solution (Fig. 12.1).

Two years later Perutz and his coworkers completed the analysis of the more complex crystal structure of haemoglobin, in which each one of the four subunits is closely analogous to a single molecule of myoglobin. In deoxygenated haemoglobin the four subunits form a stable tetrahedral array, with a lower oxygen-affinity than myoglobin, but the uptake of an

(a)

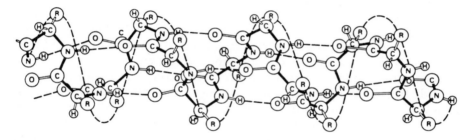

Poly-L-peptide α-helix

(b)

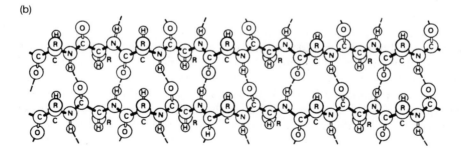

Poly-L-peptide β-sheet

Fig. 12.1 The repeat pattern of hydrogen bonding in the ordered secondary structures of L-polypeptides: (a) the right-handed α-helix conformation; (b) the pleated β-sheet conformation.

oxygen molecule by one subunit destabilizes the array and thus promotes the cooperative binding of further oxygen molecules by the other three subunits over the physiological range of oxygen partial pressure (Perutz 1964; Perutz *et al.* 1987).

The first crystal structure of an enzyme, that of lysozyme from egg-white, analysed in 1965 by David Phillips and his research associates at the Royal Institution, London, showed a smaller α-helix content than myoglobin, but the structure provided evidence for another regular polypeptide conformation predicted by Pauling and Corey, namely the antiparallel pleated β-sheet. Segments of the polypeptide chain lying side by side are linked by cross-chain hydrogen bonds between a N—H group of one segment and an adjacent C=O group of the other in a pleated β-sheet conformation, parallel if the segments run in the same direction, antiparallel if the directions are opposed (Fig. 12.1).

The X-ray crystal structure of lysozyme complexed with a trisaccharide fragment of its physiological substrate, the polysaccharide chain in a bacterial cell wall, gave support to Fischer's 'key and lock' hypothesis of enzyme–substrate interaction in the extended forms developed by Haldane (enzyme-induced steric strain of the substrate) and by Pauling (steric complementation for the reaction transition state). Lysozyme binds and then hydrolyses a hexasaccharide segment of the bacterial cell wall polysaccharide. Models based upon the structure of lysozyme complexed with the trisaccharide fragment suggest that the binding of the enzyme to the hexasaccharide segment induces a distortion of the fourth pyranoside unit from the chair to the less stable sofa (half-chair) conformation. The distortion promotes the hydrolytic cleavage of the β-glucoside $1 \to 4$-bond linking the fourth to the fifth pyranoside unit, catalysed by the adjacent hydrated carboxylate groups of an aspartate and a glutamate residue of the lysozyme main chain (Phillips 1966; Fersht 1985).

The three-dimensional molecular structure of the protein hormone insulin was finally solved in 1969 by Dorothy Crowfoot Hodgkin at Oxford, who had returned to the problem a number of times, in the intervals between solving the crystal structures of such biomolecules as cholesterol, calciferol, penicillin, and vitamin B_{12}, since her first X-ray photographs of the diffraction pattern given by an insulin crystal. With Bernal in Cambridge, she introduced in 1934 the now-standard method of maintaining the protein crystal in its mother liquor during the determination, in order to prevent crystal disorder due to dehydration, and thus obtained the first resolved X-ray diffraction patterns from protein crystals, namely those of pepsin, lactoglobulin, and insulin (Crowfoot Hodgkin and Riley 1968, Crowfoot Hodgkin 1979).

Insulin too was the first important native protein to be sequenced and then synthesized. In 1966 Bruce Merrifield at Rockefeller University reported with A. Marglin the preparation of insulin by means of a machine designed to carry out each stage of polypeptide synthesis automatically with the intermediates bound to the surface of polystyrene beads. The automated solid-phase synthesis of insulin occupied some 3 weeks, compared with the several years required by each of three other groups (in China, West Germany, and the USA) who had reported in 1963–5 the synthesis of insulin by conventional procedures.

All of these syntheses rested on the fundamental analysis of the amino acid sequences in the polypeptide chains of insulin carried out by Frederick Sanger at Cambridge over the decade 1945–55. Sanger solved the primary structure of insulin, the particular consecutive order of amino acids, by the classical method of natural product chemistry: the partial cleavage of the molecule and the characterization of the fragments, followed by the integration of the fragment structures into a coherent whole. The enterprise became feasible through the then recent developments in chromatography for the separation of polypeptide fragments, beginning with the partition chromatography of A. J. P. Martin and R. L. M. Synge (1941), and the availability of pure crystalline enzymes for the hydrolysis of specific peptide bonds.

Sanger developed the reagent, 2,4-dinitrofluorobenzene to characterize the amino acid at the amino terminus of a polypeptide chain. Two different NH_2-terminal amino acids were detected in cattle insulin, demonstrating the presence of two polypeptide chains, the A- and the B-chain made up of 21 and 30 amino acid residues, respectively. By 1953 Sanger and his coworkers had worked out the particular sequence order of the amino acids in each of the two chains and, 2 years later, they identified the specific cysteine residues which cross-linked the chains by two disulphide bridges and formed a ring within the A-chain through a third disulphide bridge.

Synthesis, the classical complement establishing the interpretation of a well-rounded degradative study in traditional natural product chemistry, followed for insulin over the next decade. The conventional synthetic procedures proved to be extended and arduous with low yields, but biologically active material was obtained. Even Merrifield's solid-phase instrumentation, automating the conventional methods, gave crystalline insulin in 3.4 per cent yield at best, although the product had 95 per cent of the specific activity of the native protein (Marglin and Merrifield 1966).

By the 1960s the nineteenth-century distinction between organic substances and organized matter had been dispelled, not only by the advances in technique but also by the reorientation from an energetic to an organizational or structural interpretation of the novel properties of macromolecules. The observation of Brownian motion in suspensions suggested to Thomas Graham in 1861 that all particles in his new category of colloids are imbued with an intrinsic self-energy. His followers among the physiologists supposed that the distinctive features of native proteins, such as the specific catalytic activity of enzymes, derived from an internal life-force, which was lost when the protein became denatured and died.

Emil Fischer's 'key and lock' hypothesis of enzyme–substrate interaction and his pioneering studies of protein analysis and polypeptide synthesis marked the beginnings of the structural reinterpretation. An early application of the hypothesis, the mechanism of enzymatic reaction kinetics proposed by Leonor Michaelis (1875–1949) and Maud Menten (1879–1960) in 1913, brought enzyme reactivity into mainstream physical chemistry, attenuating further the hold of vitalism. The vital force viewpoint had been weakened already by Fischer's use of his earlier discovery of diastereoselectivity in the ascent of the sugar series to account for the homochirality of natural products, apparent in the D-sugar and the L-amino acid monomers of the principal biopolymers.

Towards the end of the twentieth century a certain nostalgia for the lost distinction drawn earlier between the chemistry of the laboratory and the biochemistry of life lingers on, particularly among figures peripheral to the scientific tradition. The chemically proscriptive thought-style of Auguste Comte, who had declared a knowledge of the chemical composition of the Sun and stars to be impossible in principle, was resurrected a century and a half later by another philosopher, Karl Popper. In his 1986 Medawar lecture to the Royal Society, Popper assured his scientific audience, which included some of the pioneers of molecular biology, that biochemical organizations and processes can never be wholly understood in chemical terms alone. As a consequence, according to Popper, organic evolution cannot be satisfactorily explained by Darwin's theory of natural selection, which rests upon chemical reductionism: for the environment does not so much select between different organisms as the organism selects between different environments (Popper 1972).

Earlier, Popper had classified Darwinism with his twin *bête noires* of Freudism and Marxism as non-scientific theoretical constructs, since none met his demarcation criterion of falsifiability for genuine science. The demarcation criterion itself is not falsifiable, and the direct testing of a number of general scientific theories is wholly impracticable. As in the case of Darwinian natural selection, the theory of plate tectonics cannot be directly falsified (Hallam 1973), and the same may be said of much of theoretical chemistry from the time of the phlogiston theory onwards. Pauling's theory of the hybridization of atomic orbitals to account for the established stereochemistry of polyatomic molecules, e.g. the tetrahedral structure of methane in terms of the set of three mutually orthogonal 2p orbitals and the spherically symmetric 2s orbital of the carbon atom, is not falsifiable, although the theory has been scientifically fruitful and retains much pedagogic utility. Atomic orbital hybridization may be translated into the equally unfalsifiable molecular orbital description, which has come to be preferred for its fruitful applications, particularly in the field of molecular spectroscopy.

A scientific theory tends to be abandoned not so much as a consequence of direct falsification, but more as the outcome of the general adoption of an alternative proposal with more ample heuristic fruitfulness and utility. Grotthuss (1785–1822) and other German chemists of the early nineteenth century identified phlogiston with negative electricity (electrons in modern terms), lost in all oxidations (as in the contemporary view). But the new insight could not save the moribund phlogiston theory, then faced with Lavoisier's expansive vision of a new synthetic chemistry, abounding with new compounds of wholly novel properties. In the early nineteenth century there was no firm scientific basis for the immediate development of an electrochemical theory of oxidation, and the new French chemical philosophy of exploring all possible binary, tertiary, and higher combinations of the elements enjoyed a wide appeal, because of its productivity, despite the soon-evident error of the new chemistry in the postulated role of oxygen as the essential element in all acidic compounds (deriving its name from the 'acid generator') and the dubious role of the weightless element of caloric in all physical and chemical changes.

12.2 The nucleic acids

In the days when salmon still reached the Swiss head-waters of the Rhine to breed in significant numbers, Friedrich Miescher (1844–95), a physiologist at the University of Basle, investigated the massive conversion of skeletal muscle protein in the male salmon to gonadal material prior to spawning. In 1874 he reported that salmon sperm heads are largely composed of an insoluble organic salt 'nucleoprotamine', made up of a nitrogen-rich base 'protamine' and a phosphorus-rich acid 'nuclein'. The acidic nuclein was labile, requiring speedy isolation at low temperatures, and it appeared to be colloidal, failing to diffuse through a parchment dialysis membrane. The nuclein gave none of the standard colour tests for proteins, nor was it cleaved by proteolytic enzymes, such as pepsin. None the less many of Miescher's contemporaries dismissed his nuclein as an impure proteinaceous material.

A new interest in Miescher's nuclein arose when cytologists observed the distinctive dye-staining properties of the cell nucleus and the dramatic changes in the nuclear material on cell division. The development of synthetic organic dyes greatly extended the range of biological stains, traditionally vegetable colouring matters, used in cell microscopy. Paul Ehrlich (1854–1915) was inspired to pioneer chemotherapy from his early work, as a medical student, on the specificity of biological staining by synthetic dyes, published in 1877. Basic dyes like methyl green selectively stain the cell nucleus, classified as 'basophilic' by Ehrlich, in contrast to the 'oxyphilic' cytoplasm, which is preferentially stained by acid dyes, such as eosin. The selective and heavy staining of the nuclear material led in the

1880s to the coining of such terms as 'chromatin' for the network structure observed in the nucleus, and 'chromosomes' for the rod-like segments of the structure, undergoing apparent longitudinal division during cell doubling.

In 1893 Albrecht Kossel (1853–1927) at the Berlin physiological institute identified chromatin with nucleoprotamine, or nucleoprotein as it was now termed. Kossel had found in 1884 that Miescher's protamine from fish sperm is a protein, histone, containing a new basic amino acid, histidine, as well as the already-known lysine and arginine. Similar but less basic histones were isolated from other sources, such as the nucleoprotein of calf thymus or of yeast. Between 1885 and 1901 Kossel and his students discovered the new heterocyclic bases adenine, thymine, cytosine, and uracil among the degradation products of the acidic nuclein, or nucleic acid as it was renamed, together with the already-known guanine (from guano), xanthine (from kidney stones), and hypoxanthine (from pancreas tissue). The latter substances were known to be related to uric acid by giving the murexide reaction, discovered by William Prout in 1818 (Fig. 12.2). Prout found that nitric acid converts uric acid, which he isolated from the droppings of a boa constrictor, into a product forming, with ammonia, the purple–red ammonium salt of purpuric acid (murexide).

Between 1882 and the end of the century Emil Fischer established by degradation and synthesis that uric acid and the related substances are all derivatives of a new fundamental heteroaromatic bicyclic nucleus, which he termed 'purine'. During the same period Adolf Pinner (1842–1909) at Berlin discovered and named the corresponding monocyclic nucleus 'pyrimidine', to which the cleavage products thymine, cytosine, and uracil of the nucleic acids were soon related. With the new knowledge of purine and pyrimidine chemistry, Kossel and his students showed by around 1905 that the oxypurines, xanthine and hypoxanthine, are secondary derivatives from the hydrolytic deamination of the primary cleavage products, guanine and adenine, respectively, and that there appeared to be two main types of nucleic acid (DNA and RNA in present-day terms). Both types contained guanine, adenine, and cytosine, but one, plant or yeast nucleic acid, contained uracil in place of the thymine isolated from the other, animal or thymus nucleic acid. The two types appeared to differ too in their carbohydrate content. Acid treatment, Kossel showed, yielded furfural from yeast nucleic acid, characteristic of a pentose, whereas thymus nucleic acid afforded levulinic acid, more typical of a hexose.

The nucleic acid sugars and their bonding to the phosphate and the bases were characterized by Phoebus Levene (1869–1940) and his associates at the Rockefeller Institute, New York, from 1905. Levene studied chemistry under the composer Borodin, as a medical student at St Petersburg, and emigrated to the USA in 1891, returning to Europe on visits for research with Kossel and with Fischer and, later, with Pavlov. By 1912 Levene with

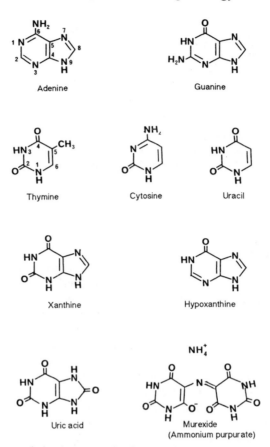

Fig. 12.2 Structures of the heterocyclic bases of the nucleic acids: the purines, adenine (A) and guanine (G); the pyrimidines, thymine (T), cytosine (C), and uracil (U); and the purine degradation products, xanthine, hypoxanthine, uric acid, and murexide (ammonium purpurate).

Walter Jacobs (1883–1967) had identified the sugar of yeast nucleic acid as D-ribose, which is linked to the bases by a glycosidic bond in what they termed 'nucleosides'—adenosine, guanosine, cytidine, and uridine—and esterified by a phosphate additionally in the 'nucleotides'—adenylic acid, guanylic acid, cytidylic acid, and uridylic acid (Fig. 12.3).

No reliable methods for the quantitative estimation of the nucleotides in a nucleic acid hydrolysate were yet available, and the approximate analytical data suggested to Levene, as to other workers in the field, that the relative molecular amounts of each of the four nucleotides must be equal. The assumption led Levene to propose in 1919 that yeast nucleic acid is a

β-D-ribonucleotide   β-D-deoxyribonucleotide

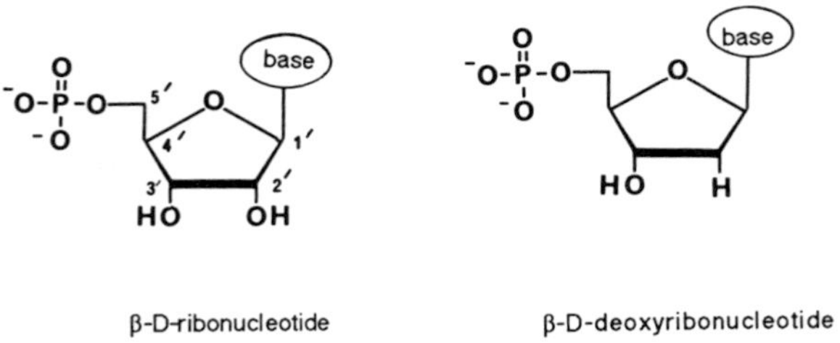

Fig. 12.3 Structures of the nucleotides: the bases are linked from the 9-position of a purine or the 1-position of a pyrimidine to the 1'-position of the sugar by a β-glycoside link (A, G, T, or C for the deoxyribonucleotides; A, G, U, or C for the ribonucleotides).

colloidal aggregate of tetranucleotide units, each unit containing one molecule of adenylic, guanylic, cytidylic, and uridylic acid linked together by phosphodiester bonds.

Thymus nucleic acid proved to be resistant to hydrolysis by aqueous ammonia to nucleosides or, under milder conditions, to nucleotides, the procedure used successfully to cleave yeast nucleic acid. Accordingly Levene employed enzymatic methods for the cleavage of thymus nucleic acid, first by passing the nucleic acid through a short section of dog's digestive tract, entering and exiting through a gastric and an intestinal fistula, respectively (a technique acquired from Pavlov), and subsequently by the use of a fractionated 'nuclease' from intestinal preparations. The first enzymatic technique served to identify not only the two purine and the two pyrimidine nucleosides but also their constituent sugar.

The sugar turned out to be, Levene reported in 1929, not the hexose surmised by Kossel, but the pentose D-2-deoxyribose, first isolated by Kiliani in 1895. Kiliani and others had devised simple colour tests to characterize the deoxysugars and, by 1931, Levene had shown, through the systematic application of these tests to the range of nucleic acids available, that the traditional distinction between animal or thymus nucleic acid and plant or yeast nucleic acid was erroneous. Plants, animals, and microbes were equally sources of both D-ribonucleic acid (RNA) and D-deoxyribonucleic acid (DNA).

The methods of carbohydrate chemistry demonstrated that both the D-ribose and the D-2-deoxyribose of the nucleic acids have a furanose structure, with a five-membered ring formed by an intramolecular semi-acetal link between the 1- and the 4-positions of the sugar. In a 2'-deoxyribonucleoside only the 3'- and the 5'-positions of the sugar are available for phosphodiester formation in a DNA tetranucleotide unit, Levene observed in 1935, and an analogous 3',5'-phosphodiester link is probable in the RNA tetranucleotide unit too. The latter proposal was supported by Levene's demonstration in 1933 that the adenylic acid isolated from muscle by Embden in 1929 is the 5'-isomer (AMP) of the adenosine-3'-phosphate obtained by the cleavage of RNA.

Initially, in 1919, the supposed tetranucleotide unit of RNA and DNA was envisaged as an open chain structure, but the demonstration in 1935 that each phosphate group in a nucleic acid is effectively monobasic implied either that the unit is cyclic or that the nucleic acids are high polymers. Levene advocated the cyclic-unit interpretation, despite new evidence. The measurement of the molecular weight and other solution properties of native DNA in forms less degraded than earlier preparations supported the high polymer formulation, and the isolation of hydrolytic 'nuclease' enzymes producing specific degradations suggested a structure for DNA less regular than a repeating simple tetramer.

Early determinations of the molecular weight of DNA gave values of the order of 10^3, consistent with the tetranucleotide hypothesis, but the DNA preparations of Rudolf Signer at Berne, by 1938, had apparent molecular weights of about 10^6 and the molecular shape, from viscosity and flow-birefringence measurements, of a thin rod with a length/width ratio of approximately 300. A 1920 report that an animal pancreas extract hydrolyses RNA but not DNA was confirmed at the Rockefeller Institute in 1937 and the purified enzyme, ribonuclease, was crystallized 2 years later. Levene too found evidence of site-selective 'nuclease' activity in fractions of animal intestinal extracts and confirmed the new molecular weight of about 10^6 for DNA. Nevertheless he maintained his long-standing hypothesis in a final conclusion (1938) that DNA and RNA are high polymers of the fundamental cyclic tetranucleotide unit.

From the 1890s, when Kossel identified nucleoprotein as the material of the chromatin network and the chromosomes of the cell nucleus, the view that nucleoprotein constituted the genetic material enjoyed wide support, but with more emphasis on the protein than the nucleic acid moiety. The 20 or so protein amino acids, through their virtually innumerable polypeptide permutations and combinations, offered far more variability to account for the diversity of genetic traits than the four DNA bases, particularly if the bases were locked into equimolecular proportions in tetranucleotide units. The nucleic acid might be nothing more than a passive anionic support linking the individual cationic carriers of genetic information, the histone proteins.

The decisive evidence that DNA itself could transfer heritable characters in microbes came in 1944 from an immunochemistry group at the Rockefeller Institute Hospital, headed by Oswald Avery (1877–1955). In 1880 Pasteur had introduced the technique of attenuating virulent bacteria, by culture at near-lethal temperatures in the case of anthrax, to produce non-virulent strains. In laboratory cultures the strains proved to be genetically stable and from 1920 on it was observed that the two types of strain formed morphologically different colonies in a number of pathogenic species. In cultures, the virulent strains of pathogenic bacteria give regular, dome-shaped, smooth (S-form) cultures, whereas the corresponding non-virulent strains produce irregular, flat, rough (R-form) cultures. The difference was shown to be due to the formation of a polysaccharide coating by the virulent S-form, the coating being absent in cultures of the non-virulent R-form.

In 1923 Frederick Griffith (1877–1941) at the Lister Institute in London reported that the non-virulent R-form of pneumococcus is transformed into the corresponding virulent S-form by the addition of heat-killed cells of the S-form to the plate culture, or to a mouse *in vivo*. In the same year Avery, who had studied the immunochemistry of pneumococci since joining the Rockefeller Hospital in 1913, identified a soluble polysaccharide from the coating of the S-forms as the substance giving rise to the specific immune reaction of each of the three types of virulent pneumococci known. Griffith then showed, in 1928, that the specific immune reaction of the S-form produced by his transformation depends solely upon the particular heat-killed virulent S-type added to any one of the R-form pneumococcus strains.

The pneumococcus transformation studies, confirmed in numerous laboratories, were extended at the Rockefeller Hospital where, by 1933, it was found that cell-free aqueous extracts of dead virulent pneumococci brought about the change from R- to S-form in cultures. Over the following decade the aqueous extracts were successively fractionated with a series of purified enzymes, each one specifically removing polysaccharide, protein, or RNA. The macromolecular material remaining after each stage of the enzymatic purification of the S-form extracts became progressively more active, and the final preparation had all the characteristics of DNA. Moreover, the material at each stage of the purification was completely deactivated by 'DNA depolymerase' enzyme fractions, or deoxyribonuclease as the enzyme was termed following its isolation and crystallization. The pneumococcus heritable transforming principle, Avery concluded in 1944, with his associates Colin MacLeod (1909–72) and Maclyn McCarty, is a 'highly polymerized and viscous form of sodium desoxyribonucleate'. The transfer of heritable properties to other microbial species by DNA preparations soon followed.

The implications of Avery's discovery were taken up more immediately by the chemical community than by geneticists and microbiologists who, in general, accepted DNA as the genetic material, instead of protein, only with caution and reserve. Avery's findings led Erwin Chargaff at Columbia University to investigate the validity of Levene's tetranucleotide theory of nucleic acid structure, and Alexander Todd, on moving from Manchester to Cambridge in 1944, reorientated his synthetic studies of nucleotides from coenzymes and vitamins to the nucleic acids (Chargaff 1979; Todd 1983).

Over the period 1940–55 Todd and his associates established by synthesis that, in the nucleosides, the sugar is linked by a β-glycoside bond to a nitrogen atom of the base in the 1-position of a pyrimidine or the 9-position of a purine (not the 7-position, as supposed earlier). Nucleotides were synthesized with the phosphate ester group at the 2'-, 3'-, or the 5'-position of the ribose sugar unit in the various nucleosides to provide compounds for comparison with the natural products of nucleic acid degradation (Fig. 12.3). The action of ribonuclease on a range of synthetic oligonucleotides established that, in RNA, the individual nucleotides are linked through phosphodiester bonds between the 3'-position of one ribose unit and the 5'-ribose position of its neighbour, as Levene had surmised from the analogy between DNA and RNA.

Levene's tetranucleotide hypothesis was effectively falsified over the period 1946–50 by the analytical studies of Chargaff, who showed that DNA from different sources varied widely in base composition, so that there could be no single DNA substance, common to all species, based on an equimolecular tetranucleotide unit. The development of partition chromatography on paper and other materials allowed the separation and isolation of minute quantities of the hydrolysis products from DNA, which was available only in small amounts from some bacterial and viral sources. The strong and characteristic light absorption of the purines and pyrimidines in the quartz ultraviolet region (about 260 nm) and the production of ultraviolet (UV) spectrophotometers from around 1945 made possible the micro-assay of the individual bases from a DNA hydrolysate.

With paper chromatography and UV spectrophotometry, Chargaff and his associates were able to show by 1950 that the molecular ratios of the four bases in DNA vary widely from one species to another and that they are rarely consistent with equimolecularity. The particular base ratios are characteristic of a species, remaining constant between one individual and another within a species and over the different tissues of an individual. From the analytical data for the relative amounts of adenine (A), guanine (G), cytosine (C), and thymine (T) in the DNA from different species, Chargaff derived three generalizations covering the molecular ratios of the bases. For the DNA of a given organic species, the 'Chargaff rules' state that: (a) the total of the purines and the total of the pyrimidines are quantitatively equimolecular, $A + G = T + C$; (b) the molecular quantities of adenine and thymine individually are equal, $A = T$; and (c) the amounts of guanine and cytosine are molecularly equivalent, $G = C$. A relation following from the three rules, (d) $A + C = G + T$, was found later to apply additionally to each individual strand of microbial DNA and, with uracil in place of thymine, to the total RNA of a cell.

By 1951 the overall features of the primary structure of DNA were clear. The DNA molecule appeared to be a highly extended deoxyribonucleotide chain, linked by 3',5'-phosphodiester bonds, with the bases in no apparent order but subject to the then-enigmatic Chargaff rules. The bond lengths and bond angles of substituted purines and pyrimidines, including the bases adenine and guanine, and the nucleoside cytidine, were available from three-dimensional X-ray crystal structure determinations, and X-ray diffraction diagrams of DNA fibres provided some indication of the arrangement of the component units in the DNA molecule.

The early X-ray fibre diagrams, obtained by William Astbury (1898–1961) and Florence Bell at Leeds in 1938, suggested that the bases in DNA are stacked face to face, like a pile of coins, in groups of eight or sixteen, providing support for Leven's tetranucleotide hypothesis. Like other biophysicists of the period, Astbury supposed that, in organic reproduction, the DNA chain provided a passive template for the replication of the genetic material, represented by the protein histones. Later X-ray patterns of hydrated DNA fibres, measured by Maurice Wilkins and by Rosalind Franklin (1920–58) at King's College London from 1950–1, showed a general superposition of two pattern types: one, dominant at a high relative humidity, was less resolved but characteristic of helical structures; while the other, pronounced at lower relative humidity, was more crystalline in form and showed a basic two-fold rotational symmetry. The DNA chains appeared to have the secondary structure of a cylindrical helix some 2 nm in diameter and with a pitch between 2.8 and 3.4 nm.

Pauling and Corey, having derived in 1950 the regular secondary structures of the polypeptides, the α-helix and the β-sheet conformation, by model building from the structural data available for the monomer and dimer units, went on to apply the technique to the problem of the DNA secondary structure in 1953. As yet unconvinced that DNA constitutes the primary genetic material, Pauling made no special allowance in their DNA model for his 1948 proposal that a pair of complementary molecular structures can serve as template moulds for the duplication of each other. The proposal was an application of Pauling's general concept of the biological importance of stereochemical complementarity, like his view of the antibody–antigen relation and his extension of Fischer's 'key and lock hypothesis' for enzyme–substrate interaction.

The secondary structure of DNA proposed by Pauling and Corey consisted of three polynucleotide chains linked together by hydrogen bonds between the phosphate groups, forming a triple-stranded helix with externally projecting nucleotide bases around an internal column of phosphate groups along the axis of the cylinder. The model was fatally flawed by the assumption that the P—O—H groups ($pK_a \sim 2$) remains undissociated under physiological conditions (pH ~ 7) in order to provide the interchain hydrogen bonding between the phosphate groups.

Earlier, John Gulland (1898–1947) and his associates at Nottingham had shown by pH titrations of DNA with acid and alkali that, after the ionization of the P—O—H groups at pH ~ 2, there are further ionic dissociations around pH ~4.5 and pH ~11.5, associated with the —NH₂ substituents of the bases and their ring amido N—H groups (or the tautomeric enol equivalents). At the latter pH values a native DNA solution lost the high viscosity characteristic of the pH range 5–11, the loss being analogous to the viscosity changes of protein solutions on denaturation with acid or alkali, owing to the breaking of hydrogen bonds. Accordingly, Gulland and his colleagues concluded, native DNA consists of aggregated micelles of polynucleotide chains, held together by hydrogen bonds between the bases of different chains, and the aggregates become dispersed into the individual chains by fission of the hydrogen bonds on denaturation with acid or alkali, leading to the observed change in hydrodynamic properties.

Model building of the DNA secondary structure by Francis Crick and James Watson at Cambridge during the period 1951–3 took into account Pauling's earlier view that a rational approach to the mechanism of organic replication lay in an analysis of the 'conditions under which complementariness and identity might coincide' in molecular structures. The approach applied to the molecular equivalencies of Chargaff's rules, in the context of Gulland's findings, led to the recognition that edge-wise hydrogen bonding between a pair of the equimolecular bases in a common plane, guanine with cytosine (G = C) and adenine with thymine (A = T), gives in the two cases heterodimers of similar shape and size overall (Fig. 12.4). An extended sequence of such heterodimers, in the form of two complementary polynucleotide chains, thus satisfies Chargaff's rules and Pauling's (1948*b*) replication condition for a two-component system, in which each part 'can serve as the mould for the production of a replica of the other part, and the complex of two complementary parts thus can serve as the mould for the production of duplicates of itself'.

The X-ray crystal structure of cytidine, carried out in 1947–9 by Sven Furberg with Bernal at Birkbeck College London, indicated that, in a nucleoside, the mean plane of the sugar ring has a near-perpendicular orientation to the molecular plane of the base. The structural dimensions and the face-to-face stacking of coplanar hydrogen-bonded base pairs into a cylindrical array, in the 1953 DNA model of Crick and Watson, required the two sugar–phosphate chains, one for each partner of a given base pair, to wrap helically round the exterior of the stacked array. The helical chains are right-handed, from the known absolute stereochemical configuration of D-deoxyribose, and the two chains of a given DNA molecule run in opposite directions (the 3',5'-phosphodiester orientation) to accommodate the two-fold rotational symmetry implied by the DNA X-ray fibre diagrams. The structural data available indicated that one complete turn of the double-stranded DNA helix, occupying a rise of 3.4 nm along the helix axis, is made up of 10 stacked base pairs, with each phosphate at a mean distance of about 1 nm from the axis (Fig. 12.5).

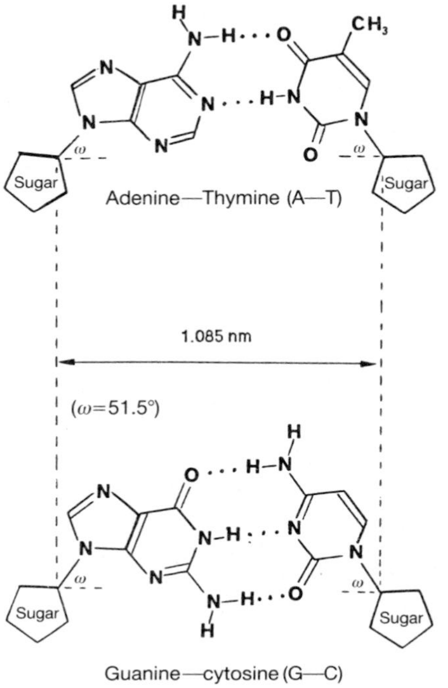

Fig. 12.4 The Watson–Crick base pairing through hydrogen bonds between adenine and thymine (A–T), or between guanine and cytosine (G–C), to give hetero-dimers of similar shape and dimensions (1.085 nm between the 1'-carbon atoms of the sugar rings in each dimer, and a common orientation of the sugar–base bonds). In the dimers and in double-stranded polynucleotides, base pairing is stronger for G–C, with three hydrogen bonds, than for A–T (DNA) or A–U (RNA), with two hydrogen bonds.

The double-stranded helix structure for DNA found immediate support in the then-limited X-ray fibre pattern data for the more hydrated B-form of DNA. A similar diffraction pattern appeared in nucleoprotein fibres from sperm head and bacteriophage preparations, suggesting that the structure had a biological significance. Over the following decade the X-ray diffraction studies of the DNA sodium salt by Wilkins and his associates confirmed and refined the structural details of the B-form (92 per cent relative humidity) and showed that the less hydrated A-form (about 75 per cent relative humidity) has the right-handed double-helix structure too, but with 11 base pairs per turn.

Subsequently the double strand RNA reoviral genetic material, and a hybrid composed of a DNA strand and its complementary RNA strand, were shown by X-ray diffraction to adopt the A-form double-stranded

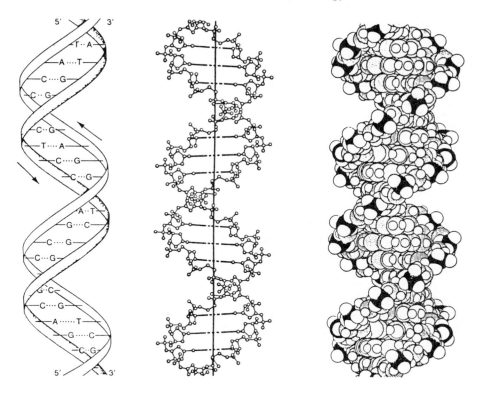

Fig. 12.5 Representations of the right-handed DNA double helix: the two strands are antiparallel with the $5' \rightarrow 3'$ phosphodiester bonding orientation running in opposite directions.

helical secondary structure. In addition a number of synthetic homopolynucleotides and heteropolynucleotides were found to associate into base-paired two-stranded helical complexes, often with the right-handed A- or B-form structures. Polydeoxyadenylic acid (poly-dA) and polydeoxythymidylic acid (poly-dT) give B-form (poly-dA.poly-dT) complexes (10 base pairs per helical turn), and so too do polydeoxyguanylic acid (poly-dG) and polydeoxycytidylic acid (poly-dC), but (poly-dG.poly-dC) at lower relative humidity (75 per cent) assumes the A-form (11 base pairs per turn). A relatively rare left-handed double-stranded helix, the Z-form with 12 base pairs per turn, is formed at low humidity (~45 per cent) by heteropolynucleotides with alternating purine and pyrimidine bases, particularly those rich in guanine and cytosine, e.g. [poly-(dG-dC)]$_2$.

The mechanism proposed for the autocatalytic reproduction of DNA whereby one strand serves as a template for the replication of the other found striking experimental support during the mid-1950s. Bacteria grown for several generations on ^{15}N (as $^{15}NH_4Cl$) as the sole source of nitrogen produced 'heavy' DNA, labelled with ^{15}N in both strands. On restoration to a normal ^{14}N nitrogen source, the bacteria gave in the first subsequent generation a DNA of 'intermediate' density, corresponding to one ^{15}N strand with ^{14}N in the other. In the second generation equal amounts of the 'intermediate' and 'light' (^{14}N in both strands) DNA were formed, and each following cell division successively halved the relative quantity of 'intermediate' density DNA.

In 1956 Arthur Kornberg at Washington University, St Louis, isolated the enzyme DNA polymerase I by fractionating the contents of rapidly dividing bacterial cells. With a single strand of DNA serving as the template and a supply of activated monomers, the $5'$-triphosphates of the deoxynucleosides (dATP, dTTP, dGTP, and dCTP), the enzyme catalysed the formation of a complementary DNA strand containing the A:T:G:C ratios expected from the composition of the template. The discovery of DNA polymerase opened up a route to the synthetic polydeoxyribonucleotides, exploited notably by Gobind Khorana and his associates at Wisconsin. Laboratory-synthesized dideoxyribonucleotides and oligodeoxyribonucleotides served equally as the primer (normally a short RNA strand) and the template for the enzymatic production of DNA high polymers from the corresponding activated monomers.

The analogous synthesis of polyribonucleotides followed from the discovery of the enzyme polynucleotide phosphorylase from bacterial extracts by Marianne Grunberg-Manago and Severo Ochoa at New York University in 1955. In the cell, polynucleotide phosphorylase normally degrades RNA to nucleoside $5'$-diphosphates (NDP), but the reaction is reversible and, in the laboratory, the enzyme catalyses the production of polymers with the $3',5'$-phosphodiester linkage from one or more of the monomers ADP, UDP, GDP, and CDP in high concentration. The polymerization is unspecific, requiring no template strand, and the composition of the polymer reflects that of the mixture of monomer starting materials.

Electron microscopy, used to follow the sequential stages of the replication of DNA *in vivo*, shows that all known forms of DNA polymerase catalyse the growth of daughter polynucleotide chains only in the $5' \rightarrow 3'$ direction of the phosphodiester bonds, these enzymes promoting the reaction of a nucleoside $5'$-triphosphate only at the free $3'$-OH end of a growing strand. The separation of the parental plus and minus strands allows the template replication of one daughter strand continuously, i.e. the leading strand complementary to the plus parental strand, while the other daughter strand is synthesized discontinuously, in the allowed $5' \rightarrow 3'$ direction, as fragments 100–1000 bases long. Subsequently the fragments, named after their discoverer R. Okazaki (1968), are linked up

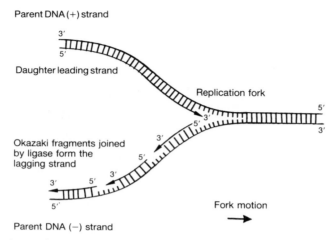

Parent DNA (+) strand

Daughter leading strand

Replication fork

Okazaki fragments joined
by ligase form the
lagging strand

Fork motion

Parent DNA (−) strand

Fig. 12.6 The replication of DNA catalysed by DNA polymerase involves the separation of the parental strands and the synthesis of both daughter strands in the $5' \to 3'$ direction, antiparallel to the orientation of the phosphodiester bonds in the paired parental strand. The leading daughter strand is synthesized continuously, following the replication fork as it moves along the parental double helix, separating the two strands. The lagging daughter strand is assembled piecemeal in Okazaki fragments 100–1000 bases long, and the fragments are subsequently linked into a continuous strand, catalysed by DNA ligase.

catalytically by the enzyme DNA ligase into the lagging daughter strand, complementary to the minus parental strand (Fig. 12.6).

The circular double-stranded DNA chromosome of the bacterium *Escherichia coli*, containing about 4×10^6 base pairs, replicates at the maximum rate of 750 base pairs per second at a given unwound replicating fork region, taking some 40 min to reproduce a complete chromosome. Under optimum growth conditons, *E. coli* undergoes division into two daughter cells every 20 min since, at any given time, replication occurs simultaneously at two or more separated-strand fork regions. The several replication sites, distanced out along the DNA double strand which measures about 1 mm if fully extended, ensure that a new round of DNA replication is initiated, and even well advanced, before the previous round is complete.

12.3 The biotechnology of the cell

The process linking the genetic material to soma substances, termed 'heterocatalysis' earlier, became divided during the 1960s into the 'transcription' of the DNA into a RNA copy, followed by the 'translation' of the RNA copy into a protein. During the previous decade a general scheme of

(a)

mRNA transcribed

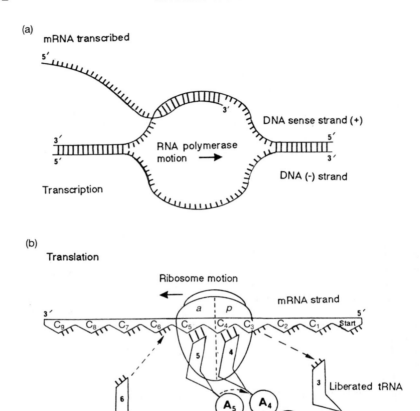

Fig. 12.7 The transcription of mRNA from DNA and the translation of an mRNA strand into a polypeptide. (a) The synthesis of an mRNA strand, catalysed by RNA polymerase, begins with the 5′-terminus and proceeds towards the 3′-end as the RNA polymerase moves along the DNA double helix. After separating the two DNA strands, the RNA polymerase matches the deoxyribonucleotide bases of the DNA sense strand with complementary ribonucleotide bases, which are then linked to form the transcribed mRNA strand. (b) The mRNA strand consists of a series of triplet codons C_n of bases pairing with the complementary triplet anticodons of a tRNA molecule specific for the nth amino acid (A_n) of the corresponding polypeptide. The translation of an mRNA strand into a polypeptide starts with the assembly of a ribosome from a small and a large subunit at the 5′-terminus, and the polypeptide is synthesized as the ribosome moves towards the 3′-end of the mRNA strand. The ribosome contains a peptidyl site (p), bonding the tRNA linked to the last amino acid residue incorporated into the growing polypeptide, and an

aminoacyl site (*a*), bonding the tRNA charged with the next amino acid to be added to the polypeptide chain. The $-NH_2$ group of the amino acid esterified to the tRNA at the *a* site attacks the ester link between the tRNA at the *p* site and forms an amide bond to the peptide chain, which is now elongated by an additional amino acid residue and attached to the tRNA at the *a* site. The uncharged tRNA remaining is liberated from the *p* site. The movement of the ribosome along the mRNA strand towards the 3'-end by one triplet codon increment places the tRNA bonded to the polypeptide chain in the *p* site and encloses in the *a* site the codon complementary to the anticodon of the tRNA charged with the next amino acid to be added to the growing polypeptide. The first codon of an mRNA strand is generally AUG for methionine (N-formyl methioine in eubacteria).

---

protein synthesis in the cell had evolved, following the isolation and the characterization of ribosomal particles by George Palade at the Rockefeller Institute. The differential sedimentation of macerated bacterial cells with the ultracentrifuge gave, after the initial deposition of the DNA chromosomal material and cell debris, the separation of small particles some 20 nm in diameter (the ribosomes) containing about 85 per cent of the RNA of the cell.

From 1953 Paul Zamecnik and colleagues at Harvard University showed that amino acids labelled with radioactive isotopes (^{3}H, ^{14}C, or ^{35}S) fed to the bacteria immediately before the fractionation gave radioactive polypeptides associated solely with the ribosomes. *In vitro* the isolated ribosomes failed to give radioactive polypeptides from the labelled amino acids unless the supernatant solution from the separation, containing the remaining 15 per cent or so of the RNA, was included; and even then polypeptide synthesis soon came to a halt, although it could be restarted briefly by the addition of fresh supernatant.

The ribosomal RNA (rRNA) and the DNA of a given organism were found generally to differ in base composition, the former usually having a higher guanine and cytosine content, independent of the particular DNA $(A + T)/(G + C)$ ratio. An exception to the apparent lack of correspondence between the RNA and DNA of an organism appeared in 1960 when Sol Spiegelman (1914–83) found that the bacteriophage T2, on infecting the bacterium *E. coli*, inhibits the production of bacterial RNA and promotes instead the generation of a new and specific T2 RNA, which matches in base composition the DNA of the bacteriophage. The matching extended to the base sequence, since the T2 RNA and a single strand of the T2 DNA formed a double-stranded hybrid helix. Labelling with radioactive isotopes and with stable heavy isotopes indicated that the T2 RNA takes over the protein manufacture of the cell; the ribosomes of *E. coli* produced before the infection become monopolized by the T2 RNA for the generation of bacteriophage protein.

The enzyme catalysing the production of T2 RNA from the bacteriophage DNA template, RNA polymerase, is a product of the *E. coli* cell, where normally it catalyses the transcription of the bacterial DNA into messenger RNA (mRNA) carrying a code which, in the ribosome, becomes translated into bacterial protein (Fig. 12.7). Evidence for messenger RNA was first presented in 1961 by Jacques Monod (1910–76) and François Jacob at the Pasteur Institute, Paris. Bacterial mRNA constitutes only a small fraction (about 4 per cent) of the total RNA of the cell, since the mRNA undergoes a rapid turnover *in vivo*. In early studies of cell-free protein synthesis, the limited duration of polypeptide production observed was due to the degradation of mRNA by enzymes, such as polynucleotide phosphorylase, present in the supernatant solution added to isolated ribosomes.

Electron microscopy mapped out the successive stages of transcription and translation, showing the appearance of the leading 5'-end of an mRNA strand from the enzyme RNA polymerase attached to a separated DNA strand in an unwound region of the double-stranded circular bacterial chromosome. The 5'-terminus of the mRNA attaches first to the smaller and then to the larger ribosomal subunit to form a translationary active ribosome which moves along the strand during the process of its transcription. The vacation of the leading 5'-end of the mRNA by the initial ribosome allows further small and large subunits to form successive ribosomes at the site and progress towards the 3'-terminus, where the two subunits of each ribosome separate once again and liberate the polypeptide formed during the translation of the mRNA (Fig. 12.7).

At 37°C the DNA of *E. coli* is transcribed at a rate of around 60 nucleotides each second into an mRNA strand with an average length of some 2000 to 3000 nucleotides. The mean half-life of an mRNA strand is only about 90 s, so that the 10–15 ribosomes translating a given mRNA strand, forming a polyribosome, become attached to the strand mainly during its period of transcription from DNA. The polypeptides formed by each ribosome of a polyribosome system are identical, the formation time of a polypeptide being of the order of 10 s under optimal conditions.

A growing *E. coli* cell contains some 15 000 ribosomes, which make up about a quarter of the cell mass. Each ribosome, about 20 nm across, has a molecular weight of approximately 2.7×10^6 and sediments in the ultracentrifuge at 70S (the Svedberg unit, $S = 10^{-13}$ s, measures the ratio of the velocity of sedimentation to the centrifugal acceleration). The ribosome dissociates reversibly, dependent upon the Mg^{2+} concentration and the ionic strength, into a large (50S) and a small (30S) subunit, each containing several proteins and approximately 55 per cent RNA. The RNA of the small subunit sediments at 16S and is composed of 1542 nucleotides, while the large subunit contains two RNA strands, one (23S) of 2904 nucleotides and the other (5S) with 120 nucleotides.

The total RNA of the cell includes, in addition to the rRNA (about 85 per cent) and the mRNA (about 4 per cent), a set of small tRNA molecules (about 11 per cent overall), each 73–93 nucleotides long, which transfer the appropriate amino acid to the site of polypeptide synthesis in a ribosome translating a strand of mRNA. The three types of *E. coli* RNA are single stranded and each is complementary in base composition and sequence to a segment of the bacterial DNA, as shown by the formation of a double-stranded hybrid helix with the parental transcript section of a single strand of that DNA.

During the 1960s the mechanism of polypeptide chain growth on a ribosome, and the relationship between the amino acid sequence of a polypeptide and the corresponding base sequence of the mRNA to which the ribosome is attached, was clarified by the use of synthetic polyribonucleotides as the mRNA in cell-free protein synthesis. In 1961 Marshall Nirenberg and Heinrich Matthaei at the National Institutes of Health, Bethesda, discovered that polypeptide synthesis in a spent preparation of bacterial ribosomes, suspended in the supernatant solution from the cell-content fractionation, could be revived by the addition of polyuridylic acid (poly-U) in small quantity (10^{-5} g). Of the various isotopically labelled amino acids added as potential feedstock, only phenylalanine was taken up and processed, the sole product being polyphenylalanine (poly-Phe).

Over the previous decade it had come to be accepted that a minimum of three nucleic acid bases are required to specify a given amino acid, since the possible combinations of the four bases taken two at a time (4^2) could not code for the complete set of 20 or so protein amino acids. The ribosomal synthesis of poly-Phe with poly-U as the mRNA implied that a sequence of three or more uracil bases coded for phenylalanine. Synthetic polyribonucleotides containing two or three different bases were found to be selective, when serving as mRNA in ribosomal protein synthesis, for a limited range of amino acids incorporated in the polypeptide formed. The study of the binding of progressively shorter synthetic oligonucleotides to a tRNA molecule charged with its specific amino acid established that a sequence of only three ribonucleotides is required to specify the incorporation of a given amino acid in ribosomal protein synthesis. The employment of a range of trimeric ribonucleotides with known sequences then identified the particular base triplets in mRNA coding for each protein amino acid (Fig. 12.8).

By 1966 the details of the genetic code had been worked out: 61 out of the total 64 triplet codons (4^3) are assigned to a particular protein amino acid and the remaining 3 signal the termination of polypeptide synthesis. The initiation signal for the synthesis is the codon AUG for methionine, or N-formyl methionine in eubacteria: thus methionine forms the initial leading amino acid (the amino-terminus) of all protein chains. Over the following decade the genetic code was found to be general, almost universal, for the mRNA transcripts of the chromosomal DNA of the three organic

(a)

Generic structure: R—$\overset{H}{\underset{CO_2^-}{\overset{|}{C}}}$—$NH_3^+$

Type		R-group or the acid
Non-polar hydrophobic	Aliphatic	Alanine (Ala) CH_3-
		Leucine (Leu) $(CH_3)_2CHCH_2-$
		Isoleucine (Ile) $CH_3CH_2CH(CH_3)-$
		Valine (Val) $(CH_3)_2CH-$
		Proline (Pro)
	Aromatic	Phenylalanine (Phe)
		Tryptophan (Trp)
	Thioether	Methionine (Met) $CH_3-S-CH_2CH_2-$

Type		R-group
Polar hydrophilic		Glycine (Gly) $H-$
		Serine (Ser) $HOCH_2-$
	Hydroxyl	Threonine (Thr) $CH_3CH(OH)-$
		Tyrosine (Tyr) $HO-\langle\rangle-CH_2-$
	Mercapto	Cysteine (Cys) $HSCH_2-$
	Amide	Asparagine (Asn) H_2NCOCH_2-
		Glutamine (Gln) $H_2NCOCH_2CH_2-$
	Anionic	Aspartic acid (Asp) $^-O_2CCH_2-$
		Glutamic acid (Glu) $^-O_2CCH_2CH_2-$
	Cationic	Arginine (Arg) $^+(H_2N)_2CNH(CH_2)_3-$
		Lysine (Lys) $H_3\overset{+}{N}(CH_2)_4-$
		Histidine (His)

(b)

Second base

First base (5'-end)	U	C	A	G	Third base (3'-end)
U	UUU UUC } Phe UUA UUG } Leu	UCU UCC UCA UCG } Ser	UAU UAC } Tyr UAA stop UAG stop	UGU UGC } Cys UGA stop UGG Trp	U C A G
C	CUU CUC CUA CUG } Leu	CCU CCC CCA CCG } Pro	CAU CAC } His CAA CAG } Gln	CGU CGC CGA CGG } Arg	U C A G
A	AUU AUC } Ile AUA AUG Met/start	ACU ACC ACA ACG } Thr	AAU AAC } Asn AAA AAG } Lys	AGU AGC } Ser AGA AGG } Arg	U C A G
G	GUU GUC GUA GUG } Val	GCU GCC GCA GCG } Ala	GAU GAC } Asp GAA GAG } Glu	GGU GGC GGA GGG } Gly	U C A G

Fig. 12.8 (a) The L-α-amino acids of the proteins, and (b) the genetic code of ribonucleotide base triplets governing the translation of an mRNA strand into a polypeptide chain.

Chemical evolution

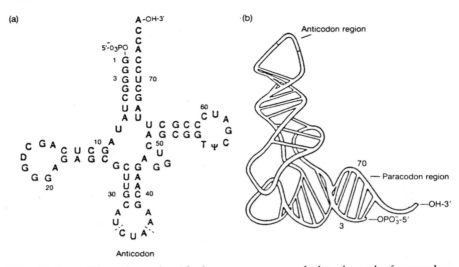

Fig. 12.9 (a) The ribonucleotide base sequence and the clover-leaf secondary structure, due to intrachain hydrogen bonding between complementary bases, of the *E. coli* alanine transfer RNA with the CUA anticodon (at the positions 34 to 36). (b) A schematic representation of the L-shaped three-dimensional conformation, the tertiary structure arising from the folding of the clover-leaf form. The upper end of the L-structure carries the anticodon, recognized by the complementary triplet codon (UAG) of an mRNA strand, normally one of the stop triplets ('amber') in the general genetic code. The lower portion ends with the CCA(OH) 3'-terminus, to which the amino acid becomes esterified in the charged tRNA. The lower portion of the L-structure includes a paracodon region, of a yet undeciphered second genetic code, recognizing the particular aminoacyl-AMP intermediate bound to its specific aminoacyl-tRNA synthetase enzyme. The 3-guanine...70-uracil base pair near the lower end forms part of the paracodon for the *E. coli* alanine-tRNA (after de Duve 1988; Hou and Schimmel 1988).

kingdoms, the eubacteria, the archaebacteria, and the nucleus of the eukaryotes. Subsequently a few deviations from the general genetic code were discovered in micro-organisms, e.g. the *E. coli* alanine-tRNA recognizing the UAG stop codon (Fig. 12.9); and, more particularly, in the autonomous translation apparatus of the eukaryote organelles, the mitochondrion and the chloroplast.

The mitochondria, like the chloroplast organelles, contain their own genetic material, a circle of double-stranded DNA, as in the prokaryotes. The human mitochondrial genome is small, consisting of only 16 569 base pairs, whereas that of the yeast mitochondrion is some five times larger. The contrast indicates a rapid evolution of mitochondrial DNA towards a smaller size in the higher organisms. The reductive evolutionary simplification has cut down to 22 the number of tRNA species required to transfer amino acids for protein synthesis from mammalian mitochondrial mRNA transcripts, compared with the 40 or more tRNA types needed for the translation of the mRNA transcribed from the eukaryote nuclear DNA. The modifications in the genetic code for mammalian mitochondria allow more latitude in the initiation and termination of polypeptide synthesis; two of the six base triplets normally coding for arginine have become *stop* signals and all four AUN codons (N = U, C, A, or G) have become *start* signals.

The tRNA molecules that transfer activated amino acids to ribosomes translating mRNA were first detected as their amino acyl esters by the use of isotopically labelled amino acids in cell-free ribosomal protein synthesis. By 1965 Robert Holley and his associates at Cornell University had identified the particular bases and their sequential order in an alanine tRNA molecule, composed of a single chain of 77 ribonucleotides. A number of the constituent nucleosides were found to be methylated (N-Me purines, C-Me pyrimidines), or reduced (dihydrouridine), or non-standard in other ways, such as inosine (I, deaminated adenosine) or pseudouridine (Ψ) with the uracil C-5 bonded to the ribose C-1'. Optimum base pairing by hydrogen bonding between complementary segments of the tRNA chain suggested a possible clover-leaf secondary structure made up of three unpaired loops and four base-paired stem regions.

Subsequently the clover-leaf secondary structure was confirmed by several tRNA X-ray crystal structure determinations, which show an overall tertiary L-shaped conformation from the folding of the clover-leaf hydrogen-bonded assembly. One end of the L-form of the tRNA tertiary conformation, the stalk of the clover leaf, is made up of the base-paired terminal segments of the RNA chain, except for four nucleotides at the 3'-terminus, which ends with the sequence CCA(OH). The free 3'-OH group of the last adenosine is esterified in a tRNA charged with its specific amino acid. The other end of the tRNA L-shaped conformation is formed by the middle loop of unpaired bases in the clover-leaf structure and contains a sequence of three bases, the anticodon, complementary to the mRNA codon for the particular amino acid carried by the charged tRNA (Fig. 12.9).

With 61 base triplets coding for 20 amino acids, there is substantial redundancy or degeneracy in the genetic code. Eight of the amino acids are coded by a set of four base triplets that differ only in the third position (the 3'-end), and three of these, leucine, serine, and arginine, are each coded by a further two base triplets (Fig. 12.8). The redundancy relates to the finding that the number of distinct tRNA molecular species is invariably less than 61 and that a given tRNA anticodon may recognize two or more codons for the amino acid with which it is charged. The mRNA transcripts of the nuclear DNA in yeast are translated by only 46 tRNA species, which make use of 56 out of the 61 sense codons. Only 20 of the yeast tRNA species are present in relatively large amounts, one for each amino acid. Several of these tRNA recognize and accept two or even three codons for the amino acid with

Chemical evolution

which the tRNA molecule is charged in the cases where the codons differ
only in the third position (3'-end). The flexibility of the tRNA response is
ascribed to a lack of steric constraint on the complementary base of the
anticodon (the 5'-end), allowing some latitude of hydrogen bond pairing to
the base in the third position of the codon (Crick's 'wobble hypothesis').

The enzymes which catalyse the esterification of the free 3'-OH group of
the terminal adenosine residue in a tRNA chain with a unique amino acid,
the aminoacyl-transfer RNA synthetases, have a double and more restricted discrimination. There are only 20 such enzymes, one for each
protein amino acid, and each aminoacyl-tRNA synthetase recognizes and
charges all the tRNA species which pair with the several codons for the
amino acid singular to the enzyme. A given aminoacyl-tRNA synthetase
recognizes two very different molecular types, the amino acid and the
tRNA polyribonucleotide, and the overall charging by activation and
esterification occupies two distinct stages, one for each type of molecular
recognition (Fig. 12.10). First, the enzyme (E) selectively catalyses the
reaction of its specific amino acid with ATP to form pyrophosphate and

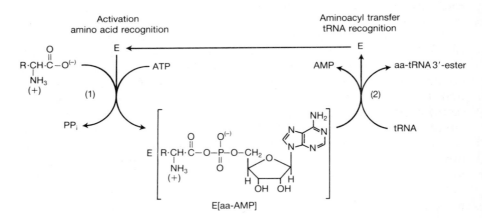

Fig. 12.10 The two recognition steps in the charging of a tRNA molecule by its
specific amino acid, catalysed by a given aminoacyl-tRNA synthetase enzyme (E).
The enzyme first identifies the particular amino acid (aa) and catalyses the formation, from ATP, of the aminoacyl-AMP intermediate [aa-AMP], which is bound by
the enzyme in the complex E[aa-AMP]. In the second step the complex E[aa-AMP]
recognizes a tRNA molecule appropriate for the amino acid bound as the
[aa-AMP] intermediate through structural features (the paracodon region) near to
the CCA(OH) 3'-terminus of the tRNA. Finally the enzyme catalyses the esterification of the terminal 3'-OH of the tRNA with the bound amino acid, releasing
AMP, to give the charged tRNA 3'-ester (aa-tRNA). Participation of the specific
amino acid in the second recognition step, as the complexed [aa-AMP] intermediate, is probable (after de Duve 1988).

aminoacyl-adenylate (aa-AMP), which remains bound to the enzyme as the complex E[aa-AMP]. Second, the complex E[aa-AMP] recognizes one of the tRNA specific for the amino acid, which is then transferred to the CCA(OH) terminus of the tRNA molecule to form the 3′-ester (aa-tRNA), with the liberation of the enzyme and AMP.

The second stage implies a possible 'second genetic code' of base groupings (paracodons) in the tRNA molecule recognized by one, and one only, of the 20 complexes of an aminoacyl-adenylate with its specific enzyme E[aa-AMP] (de Duve 1988). Studies of nucleotide substitutions in several tRNA species suggest that the base groups identified by a complex E[aa-AMP] lie remote from the anticodon and close to the CCA(OH) terminus, to which the amino acid is transferred. A hydrogen-bonded pair of bases in the clover-leaf stem of an alanine-tRNA from *E. coli*, guanine at the 3-position and uracil at the 70-position (out of 76 bases in all), govern the aminoacylation (Fig. 12.9). The substitution of these two bases into the corresponding positions of either a phenylalanine-tRNA or a cysteine-tRNA confers upon each the capacity for aminoacylation with alanine. The replacement of the two bases, 3-guanine and 70-uracil, in the alanine-tRNA molecule eliminates its capacity for alanine aminocylation (Hou and Schimmel 1988).

The two-stage character of the charging of a tRNA molecule with its specific amino acid promotes the accuracy of aa-tRNA formation by providing for the correction of a first-stage error at the second stage. Competition experiments show that, at the second stage, the recognition and charging of a tRNA molecule is sensitive to the particular amino acid complexed as its [aa-AMP] intermediate with the tRNA synthetase enzyme during the first stage. The amino acids ($RCH(NH_2)COOH$) valine (val, $R = (CH_3)_2CH—$) and isoleucine (ile, $R = C_2H_5(CH_3)CH—$), differing only by a methylene group, compete at equimolar concentration for isoleucyl-tRNA synthetase to give, at the first stage, one incorrect activation, forming val-AMP, for every 100 or so correct activations to ile-AMP. There is a discrimination of the same order of magnitude at the second stage, where most of the val-AMP molecules complexed to the isoleucyl-tRNA synthetase are hydrolysed to valine and AMP, so that the overall error level is reduced to approximately one part in 10^4. The competitive discrimination is achieved under conditions far removed from equilibrium with much dissipation of free energy by the coupled hydrolysis of ATP to AMP. One incorrect charging to give the product val-tRNA^{ile} is accompanied by the formation of approximately 10^2 to 10^3 AMP molecules, dependent upon pH and other conditions. These AMP coproducts measure the numerous successful eliminations of errors for each surviving mistake, and show that a considerable expenditure of free energy is required for sensitive discrimination (Cramer and Freist 1987).

Further proof-reading stages in protein synthesis appear to be absent, apart from a ribosomal check that the anticodon of the charged tRNA correctly matches the mRNA codon at the initial binding site. If cysteine (cys, $R = HSCH_2-$) bonded to one of its own tRNA carriers (cys-tRNA^{cys}) is converted to alanine (ala, $R = CH_3-$) by removal of the sulphur atom, the modified ester ala-tRNA^{cys}) is recognized by the cysteine codons of the mRNA in cell-free ribosomal protein synthesis, and the polypeptide produced has alanine residues in the expected cysteine positions.

The two binding sites of a ribosome consist of cavities with a shape governed by the mRNA codon enclosed by the two ribosomal subunits at each site. The shape of both the *p* (peptidyl) and the *a* (aminoacyl) site are specific for a charged aminoacyl-tRNA molecule with an anticodon matching the enclosed mRNA codon by base-pairing. In eubacterial protein synthesis the initial peptide bond is formed by the transformation of the N-formylmethionine ester link to its specific tRNA, which is bound at the *p* site, into an amide link with the second amino acid, similarly esterified to one of its specific tRNA, which is bound at the *a* site. The free methioine-tRNA is then liberated for a further round of charging with N-formylmethioine and peptide initiation on another ribosome.

At the same time the assembly of the dipeptide esterified to the second tRNA, still hydrogen bonded by the matching triplet codon to the mRNA strand, is translocated as a whole from the *a* site to the *p* site by a movement of the ribosome towards the 3′-end of the mRNA strand through a codon increment. The translocation moves a new mRNA codon into the vacated *a* site which then assumes the shape appropriate for the tRNA charged with the third amino acid of the growing polypeptide chain. After attachment to site *a*, the free amino group of the third residue attacks the ester link of the dipeptide to the tRNA at site *p*, from which the tRNA corresponding to the second amino acid residue of the chain is liberated for subsequent recycling. In turn the tripeptide ester formed, with its tRNA hydrogen bonded to the matching triplet section of the mRNA, translocate as a whole from the *a* site to the *p* site by another ribosomal movement through a codon increment, allowing for a further addition to the polypeptide sequence and the subsequent elongation (Fig. 12.7(b)).

As the 5′-end of an mRNA strand emerges from a ribosome through the successive translocations, a small ribosomal subunit complexed with three protein initiation factors and a molecule of guanosine 5′-triphosphate (GTP) bind to the strand. One of the initiation factors is liberated when an initiator tRNA charged with N-formylmethionine hydrogen-bonds to the first coding mRNA triplet, and the other two when the large ribosomal subunit completes the assembly of a new ribosome. The formation of the ribosome is driven by the free energy from the coupled hydrolytic cleavage of GTP to GDP and inorganic phosphate.

Each amide bond formed in the growing polypeptide chain involves the coupled hydrolysis of two further GTP molecules to GDP and inorganic phosphate, and the catalytic mediation of three protein elongation factors. The free energy from the cleavage of one GTP molecule is linked to the transport and the location of an appropriate charged tRNA molecule into a vacant *a* site, and that of the second GTP molecule is coupled to the translocation of the ribosome, shifting the peptidyl-tRNA-mRNA-codon assembly from site *a* to site *p*. The formation of a new amide bond between the peptide esterified to the tRNA at the *p* site and the amino acid bound to the tRNA at the *a* site is catalysed by a peptidyl transferase centre located in the large subunit of the ribosome.

The free energy of hydrolysis of the first GTP molecule normally involved in amide bond formation is diverted to editing out the error if an incorrect aminoacyl-tRNA ester becomes located at the *a* site. The specificity of the hydrogen bonding between the mRNA codon and the tRNA anticodon has an error level of about one part in a hundred, and the ribosomal proof-reading dependent upon the hydrolysis of the GTP molecule has a similar error level, so that the overall fidelity of the translation of an mRNA strand into a polypeptide chain is prone to error at the approximate level of one part in 10^4. With up to 20 ribosomes on a polyribosome translating a single mRNA molecule simultaneously, each producing a polypeptide some 500 amino acid residues long, there is an even chance that one of the polypeptides contains a single incorrect residue or that all are error free.

The appearance of an erroneous codon in the mRNA undergoing translation to a polypeptide has a comparable or somewhat lower probability. Such errors arise mainly during the transcription of the mRNA sequence from a DNA strand catalysed by the enzyme RNA polymerase, which lacks a proof-reading capacity. The analogous enzyme RNA replicase, which copies viral RNA strands, has an error level near to one part in 10^4, and it appears likely that the error rate of RNA polymerase transcription is not much lower. In contrast the fidelity of DNA replication is substantially higher, with error levels in the range from one part in 10^7 (prokaryotes) to 10^{11} (eukaryotes). The enhanced accuracy arises from proof-reading by the enzyme DNA polymerase during DNA replication and a number of post-replication DNA repair and error-correction enzyme systems.

The structure of the genetic code suggests an evolutionary selection and organization of codon base triplets such that base substitutions in an mRNA strand conserve in large measure the type of amino acid at a given position in the corresponding polypeptide. The most common type of point mutation, by an order of magnitude, is a transition, the interchange of either the two pyrimidine bases or the two purine bases, as opposed to a transversion, the change from a purine to a pyrimidine base or vice versa. The interchange of the two pyrimidine bases (U/C) at the third position of a codon (the 3'-end) leaves the coding of the amino acids unaltered, and the same holds for the majority of the amino acids if a purine base (A/G) at the corresponding position is interchanged with the other. Codons with a pyrimidine in the second position, particularly U, code mainly for hydrophobic amino acids, whereas the hydrophilic amino acids are coded principally by triplets with a purine, especially A, in the second place. Thus a transition in the second position of a codon tends to replace one amino acid with another of a similar type, as do transitions and transversions in the first position if either U or A occupy the second place (Fig. 12.8).

The amino acid sequences of proteins serving the same role in different organisms, such as the α- and β-polypeptide chains of the haemoglobin tetramer $(\alpha_2\beta_2)$ in the vertebrates, usually show significant differences of primary structure with no major changes in the secondary conformation and the tertiary folding, nor a modification of function. In mammals the haemoglobin α-chain contains 141 amino acid residues and, of those found in the human species, 19 per cent are changed by substitutions in the dog. The α-chain residue differences rise to 25 per cent in the chicken and to 49 per cent in the carp. For most of the substitutions, one hydrophobic amino acid is replaced by another and, of the acidic or basic hydrophilic types, the replacement generally conserves the type.

The observation of a wide variation from one organic species to another in the primary structure of a given protein without appreciable alterations in secondary and tertiary structure, or in the function, has led to the conjecture that most surviving DNA mutations, and their consequent mRNA codon changes, are 'neutral' to Darwinian natural selection pressure, and that organic evolution is largely a random drift process (Kimura 1983). The evolutionary hypothesis of random drift deriving from neutral mutations applies principally to codon changes conserving the amino acid type in organisms confined to a uniform and indifferent environment. Other surviving codon changes tend to be milieu sensitive. The haemoglobin mutation responsible for sickle-cell anaemia involves a change of amino acid type from the acidic hydrophile (glutamic acid) in the 6-position of the 146 residues in the β-polypeptide chain to a hydrophobe (valine), and, while 'favourable' for the resistance conferred in regions where malaria is endemic, the mutation becomes 'deleterious' elsewhere as a consequence of the ensuing anaemia.

12.4 Biopolymer sequences and evolutionary trees

From the late 1950s the sequence of amino acids in proteins with a specific function, such as insulin, haemoglobin, or cytochrome *c*, were determined and catalogued for a range of organisms (Dayhoff 1968–78), followed a decade later by corresponding determinations of nucleic acid sequences. Commenting on the sequences then available, Zuckerkandl and Pauling (1965a) pointed out that a present-day protein embodies its own ancestral history in its organization, and that comparisons of homologous polypeptide chains, together with the dated fossil record, provide evolutionary information of three types. First, the probable amino acid sequence of the ancestral polypeptide from which the chains compared had diverged. Second, the approximate epoch at which the divergence had begun. Third, the lines of descent along which the changes in amino acid sequence had occurred.

In an application of the approach, Zuckerkandl and Pauling (1962) estimated a mean substitution rate for a haemoglobin polypeptide of one amino acid every 14.5 million years from the 18 residue differences found between the horse and the human α-chain. Since the α- and the β-chains of the human haemoglobin tetramer ($\alpha_2\beta_2$) show 78 differences, it was concluded that the two polypeptides had diverged from a common origin, by gene duplication, some 565 million years ago. The common origin, a monomeric haemoglobin consisting of a single polypeptide chain, has a modern representative in the blood of primitive jawless fishes, such as the lamprey and the hagfish, which lack the advantage of cooperative oxygen uptake and release that emerged with the evolution of the haemoglobin tetramer.

The approach of Zuckerkandl and Pauling with their concept of a 'molecular evolutionary clock' was developed by Kimura (1969), who employed a Poisson distribution, which is appropriate for events of low probability in a vast population, like the decay of radioactive isotopes already used to calibrate the geological evolutionary clock. Kimura showed that, for two homologous polypeptides composed of n residues with different amino acids at d sites due to divergence over a time t, the rate of evolutionary amino acid substitution per site per year, k_{aa}, is given by the relation

$$2tk_{aa} = -\ln(1 - d/n) \sim -2.3\log(1 - d/n). \qquad (12.1)$$

The application of relation (12.1) to a series of sequenced polypeptides with a common biochemical role, such as the α-chain of vertebrate haemoglobin, gives an approximate constant value of k_{aa} of the order of 10^{-9} residue replacements per site per year, a unit Kimura termed the *pauling* or the 'molecular evolutionary unit' (MEU). While roughly constant for a given type of protein, the value of k_{aa} varies widely for proteins of different functional types by three orders of magnitude between the rapidly evolving fibrinopeptides (8.3 MEU) and the highly conserved histone proteins (0.008 MEU). The fibrinopeptides are involved in the formation of blood clots, and only a small segment of the polypeptide has an essential role. The amino acids of the other segments are free to undergo substitution without selectional control, since no drastic functional changes are involved. In contrast, the histone polypeptides are vital as a whole to the conservation of the genetic material, making up the nucleosome particles round which the double-stranded DNA supercoils in the eukaryote nucleus to give the nucleoprotein chromatin a 'string of beads' form.

Proteins in which amino acid substitution leads to gross impairment of function have low evolutionary rates, since the rate estimated refers to surviving residue replacements. In general the several segments of a polypeptide chain undergo evolutionary substitution at different rates, dependent upon the role of the segment in the overall biochemical function. In both the α- and the β-chain of haemoglobin the segment forming the functionally rigid pocket to hold the iron–porphyrin oxygen carrier has a lower evolutionary rate (α 0.165, β 0.236 MEU) than the segments forming the globin surface (α 1.35, β 2.73 MEU). Amino acid substitutions are considered to occur as frequently in the haem pocket regions as in the surface regions of the haemoglobin chains, but those in the surface segments tend to be less lethal, surviving to register the higher estimated evolutionary rate. The substitution leading to sickle-cell anaemia, valine in place of glutamic acid in the 6-position from the N-terminal of the β-chain, is a surface replacement.

As amino acid replacements in a protein are the tertiary product (through transcription and translation) of nucleotide substitutions in DNA, more detailed evolutionary information became available from comparisons of nucleic acid sequences. Approximately one-third of the mutations in coding DNA result in no change of the amino acid residues in the polypeptide coded, owing to the degeneracy of the genetic code in which the 20 protein amino acids correspond to 61 base triplets. Coding DNA mutations which are 'silent' or 'synonymous', owing to the code degeneracy, do not lead to any amino acid replacement in the corresponding protein. These synonymous mutations are not edited out by natural selection and are expected to appear with the highest frequency in a set of comparable DNA or transcribed mRNA sequences. The synonymous mutations provide a 'silent molecular clock' which keeps an absolute evolutionary time, as opposed to the variable 'protein clocks' that run at different rates for proteins with different functions. Studies of the silent clock of synonymous mutations established that molecular evolution depends upon the succession of elapsed years, rather than the number of successive organic generations.

The histone proteins are highly conserved, with only two differences out of about 100 amino acids in the H4 polypeptide between calf thymus and the pea plant, although animals and plants diverged some 1.2 billion years ago. A comparison of the mRNA sequences coding for the histone H4 polypeptide in two sea urchin species shows, however, that the synonymous mutation rate in the DNA transcribed is as high as that of the fast-evolving fibrinopeptides. In general the degeneracy of the genetic code is prominent for the third position (3'-end) of the coding base triplets where the transitional mutations (U/C) or (A/G) are mostly silent. Comparisons of the mRNA sequences coding for the α- and the β-polypeptide chains of mammalian haemoglobin indicate that natural selection has pruned down the rates of surviving base changes in the first and second positions of the triplet codons, which are approximately equal, to one-quarter or less of the mutational rate found for the third codon position. The estimated rate of base change in the third codon position is closely comparable to the synonymous mutation rate for coding DNA sequences (exons) and the general mutation rate for non-coding intervening sequences (introns).

Comparisons of the monomer sequences in the proteins and the nucleic acids of different organisms served to distinguish between alternative general forms of the evolutionary relationships between the enormous variety of organic species, and to identify the connections of the optimum genealogical tree. General alternative relationships consist of the Big Bang starburst in which the species emerge simultaneously from a common origin, as in early cosmogonies; or the monotonically linear concept of the Great Chain of Beings, current in the classical Greek, medieval, and early modern period (Lovejoy 1950); or the successive divergence of the species in a tree-like reticulation, advocated from the eighteenth century on. The application of the razor proposed by William of Occam (1280–1349), the principle of parsimony, to sets of homologous sequences confirmed and extended the classical phylogenetic trees, constructed over the previous century or more from the morphological characters of the organic species and the fossil evidence. The optimum genealogical tree, according to the principle of parsimony, embodies the minimum number of mutations required to account for the evolution of a set of present-day homologous sequences from a common ancestral sequence. The time-scale for the successive branchings of the evolutionary trees comes from the molecular clocks, the rate of amino acid substitutions or of base replacements in the particular protein or nucleic acid type employed to construct the tree relationship. Proteins of different functional types and the nucleic acids coding for them, or non-coding but functional, like ribosomal RNA, give genealogical trees which are in general agreement where they overlap and serve to complement one another elsewhere.

The tree based upon the amino acid sequences of the cytochrome *c* protein, a general but not universal redox carrier, is confined largely to aerobic organisms. The corresponding ferredoxin tree has branches covering both the anaerobes and the aerobes, the latter branches providing support and the former a complementation for the cytochrome *c* connections. In turn the relationships found between the base sequences of the 5S rRNA (~120 nucleotides) in the larger ribosomal subunit overlap and reinforce both the ferredoxin and the cytochrome *c* protein trees and, combined with the latter, give an even more comprehensive evolutionary tree (Schwartz and Dayhoff 1978).

The classification and phylogeny of microbes had remained a grey area since the recognition in 1866 by Ernst Haeckel (1834–1919) at Jena of the 'protista' as a unique organic kingdom, distinct from the plant and animal kingdoms, until the nucleic acid base sequences of microbes were analysed and compared a century later. Bacteria had been classified morphologically from the shapes first observed by van Leeuwenhoek around the turn of the seventeenth century and by staining based upon a procedure introduced by Christian Gram (1853–1938) at Copenhagen in 1884. Bacteria divide into Gram positive or Gram negative classes, dependent on the respective retention or loss of the dye after treatment with Ehrlich's gentian–violet stain and a potassium tri-iodide (KI_3) solution, followed by an alcohol wash.

Cell microscopy during the 1930s afforded the broad division of microbes into the prokaryotes with no apparent nucleus, and the eukaryotes containing a membrane-enclosed nucleus and organelles enclosed by a double membrane (mitochondria, chloroplasts). From the 1950s authentic microfossils were discovered in increasing numbers and variety, the remains of prokaryotes dating back to more than 3 billion years. A size increase in some microfossils ($\sim 1\,\mu$m to $\sim 10\,\mu$m) around 1.4 billion years ago suggests the appearance of eukaryotes around that time. Unlike the macrofossils of shells and bones, which first appeared some 600 million years ago, the microfossils lack detailed characteristics allowing comparisons with present-day species.

The ribosomal and transfer RNAs are essential to all living creatures for protein manufacture, and their well-conserved sequences permit correlations between distantly related species. The nucleotide chains of tRNA and 5S rRNA are too short for detailed comparisons among the bacteria, although they serve as well as the protein sequences to relate the prokaryotes to the algae, fungi, animals, and higher plants. From the late 1960s, Carl Woese at Urbana, Illinois, and his colleagues sequenced the 16S rRNA of the smaller ribosome subunit in bacteria, some 1500 nucleotides long, to obtain correlations of greater resolution among the prokaryotes. Over the following two decades the 16S rRNA of some 500 prokaryote species were sequenced, together with the corresponding 18S rRNA (~ 1800 nucleotides) of a number of eukaryotes (Woese 1987).

Initially the sequencing of only short lengths of RNA was practicable, so that the 16S rRNA was cleaved enzymatically at the $3'$-phosphodiester bond of each guanosine residue with a specific endonuclease (T_1) to give a mixture of oligonucleotides which were separated, sequenced, and catalogued for each organism. Subsequent technical developments made possible the complete serial identification of each base in a 16S (or 18S) rRNA strand. Comparisons between the catalogues of oligonucleotides containing six or more bases were made, for each pair of species, in terms of the traditional quantitative ratio of like characteristics to the sum of the like and unlike characteristics, the similarity coefficient, due to the French naturalist Michel Adanson (1727–1806).

The similarity coefficient S_{AB} for two different 16S rRNA sequences, containing N_A and N_B bases, is defined by the ratio of the number of bases common to the two sequences, N_{AB}, to the mean sequence length:

$$S_{AB} = 2N_{AB}/(N_A + N_B). \tag{12.2}$$

The similarity coefficient is a sensitive function of the percentage homology P between two strands which have been completely sequenced, S_{AB} being proportional to a power between P^5 and P^6. Over the range of significant discrimination, $P \sim 80$ per cent for $S_{AB} = 0.3$ while $P \sim 60$ per cent for $S_{AB} = 0.05$.

As expected, the similarity coefficient between typical prokaryotes and eukaryotes was found to be small. While S_{AB} values of about 0.3 are observed for diverse unicellular eukaryotes (fungi, algae, and protozoa) among themselves, their rRNA sequences overlap those of prokaryotes at the level of $S_{AB} \sim 0.1$ or less ($S_{AB} = 0.05$ for yeast and *E. coli*). Unexpectedly the prokaryotes divided into two kingdoms, the archaebacteria and the eubacteria, as remotely related to one another as to the kingdom of the eukaryotes individually (for *Thermoplasma*, $S_{AB} = 0.08$ with yeast and 0.09 with *E. coli*).

The archaebacteria characteristically thrive in extreme environments, hot sulphurous vents, salt marches, or anaerobic organic-rich muds. They are considered to be ancient, as the term archaebacteria implies, from the heterogeneity of their 16S rRNA sequences, due to divergence over a long period of time. The similarity indices of those sequences distinguish three major divisions: first, the methanogens and halophiles; second, *Thermoplasma*, a group containing a single organism; and third, a heterogeneous group of sulphur-dependent thermoacidophiles, possibly the most ancient of the prokaryotes. The latter grow under acidic conditions (pH ~ 1–6) and at elevated temperatures (~ 60–95°C). Some, such as *Sulfolobus*, oxidize sulphur to sulphate while others reduce sulphate to sulphide, like *Archaeoglobus*, which generates small quantities of methane as well.

The 16S rRNA sequences of the eubacteria indicated that, while the Gram positive organisms form a distinct group of related species, the Gram negative organisms are much more diverse, falling into nine groups defined by a similarity coefficient of 0.2–0.3 within a group. The largest Gram negative group, the purple bacteria, contains not only the coloured photosynthetic organisms but also many colourless analogues, such as *E. coli* and other enteric bacteria. The green photosynthetic bacteria fall into two distinct groups, the sulphur (hydrogen sulphide substrate) and non-sulphur (organic substrate) types, which are only distantly related ($S_{AB} \sim 0.18$). The cyanobacteria make up a coherent group, and they are related to the chloroplast organelles of eukaryotic red and green algae, and green plants (S_{AB} values of 0.25–0.30).

The sulphate-reducing bacteria, which use the oxygen of the sulphate for a primitive form of respiration, comprise a Gram negative eubacterial group. One member of the group, *Desulfovibrio sulfodismutans*, obtains free energy through the disproportionation of sulphite or thiosulphate to sulphide and sulphate, an 'inorganic fermentation' analogous to the redox disproportionation of glucose to ethanol and carbon dioxide (Bak and Cypionka 1987).

The eukaryote organelles, the mitochondrion and the chloroplast, have a prokaryote size (~ 1 μm) and contain their own DNA, a circular double strand as in bacteria, and ribosomes of the bacterial type. The 16S rRNA sequences of the organelles relate the mitochondria to the purple group of eubacteria and the chloroplasts to the group of cyanobacteria. The discovery of indigenous DNA in the organelles and the similarities of their 16S rRNA sequences to those of extant eubacteria supported the long-standing conjecture that mitochondria and chloroplasts evolved by a process of endosymbiosis. It is surmised that an ancestral eukaryote, with a lifestyle based upon the enclosure of nutrients into a membrane-bounded vacuole, engulfed a purple-group aerobic bacterium generating ATP through the respiratory chain and, in return for a supply of ATP, the host archae-eukaryote provided the guest proto-mitochondrion with the materials required for its maintenance and ATP production. In a similar way, the engulfment of a cyanobacterium into a vacuole led to the evolution of the chloroplast, which supplied carbohydrate to its host for the return provision of a life-support system. The two membranes surrounding an organelle are vestiges of the endosymbiotic origin: the inner membrane represents the cytoplasmic membrane of the eubacterial guest while the outer membrane corresponds to that lining the vacuole of the host eukaryote.

Possible living relatives of the ancestral eukaryote, as many as a thousand species, contain no mitochondria or other organelles. These species form the *archaezoa* subdivision of the eukaryotes, the other being the *metakaryota*, which contain mitochondria. Sequenced representatives of the archaezoa are minute parasites, the *metamonada* and the *microsporidia*, which contain bacterial-sized ribosomes with 16S rRNA in the smaller subunit, rather than the 18S rRNA typical of the higher eukaryotes. The substantial difference between the 16S rRNA sequences of a microsporidia and a metamonada species and those of mitochondrion-containing eukaryotes, the metakaryota, suggests an early evolutionary divergence (Fig. 12.11). The archaezoa and the metakaryota diverged perhaps before the age (~ 1.4 billion years) of the larger microfossils (~ 10 μm) which, on account of their size, probably represent the remains of organelle-equipped eukaryotes (Cavalier-Smith 1989).

Biochemically the eukaryotes have resemblances to both the eubacteria and the archaebacteria. The translation of mRNA into a polypeptide begins with methionine in the eukaryotes and in the archaebacteria, but with N-formylmethionine in the eubacteria. The subsequent elongation of the polypeptide chain is inhibited by dipthera toxin in the eukaryotes and

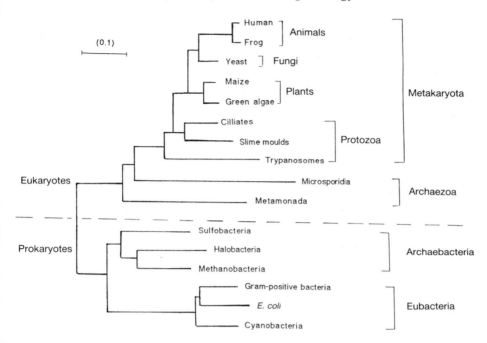

Fig. 12.11 Universal evolutionary tree based on the nucleotide sequences of the rRNA in the small subunit of the ribosome. The scale length (0.1) corresponds to 10 differences for each set of 100 nucleotides (after Cavalier-Smith 1989).

the archaebacteria but not in the eubacteria; in addition, chloramphenicol inhibits the elongation in eubacteria but not in eukaryotes or archaebacteria. But, in contrast, the lipids of the cytoplasmic membrane in both eubacteria and eukaryotes consist of the diesters of L-glycerophosphate with straight-chain fatty acids, whereas the archaebacterial lipids are composed of the diethers of D-glycerophosphate with phytanol and other branched-chain alcohols.

Whether the ancestral organisms of the three kingdoms emerged around the same time from a common universal ancestor, or evolved serially one from the other, and, if so, in what particular order, remain open questions. The evidence from the rRNA sequences indicates that the eubacteria and the eukaryotes are related more closely to the archaebacteria than they are to one another, so that the latter may have contained the common ancestor of all three of the modern kingdoms. The diversity of the present-day sulphur-dependent thermoacidophiles is consistent with the assumption that they derive from the most ancient lines of descent. The lifestyle of the modern representatives corresponds to an environment of volcanic activity and elevated temperatures more widespread globally about 4 billion years ago than today, with then ample supplies of organic substrates, continuously generated by prebiotic organic geochemical processes and augmented episodically by the organic substances in the carbonaceous chondrite fraction of the frequent bolide impacts.

13
Prebiotic chemistry

13.1 Abiotic synthesis of biomonomers

Charles Darwin and his followers took the view that Pasteur's experiments refuting the present-day spontaneous generation of organisms were not strictly relevant to the question of the origin of life aeons ago before the mechanism of natural selection came into play. Writing in 1871 to his long-standing associate, Joseph Hooker (1817–1911), director of the Kew botanic gardens, Darwin pointed out that if,

in some warm little pond with all sorts of ammonia and phosphoric,—light, heat, electricity, etc. present, that a protein compound was chemically formed, ready to undergo still more complex changes, at the present day such matter would be instantly devoured, or absorbed, which would not have been the case before living creatures were formed.

A number of chemical reactions later recognized to have prebiotic significance were already known, but the chemical sciences were too un-developed at the time for any realistic articulation of the warm-pond model. The synthesis of new aromatic dyestuffs, guided by Kekulé's flat-land molecular structures, dominated the chemical scene. Darwin's model became significant only after the investigation of the sugars, amino acids, and polypeptides (1884–1919) by Emil Fischer, and related studies by his contempories based upon the three-dimensional stereochemistry of Le Bel and van't Hoff. Then the warm-pond conjecture was taken up and extended in the hypothesis that postulated the origin of terrestrial life from a plenitude of prebiotic biomolecules in the early oceans.

The Soviet biochemist A. I. Oparin (1894–1980) reviewed, briefly in 1924 and at book length in 1936, a number of known chemical reactions which could have produced molecules of biological importance during the Precambrian period, based upon what was then known of the chemistry of the Earth and the solar system at large, the biochemistry of fermentation and glycolysis, and the metabolism of the chemoautotrophic bacteria that fix carbon dioxide using inorganic electron donors (sulphur, nitrite, ferrous iron). The potentially prebiotic reactions included the hydrolysis of metallic carbides and nitrides to give hydrocarbons and ammonia, respectively; the formation of cyanamide from calcium carbide and nitrogen; the base-catalysed polymerization of a formaldehyde solution to a mixture of sugars, discovered by Butlerov in 1861; the deposition of polyglycine from a solution of glycine ethyl ester on standing, observed by Curtius in 1904;

and the reactions of aldehydes giving a redox disproportionation (Cannizzaro in 1853) or an aldol condensation (Wurtz in 1872). Following the
appearance of an English translation in 1938, Oparin's book, *Origin of
life*, became widely influential and was much quoted, like the independent
and similar view of the rise of living systems for prebiotic biomolecules,
published in 1929 by the biochemist and geneticist, J. B. S. Haldane (1892–
1964), then at Cambridge.

Haldane proposed that the primordial terrestrial seas had been a vast
chemical laboratory, powered by solar radiation. The early atmosphere
was oxygen-free, since there is now less atmospheric oxygen than would be
required to burn up all the coal and petroleum reserves; and metabolically
primitive bacteria are anaerobic, living by fermentation, like the mammalian
embryo for the first few days after conception. With no ozone layer the
ultraviolet radiation of the Sun penetrated to the seas, which contained
dissolved carbon dioxide and ammonia, to generate an immense variety of
organic compounds (as the 1927 photochemical studies of E. C. C. Baly
and his associates at Liverpool appeared to demonstrate). Consequently
the early oceans became a warm dilute soup in which a range of biopolymers were elaborated, some of virus-like proportions and capable of replication by a process analogous to crystallization. The stage was then set for
the emergence of the first cell, by the acquisition of a lipid membrane and
other supporting elements, doubtless after numerous failures. As the
nutrient reserves in the oceans became exhausted by fermentation, the early
organisms turned to direct capture of the solar energy by photosynthesis.
There could be only one origin, a single common ancestor for the entire
organic world, since the biomolecules are uniquely handed, derived solely
from the D-sugars and the L-amino acids.

The Oparin–Haldane hypothesis of the emergence of organized systems
with autocatalytic and heterocatalytic properties in the primordial soup-
like oceans was first taken up in the USA by Norman Horowitz at Stanford
in 1945. At Stanford, George Beadle (1903–89) and Edward Tatum had
developed the then new field of biochemical genetics, based upon the
hypothesis that each enzyme catalysing an individual biosynthetic step is
controlled by a single gene. A mutation in the gene eliminated the related
enzyme and the corresponding biosynthetic intermediate, which was then
required by the organism for survival. Studies of the nutrient requirements
in the mutants of the mould *Neurospora* (induced by X-rays) revealed by
1945 seven stages in the biosynthesis of arginine by the organism. While
arginine is an essential constituent of proteins, two of the intermediates,
ornithine and citrulline, have no function other than the role of precursors
for arginine in *Neurospora* (Fig. 13.1). From these and analogous results,
Horowitz concluded that biochemical synthesis had evolved backwards
from complex products to simpler precursors. Initially the primordial

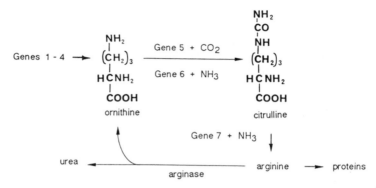

Fig. 13.1 The interpretation, based on the 'one gene–one enzyme' theory, of the nutrient requirements, for ornithine and citrulline on the biosynthetic pathway to the essential amino acid, arginine, found in *Neurospora* mutants (after Beadle 1946).

oceans had abounded with complex molecules, such as arginine, which became progressively depleted by incorporation into the proteins and other materials of the early organisms. The organisms that developed the capacity for the synthesis of the essential complex molecules from precursors survived as long as the precursors themselves remained abundant. Thereafter, by a progressive recourse to ever more simple intermediates, intricate biosynthetic pathways were developed, ultimately dependent upon inorganic substances alone in the case of the autotrophic bacteria.

Oparin's theory prompted Melvin Calvin and his associates at Berkeley in 1950 to bombard a trial primitive atmosphere of hydrogen and carbon dioxide equilibrated with an aqueous solution of ferrous iron with 40 MeV α-particles, generated from helium in a cyclotron, on the assumption that natural radioactivity had been a significant energy source for prebiotic reactions. Up to 22 per cent of the dissolved carbon dioxide was transformed into formic acid and a small fraction (0.1 per cent) into formaldehyde. In subsequent experiments, with acetic acid added initially, multi-carbon acids were generated (lactic, malic, succinic, fumaric, etc.).

At Chicago, Harold Urey in 1952 gave reasons for doubting that natural radioactivity had been an important energy source for early geochemical and prebiotic reactions, compared with the enormous input of solar ultraviolet radiation, the kinetic energy of bolide impacts, and powers of other provenance. Moreover, Urey argued, the primordial atmosphere was highly reducing, composed of hydrogen and the hydrides of the lighter elements (CH_4, NH_3, H_2O, etc.), like the present atmospheres of the outer planets. The ultraviolet photolysis of the hydrides and the gradual loss of hydrogen from the Earth's gravitational field produced a secondary atmosphere of nitrogen and carbon dioxide, and much of the carbon dioxide was taken up in the weathering reaction with the silicate rocks to give the limestones and silica. Reducing conditions were possibly maintained until some 800 million years ago, by which time the organic substances in the oceans had been much depleted by the anaerobic fermenting heterotrophs and the photosynthetic autotrophs had evolved. Urey estimated that the prebiotic fraction of organic substances in the oceans was about 1 per cent, a concentration rather small for appreciable product formation by known chemical reactions conducted at Earth-surface temperatures over a finite period, however extended.

The problem of the small fraction of organic compounds in the primordial soup had been addressed in 1951 by J. D. Bernal in London. Bernal proposed that the dissolved organic substances had been concentrated by adsorption on the surfaces of clay mineral particles suspended in the early oceans, and that the subsequent prebiotic reactions elaborating the organic substrates were catalysed at active sites on the clay mineral surfaces or located within the layered structures.

Definitive experiments on the prebiotic synthesis of the acidic biomonomers were reported by Stanley Miller in Urey's laboratory from 1953. Gas mixtures corresponding to the primitive terrestrial atmosphere according to Urey's analysis (CH_4, NH_3, H_2, H_2O) in a large flask equipped with tungsten electrodes were subjected to a high-frequency electric discharge from a Tesla coil, simulating lightning discharges. Steam from a subsidiary flask of boiling water continuously entered the system and left as condensed water for return to the liquid reservoir, modelling the evaporation from the oceans and the back-condensation in rain. After a week of continuous spark discharges, the accumulated products in the gaseous, liquid, and solid phases were analysed. Carbon monoxide and nitrogen turned out to be the main gaseous products; a little methane remained but most of the ammonia was consumed. A substantial quantity of insoluble material was generated, probably a cyanide, aldehyde, or cyanide–aldehyde polymer. Chromatographic analysis of the residual aqueous solution showed that some 25 amino acids had been synthesized, together with several hydroxy acids, fatty acids, and amide products. The compounds identified accounted for 15 per cent of the carbon in the original carbon source, methane.

Subsequent studies of the reaction mechanisms involved in the electric discharge syntheses demonstrated, from samples withdrawn during the run, a rapid initial formation of hydrogen cyanide and of aldehydes, followed by a decrease in the concentration of these products as the amino acids were more slowly formed. The result indicated that the long-known Strecker synthesis (1850) occurred in the condensate from the discharge, and explained why specifically α-amino acids and α-hydroxy acids were formed predominantly, with only minor yields of acids substituted in other positions:

$$
\begin{array}{ccc}
\text{RCHO} + \text{HCN} \longrightarrow \text{RCH(OH)CN} & \xrightarrow{\text{NH}_3} & \text{RCH(NH}_2\text{)CN} \\
\downarrow \text{H}_2\text{O} & & \downarrow \text{H}_2\text{O} \\
\text{RCH(OH)COOH} & & \text{RCH(NH}_2\text{)COOH.}
\end{array}
$$

During the 1950s it became clear, from the depleted terrestrial abundance of the non-radiogenic noble gas isotopes, that the primary atmosphere of the Earth had been formed mainly by degassing from the molten interior, rather than by accretion from the solar nebula, so that a close correspondence in composition between the early terrestrial atmosphere and the present-day atmospheres of the outer planets is not expected. The revised view held that, while the primary atmosphere was reducing, nitrogen appeared mainly in molecular form (N_2) already, and carbon was represented principally by the oxides (CO, CO_2) rather than the hydrides. Repetitions of the spark discharge experiments, with the initial gas phase made up of carbon monoxide or carbon dioxide in place of methane, and molecular nitrogen instead of ammonia, were found to give essentially the same range of amino, hydroxy, and fatty acids as the original prebiotic synthesis experiments, provided that the initial gas mixture was reducing overall ($H_2/CO > 1$; $H_2/CO_2 > 2$). The yields proved to be lower, the identified products making up some 2.7 per cent of the initial carbon at an optimum. The crucial component turned out to be ammonia which, if added to the initial carbon oxide gas mixture, raised the yield of characterized products to the 15 per cent level of initial carbon typical of the original spark discharge experiments using methane, ammonia, and water mixtures (Miller 1986).

Similar ranges of amino acids, generally in smaller yields, were obtained from gas mixtures simulating the early terrestrial atmosphere using other energy sources, such as ultraviolet radiation, electron bombardment or γ-radiation (modelling natural β- and γ-radioactivity), and thermally at about 1000°C on quartz sand or silica gel surfaces (Lemmon 1970). Parallel analyses of the carbonaceous chondrites showed that the range of amino acids characterized in the organic matter of these meteorites overlapped, in both identity and relative abundance, the particular amino acids produced by the spark discharge and other prebiotic syntheses.

The observed fall of a large carbonaceous chondrite at Murchison, Australia, in 1969, followed by the early collection of some 83 kg of samples, provided a stimulus and wider opportunities for comparative studies. The comparisons suggested that no single mode of prebiotic synthesis could account for the wide variety of organic compounds identified in the carbonaceous chondrites, the two most important processes being the electric discharge synthesis and the Fischer–Tropsch-type (FTT) synthesis.

The FTT reaction (Section 8.3), discovered by Döbereiner in 1818 and patented by Fischer and Tropsch in 1922, affords a variety of alcohols, acids, and hydrocarbons from water–gas mixtures (CO/H_2O) passed over metal oxide and silicate catalysts at relatively moderate temperatures (100–300°C). The addition of ammonia to the water–gas feedstock gives numerous amines, amino acids, and N-heterocyclic compounds as well. Hydrogen and carbon monoxide are the most abundant diatomic gases in the dark molecular clouds, and ammonia is a predominant tetra-atomic molecule. These gases in the solar nebular provided a feedstock for the production of meteoritic organic materials, additional to the more complex organic molecules of the parent cloud, which were extensively reformed, where not decomposed. During the accretion of the parent bodies of the meteorites, FTT catalysis by grains of magnetite (Fe_3O_4) or of hydrated silicates would have led to the formation of the variety of organic compounds found in the carbonaceous chondrites (Anders *et al.* 1973).

The organic substances may have formed in the atmospheres of the meteoritic parent bodies, since some of the organic matter in the carbonaceous chondrites has a globular form, 1–3 μm in diameter, with a magnetite or silicate core (Anders and Hayatsu 1981). Similarly, in the early atmosphere of the Earth, meteoroid grains of magnetite or silicate in the 10^{-6}–10^{-9} g mass range, which only warm up without melting during their fall, could have served as FTT catalysts for the prebiotic terrestrial production of organic substances in their descent through the reducing gas mixture. The short contact time between the catalytic particles and the reactant gases would favour the formation of unbranched chains in the aliphatic compounds, such as the normal paraffins, and the straight-chain fatty acids of the eubacterial and the eukaryote lipids. Meteorites with a mass above about 100 g form an ablation crust, owing to surface melting during the deceleration, but the internal materials reach the Earth's surface intact, and the falls of carbonaceous chondrites, much more frequent in the early Precambrian than at later times, would have enriched the organic content of the primordial oceanic soup.

The ranges of amino acids produced by the electric discharge and by the FTT syntheses overlap, but, in part, they are complementary. The aromatic and the basic amino acids (tyrosine, histidine, lysine, arginine) are formed in the FTT reaction but not in the spark discharge synthesis, whereas the latter affords the hydroxy amino acids (serine, threonine) which remain unreported among the products of the former reaction. Of the 20 protein amino acids, 13 are formed by the spark discharge synthesis and an additional 4 by the FTT reaction, while the remaining 3 are available by photolytic and pyrolytic reactions. Many non-protein amino acids are formed as well. Some 35 amino acids have been identified among the products of the spark discharge synthesis (Miller 1986), compared with the 55 amino acids characterized in the organic material of the Murchison meteorite (Cronin and Pizzarello 1986).

Both the purine and the pyrimidine nucleotide bases have been identified among the variety of N-heterocyclic compounds produced by the FTT reaction, and these bases are readily formed by the further reaction of intermediates generated during the spark discharge synthesis. Adenine is essentially a pentamer of hydrogen cyanide, which is produced in quantity early in the discharge synthesis. Dilute solutions of hydrogen cyanide hydrolyse to ammonium formate and only relatively concentrated solutions (>1.0 M) give oligomers, including adenine, and polymers.

In a remarkable series of experiments it was shown that the freezing of a dilute solution of ammonium cyanide (10^{-3} M) resulted, through the separation of ice, in the concentration of the liquid phase to a eutectic mixture (75 per cent HCN) at $-21°C$, and that adenine in 0.5 per cent yield with other products was formed at a measurable rate over the temperature range 0 to $-20°C$. The intermediates identified included the hydrogen cyanide trimer and tetramer, aminomalonitrile and diaminomalonitrile, respectively (Fig. 13.2). In sunlight the hydrogen cyanide tetramer photoisomerizes to 4-aminoimidazole-5-carbonitrile, which is formed additionally by the reaction of the hydrogen cyanide trimer with formamidine (from ammonia and hydrogen cyanide). The carbonitrile product serves both as the immediate precursor of adenine, by reaction with a further molecule of hydrogen cyanide, and as an intermediate in the production of guanine and other purines (Orgel and Lohrmann 1974).

The pyrimidine nucleotide bases are formed in aqueous solution under mild conditions from cyanoacetylene and the cyanate ion. A spark discharge through mixtures of nitrogen and methane generates cyanoacetylene in quantity; hydrogen cyanide alone is the more abundant product. In the interstellar dark molecular clouds cyanoacetylene and hydrogen cyanide are equiabundant ($\sim 10^{-9}$ relative to unit hydrogen abundance). Cyanogen, also formed during the electric discharge experiments, hydrolyses to cyanide and cyanate, the latter reacting with cyanoacetylene to give cytosine, and thence uracil by hydrolytic deamination (Fig. 13.3). In addition cyanoacetaldehyde, from the hydration of cyanoacetylene, reacts with guanidine to form 2:4-diaminopyrimidine, which hydrolyses to cytosine and then to uracil (Miller 1986).

The prebiotic source of the sugars is generally considered to be formaldehyde, through the base-catalysed formose reaction of Butlerov (1861). Formaldehyde and ammonia are the main four-atom molecules present in the dark molecular clouds, each some 10^{-8} abundant relative to hydrogen, although neither species could have survived as such within the inner solar system. It is estimated from the rate coefficients of the reactions involved

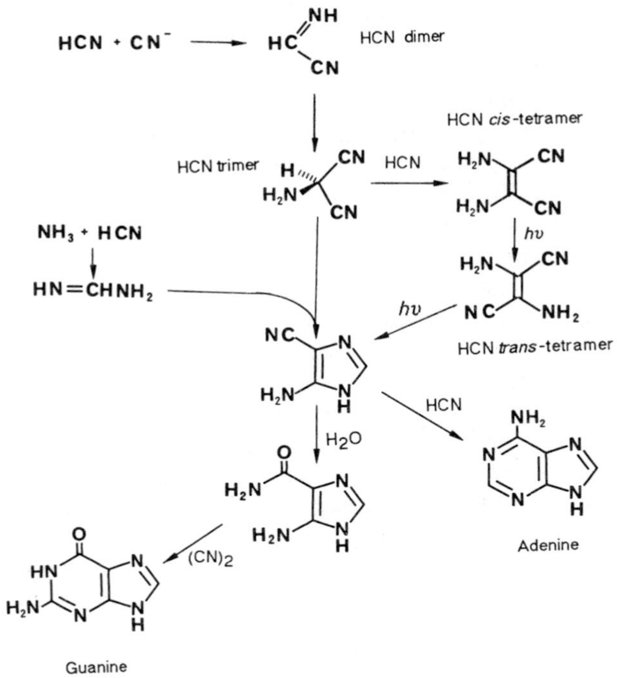

Fig. 13.2 The prebiotic synthesis of the purine nucleic acid bases from hydrogen cyanide through the dimer and oligomers, with related substances (formamidine and cyanogen) produced from a reducing gas mixture by an electric discharge.

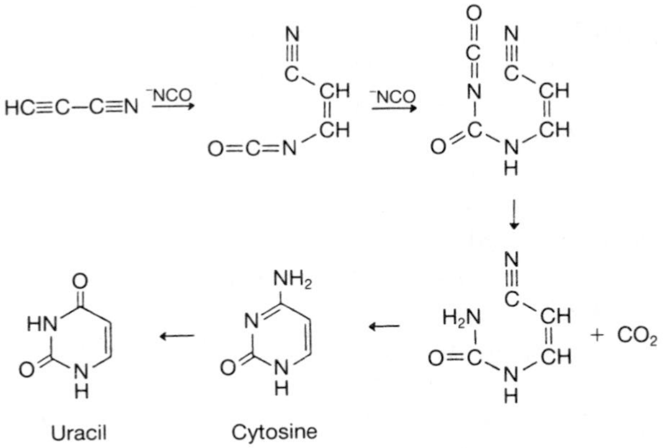

Fig. 13.3 The prebiotic synthesis of the pyrimidine nucleic acid bases from cyanoacetylene with isocyanate produced by the hydrolysis of cyanogen.

that formaldehyde was produced in quantity by the solar ultraviolet photolysis of water vapour and carbon dioxide in the early terrestrial atmosphere. Most of the formaldehyde formed ($\sim$99 per cent) soon decomposed photolytically, but the remaining fraction rained out at a rate of 10^{11} moles per year to give the oceans, at their present volume, an effective formaldehyde concentration of 10^{-3} M over 10 million years (Pinto *et al.* 1980). The linear polymer of formaldehyde, polyoxymethylene, readily formed in neutral solution or on mineral surfaces from the gas phase, is considerably more stable to ultraviolet radiation than the monomer, and the polymer serves equally well as the substrate for the formose reaction.

Following the work of Butlerov in 1861, in which he obtained a sweet-tasting substance from the addition of slaked lime to aqueous formaldehyde, Loew (1886) obtained a fermentable hexose, termed formose, by the action of gaseous formaldehyde on calcium hydroxide. Fischer and his contemporaries identified (1888–1906) DL-glucose, DL-fructose, DL-arabinose, and other sugars among the products of the formose reaction, and characterized the formaldehyde dimer, glycolaldehyde ($HO–CH_2–CHO$), as the initial product formed by the aldol condensation of Wurtz. The initial rate of reaction is low, glycolaldehyde being generated slowly during the 'induction period', after which the reaction becomes autocatalytic and speeds up to give mainly a pentose and hexose mixture with some of the shorter- and a little of the longer-chain-length sugars (Fig. 13.4). The induction period is eliminated if glycolaldehyde or the triose mixture (glyceraldehyde and dihydroxyacetone) is added to the reaction medium.

With the introduction of chromatographic methods during the 1950s, additional pentose and hexose sugars in the formose mixture were

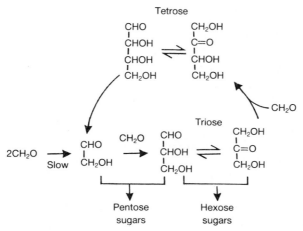

Fig. 13.4 An autocatalytic cycle proposed for the formose reaction, generating monosaccharides (mainly pentose and hexose sugars) from formaldehyde.

characterized, together with their precursors and products. A quantitative estimate of the sugar composition of formose by gas–liquid chromatography in terms of the carbon chain length gives: C_4 10 per cent; C_5 30 per cent; C_6 55 per cent; with C_7 and C_8 5 per cent. The facile and efficient condensation of the triose mixture (yields about 90 per cent) accounts for the dominance of the C_6 sugars in formose, and the stable all-equatorial orientation of the bulky ring substituents in the pyranose chair conformation of β-glucose (Fig. 11.3) explains the relatively high abundance of glucose in the hexose mixture formed.

The range of catalysts for the formose reaction has been extended from the original alkaline earth hydroxides to the corresponding carbonates, the hydrated heavy metal and lanthanide oxides, alumina and the aluminosilicate clay minerals (kaolinite and illite), and some organic bases. The caustic alkali metal hydroxides induce the redox disproportionation of formaldehyde to methanol and formic acid (the Cannizzaro reaction), which is competitive with the formose aldol condensation, giving side products such as glycerol and glyceric acid from glyceraldehyde. The calcium ion, and other divalent and polyvalent metal ions, favour the formose reaction by complexing with the sugars formed, and the coordination complex of the metal ion itself catalyses the reaction (Mizuno and Weiss 1974).

The formose condensation in aqueous solution typically requires formaldehyde concentrations of about 1.0 M, yields being modest at 10^{-2} M and negligible at the concentration estimated for the prebiotic oceans ($\sim 10^{-3}$ M). Formaldehyde in concentrated form as the polymer, polyoxymethylene, forms much of the envelope of silicate grains in the circumstellar dust belts of some cool stars and, in the laboratory, it is found that formaldehyde polymerizes even at approximately 4 K, predominantly by a non-Arrhenius quantum tunnelling process at low temperature (Goldanskii 1986). Any surviving formaldehyde polymer in the dust grains accreted in the cool outer reaches of the solar nebula could provide the early Earth, from the then high impact rate of carbonaceous meteorites, only with a priming supply of sugars at best. The free sugars are unstable in aqueous solution and, while stabilized by glycoside formation, e.g. as nucleosides with the purine and pyrimidine bases, their ultimate decomposition is only postponed. The more significant source of sugars on the early Earth was the probable continuous supply of formaldehyde, generated photochemically in the anoxic atmosphere, augmented by the concurrent electric discharge production, and concentrated by polymerization or by direct aldol condensation from the gas phase on catalytic mineral surfaces.

Already in 1909 William Huggins, reviewing the spectroscopic studies of the stars over the previous half-century, judged it 'remarkable that the elements most widely diffused through the host of stars are some of those most closely connected with the constitution of the living organisms of our globe'. By 1974 Leslie Orgel and Rolf Lohrmann in a similar vein could point to the

striking fact that biologically important sugars, amino acids, purine bases, and pyrimidine bases can all be obtained in aqueous solution, under mild conditions, from the small family of molecules ... [which] has recently been shown to be abundant in extraterrestrial dust clouds. It is hard to believe that this web of connections is coincidental; it is far more likely that the aminoacids, sugars, and nucleotide bases are important in contemporary biochemistry because they were prominent among the organic compounds that formed on the primitive earth.

13.2 Abiotic condensations and polymerizations

As guidelines for the synthesis of biomolecules under prebiotic conditions, Orgel and Lohrmann (1974) laid down the following four conditions for realistic simulations:

(a) All primary organic reagents must be derivable as significant products from a reducing atmosphere composed of a selection of the following simple gases: CH_4, CO, CO_2, NH_3, N_2, H_2O. Ultraviolet light, heat or electric discharges may be used as sources of the energy involved in the synthesis of primary reagents from these elementary gases.

(b) No solvent other than water may be used. Reactions in aqueous solution must be carried out at moderate pH's, preferably between 7 and 9.

(c) Solid-state reations must occur without excessive drying of the reactants. Solid-state reaction mixtures are preferably obtained by evaporating aqueous solutions that are initially at pH's between 7 and 9.

(d) All reactions must occur under conditions of temperature and pressure that occur on the surface of the earth today. We doubt that volcanoes or thermal springs contribute much to the origins of life, so we prefer to carry out solid-state reactions at temperatures below 80°C...

Syntheses under conditions approximating to those of the protocol are available for most of the biomonomers, but the elimination of the elements of water from the monomers to form nucleosides, nucleotides, nucleic acids, polypeptides, and polysaccharides turned out to be more problematic, particularly using aqueous reaction media to simulate the proto-biochemistry of early replicating systems. The two general methods employed consist in either the solid-state thermal dehydration of single or mixed monomers, usually with catalytic substances, or the coupling of the water-eliminating condensation in aqueous solution to the reaction of a substance with a large negative free energy of hydrolysis, modelled on the coupling of ATP hydrolysis to biomonomer condensations *in vivo*. A third method, restricted to the production of polypeptide-like materials, involves the partial hydrolysis of the high polymer deposited from concentrated ammonium cyanide solutions, or the polymer of the α-amino nitrile intermediate of the

Strecker amino acid synthesis, $RCH(NH_2)CN$, which is a probable constituent of the tars generated in the spark discharge prebiotic experiments.

Polypeptide-like substances, termed 'thermal proteinoids', result from heating mixtures of anhydrous amino acids containing some 50 per cent of the dicarboxylic acids, aspartic and glutamic acid, at 100–200°C for a week, or at lower temperature if phosphoric acid is added to the mixture. The dicarboxylic acids are essential and form a melt in which the other protein amino acids dissolve and react. Glutamic acid dehydrates to the lactam, 2-pyrrolidone-5-carboxylic acid (pyroglutamic acid), which ring-opens to form a dipeptide with a second amino acid molecule. The proteinoids have molecular weights of several thousand, and their hydrolysis back to amino acids is catalysed by proteolytic enzymes. Warm saturated solutions of proteinoids have the curious property that, on cooling, 'microspheres' some $2\,\mu m$ in diameter are deposited with 'proto-cellular' attributes (Fox and Dose 1977).

In the presence of a variety of inorganic salts, the purine bases condense with ribose in the solid state at about 100°C to give a mixture of purine ribosides in 2–20 per cent yield. Magnesium salts favour the formation of the β-nucleosides used by living organisms in yields of up to 8 per cent. It is possibly significant that the salt mixture obtained by evaporating sea water is a particularly effective catalyst. Pyrimidine nucleosides are not formed by an analogous thermal solid-state condensation, but they have been obtained by building up the pyrimidine ring system on the 1-position of ribose from cyanoacetylene, analogous to the synthesis of cytosine itself.

The first-investigated reagents of possible prebiotic significance with a large negative free energy of hydration and the potential of coupling to water-eliminating condensations in aqueous solution were typically reactive nitriles: hydrogen cyanide, cyanamide (H_2N-CN), dicyanamide ($NC-NH-CN$), cyanogen ($NC-CN$), and their dimers or oligomers and derivatives such as isocyanic acid ($HNCO$). These reagents react in aqueous solution with both inorganic and organic acids. Thus cyanamide forms amidine derivatives, e.g. $^+(NH_2)_2C-OPO_3H^-$ from orthophosphates or $^+(NH_2)_2C-O-CO-CHRR'$ from carboxylic acids, which react further with alcohols to form esters, or with amines to give amides, liberating the amidine moiety in hydrated form as urea $(NH_2)_2CO$. The phosphate derivative from cyanogen efficiently phosphorylates aldose sugars in aqueous solution but not nucleosides, since the reaction is specific for the 1-position of the sugar, to which the base is bonded in the nucleoside. The analogous peptide synthesis has a limited scope. The amidine derivatives of the amino acids condense with further amino acid molecules in aqueous media to form mainly dipeptides together with a little trimer and tetramer but no high polymer, owing to the competitive reaction of the cyanamide with the amino group, to give the guanidine derivative of the amino acid $^+(NH_2)_2CNH-CHR-COO^-$.

As in the production of polypeptides from amino acids, thermal solid-state reactions for the conversion of nucleosides to nucleotides presented no major problems. Urea was found to catalyse at moderate temperatures and in good yield the long-known thermal conversion of inorganic phosphates to pyrophosphate and polyphosphates. Nucleosides added to the reaction mixture are converted initially to the nucleoside 5′-phosphate with smaller amounts of the 2′- and 3′-isomer and then, on prolonged heating, to the diesters, mainly the intramolecular nucleoside cyclic 2′,3′-phosphate with some of the intermolecular dinucleoside phosphate. Magnesium phosphates are thermally transformed specifically to pyrophosphates, and the use of $Mg(NH_4)PO_4$ as the phosphate source with the urea catalyst was found to convert nucleosides at moderate temperature mainly to the nucleoside 5′-diphosphate, i.e. the 5′-pyrophosphate, rather than the diesters formed in the absence of magnesium ions. The mechanism of the catalysis by urea, and by analogues containing a primary amide group ($-CONH_2$), remains as yet uncertain, like the mechanism of pyrophosphate formation promoted by the magnesium ion (Orgel and Lohrmann 1974).

The evolution of self-replicating nucleic acids, generally taken to have been the ribose type (RNA) initially, involves two types of polymerization. First, activated nucleotides combined directly, either in the solid state or in solution, to form the primary polynucleotides. Second, the primary strands served as templates directing the synthesis of secondary complementary polynucleotides from activated monomers in aqueous solution, governed by specific base-pair hydrogen bonding (A–U and G–C), to give an ordered double-stranded conformation. The prebiotic chemistry (non-enzymatic) of both polymerizations, particularly the template reactions, have been extensively studied by L. E. Orgel and his associates at the Salk Institute, San Diego (Orgel 1986b).

The thermal solid-state synthesis of nucleotides produces two types of activated monomer; the nucleoside cyclic 2′,3′-phosphate and, if magnesium ions are present, the nucleoside 5′-diphosphates and polyphosphates. Heated alone above 100°C the nucleoside cyclic 2′,3′-phosphates form oligonucleotides. The reaction is catalysed by amines, which promote a transformation from the intramolecular to the intermolecular phosphate diester bonding at a moderate temperature. The natural 3′−5′ phosphate diester internucleoside linkage predominates in the oligomers, which are formed in some 25–68 per cent yield.

Following the limited success of the activated nitrile condensing agents, cyanamide and the like, for the production of biopolymers, attention turned to the molecules involved in the biochemical activation steps leading to polypeptide and polynucleotide synthesis, namely the nucleoside 5′-polyphosphates, represented mainly by ATP and, to a lesser degree, by GTP. In protein synthesis, ATP activates the amino acids to the aminoacyl adenylate intermediates ($^+NH_3$–CHR–CO–AMP). These activated intermediates are reported to condense to polypeptides up to 50 or more residues long, following absorption between the layers of clay minerals, such as montmorillonite.

In the solid state, heating an amino acid at moderate temperature (40–100°C) with the magnesium salt of ATP, or of an adenosine pyrophosphate, affords the phosphoramidate ^-OOC–CHR–NH–AMP, which does not polymerize directly but will form an oligopeptide product indirectly if heated with imidazole, either in solution or in the solid state. The treatment with imidazole rearranges the phosphoramidate to the aminoacyl adenylate and, in addition, forms the aminoacyl) imidazolide ($^+NH_3$–CHR–CO–imidazole). Both of these activated aminoacyl derivatives condense thermally to form oligopeptides (Fig. 13.5). At the same time the imidazolide of AMP is generated (adenosine-5'-phosphorimidazolide), which serves as an activated monomer for the efficient template-directed synthesis of oligoadenylic acid in aqueous solution (Fig. 13.6). This set of reactions

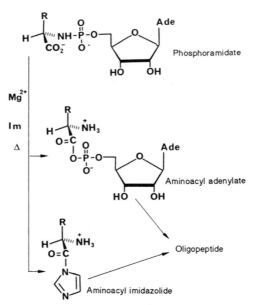

Fig. 13.5 The thermal solid-state conversion of the phosphoramidate, formed by heating an amino acid with ATP or ADP and a magnesium salt, with imidazole (Im) to the corresponding aminoacyl adenylate [aa-AMP] and aminoacyl imidazole [aa-Im]: both [aa-AMP] and [aa-Im] condense in solution to form an oligopeptide from the aminoacyl moiety. Additionally the conversion produces adenosine 5'-phosphoimidazolide (ImpA) which, like the phosphoramidate, condenses in solution to give oligonucleotides.

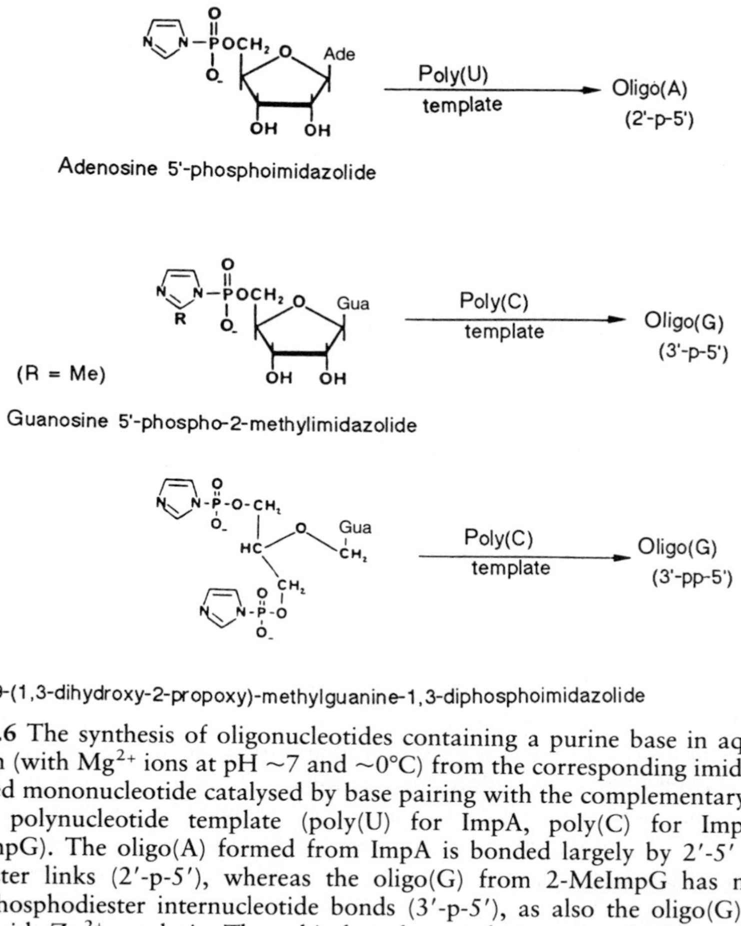

Adenosine 5'-phosphoimidazolide

(R = Me)

Guanosine 5'-phospho-2-methylimidazolide

9-(1,3-dihydroxy-2-propoxy)-methylguanine-1,3-diphosphoimidazolide

Fig. 13.6 The synthesis of oligonucleotides containing a purine base in aqueous solution (with Mg^{2+} ions at pH ~7 and ~0°C) from the corresponding imidazole-activated mononucleotide catalysed by base pairing with the complementary pyrimidine polynucleotide template (poly(U) for ImpA, poly(C) for ImpG or 2-MeImpG). The oligo(A) formed from ImpA is bonded largely by 2'-5' phosphodiester links (2'-p-5'), whereas the oligo(G) from 2-MeImpG has mainly 3'-5' phosphodiester internucleotide bonds (3'-p-5'), as also the oligo(G) from ImpG with Zn^{2+} catalysis. The achiral analogue of guanosine doubly activated with phosphoimidazole groups at the 3'- and the 5'-positions, schematically formed by removing the 2'-CHOH group from the ribose ring, condenses under catalysis with a poly(C) template to give an oligo(G) with pyrophosphate inter-nucleotide bonds between the positions corresponding to 3'- and 5'- in guanosine (3'-pp-5').

models and links together the prebiotic production of oligopeptides and oligonucleotides (Lohrmann and Orgel 1973).

Organized helical conformations are adopted in solution at ambient or low temperature by the monomeric purine nucleosides or nucleotides complexed by base-pair hydrogen bonding with the complementary pyrimidine polynucleotide. This observation provided a basis for studies of the non-enzymatic template-directed synthesis of oligonucleotides. Thus

polyuridylic acid, poly(U), forms a triple-stranded helical secondary structure with adenosine or its 5'-monophosphate (AMP) in which two poly(U)
strands hydrogen-bond to a single stacked column of the adenine-containing
monomers. Guanine derivatives form similar triple helices with two strands
of polycytidylic acid, poly(C), under most conditions, although an excess
of guanosine and a single strand of poly(C) gives a double helix at pH 8.
The pyrimidine nucleoside or nucleotide monomers do not form organized
secondary structures with the complementary purine polynucleotide
strands, owing to the limited stacking propensity of pyrimidine derivatives.
The attractive intermolecular forces between the monocyclic pyrimidines
are substantially smaller than the corresponding forces between the bicyclic
purines.

The triple helices formed suggested that the stacking of preactivated
monomers should promote the self-condensation of the latter to form a
complementary polynucleotides strand, but the biochemically activated
monomer, ATP, failed to condense on a poly(U) template and slowly
hydrolysed to ADP and AMP. In contrast the imidazolide of adenosine
5'-phosphate (ImpA) with a poly(U) template and a magnesium ion
catalyst slowly oligomerized at 0°C and neutral pH over a period of a
week. Oligomers up to the hexanucleotide were characterized in the product, although the phosphate diester linkage turned out to be almost
exclusively 2'–5', rather than the natural 3'–5' internucleotide bonding.
The internucleotide 2'–5' linkage remained unaffected by metal ion catalysis (Pb^{2+}, Zn^{2+}) of the reaction of ImpA on a poly(U) template, but in the
analogous case of the oligomerization of guanosine 5'-phosphorimidazolide
(ImpG) on a poly(C) template the phosphate diester bonding was changed
largely to 3'–5' specifically by the Zn^{2+} catalyst. The formation of the
natural 3'–5' phosphate diester internucleotide linkage was similarly
promoted by changing the activating group to 2-methylimidazole from the
parent nucleus (Fig. 13.6). Additionally the new activated monomer,
guanosine 5'-phospho-2-methylimidazolide (2-MeImpG), extended the degree of oligomerization on a poly(C) template. Products with a chain length
in excess of 30 nucleotides were detected in oligomer mixtures with a mean
length of about 15 residues.

The oligomerizations are remarkably selective. The products from the
reaction of an equimolecular mixture of ImpA and ImpG on a poly(C)
template were found to contain only one adenosine for every 200 guanosine
residues incorporated into the oligomer. The fidelity of this non-enzymatic
replication is only an order of magnitude less than that of the RNA
polymerases (1 in about 4000). The discrimination extends to the incorporation of the pyrimidine nucleotides into the oligomers. Neither ImpU nor
ImpC polymerize on the corresponding complementary polynucleotide
template, poly(A) or poly(G) respectively, since no initial helical complex
with the template forms. The use of mixed copolymers as templates allows

the significant incorporation of a pyrimidine nucleotide into the oligomer product if, and only if, the copolymer template includes the complementary purine nucleotide.

Mixed polynucleotide templates containing cytidine as the major component select from an equimolecular solution of activated mononucleotides, 2-MeImpN, where N is adenine, guanine, uracil, or cytosine, only the complementary monomers for oligomerization, with a discrimination factor lying between 100 and 500, dependent upon the particular conditions. The poly(C,U) template afforded principally oligo(G,A) products, while poly(C,A) gave mainly oligo(G,U), and the complementary oligo(G,C) strands resulted from the poly(C,G) template. The three-component template poly(C,U,A) selected the complementary three out of the four activated monomers, giving oligo(G,A,U), and all four monomers were incorporated with the four-component template poly(C,G,U,A) (Inoue and Orgel 1983).

The template-directed oligomerizations exhibited additionally a marked stereoselective chiral discrimination. The enantiomeric adenosine molecules A_D, composed of the natural D-ribose, and A_L, containing the synthetic mirror-image sugar L-ribose, form triple helices of comparable stability with poly-D-uridylic acid, poly (U_D), made up of D-ribose. The activated monomer from D-adenylic acid, $ImpA_D$, reacts much more rapidly with D-adenosine than with L-adenosine in the triple helix with poly(U_D), so much so that racemic DL-adenosine is optically resolved by the oligomerization reaction. The enantiomeric stereoisomerism has the consequence that the reactive groups for oligomerization, appropriately aligned for the reaction of $ImpA_D$ with A_D, have the incorrect relative orientation in the corresponding reaction with A_L, and the A_L remains uncombined after the oligomerization of DL-adenosine, with only the D-isomer undergoing template-directed combination.

The oligomerization reactions of activated D- and L-guanosine mononucleotides on a poly-D-cytidylic acid (poly(C_D)) template had an even more striking outcome. The natural D-ribose activated monomer, D-guanosine 5′-phospho-2-methylimidazole, D-2-MeImpG, with the poly(C_D) template reacts to give all 3′–5′ phosphate diester-linked oligomers with lengths up to more than 20 residues. The oligomerization is not affected by the addition of other activated D-nucleotide monomers, from D-adenosine, D-cytidine, or D-uridine, but the chain elongation is specifically inhibited by the enantiomeric L-guanosine derivative, L-2-MeImpG. In the template-directed poly(C_D) oligomerization of the racemic monomer, DL-2-MeImpG, the L-enantiomer serves as a selective chain terminator to the elongation of oligo-D-guanylic acid in the normal 5′→3′ direction by forming a 2′–5′ or 3′–5′ phosphate diester linkage to L-guanosine at the growing end of the oligomer. At the same time the 5′-end of the oligomer is capped by 5′-pyrophosphate bonding to L-guanosine, giving a mixture of isomers for each n-mer set in the limited range of oligomers formed. By itself the L-2-MeImpG monomer with a poly(C_D) template has only a limited condensation capacity, forming a pentamer at most, with much 5'-pyrophosphate and 2'–5' linked dimer and trimer (Joyce *et al.* 1984).

The normal formose condensation of formaldehyde affords only a mixture of racemic sugars, including DL-ribose. The specific inhibition of a template-directed D-ribonucleotide oligomerization by the L-enantiomer suggests that the original prebiotic structural support system for protogenetic replication by specific A–U and G–C base pairing was not the D-ribose-phosphate diester chain, nor any other chiral backbone.

The removal of the ribose 2'-carbon atom of a nucleoside, and the replacement of its bonds to the 1'- and the 3'-carbons by hydrogen atoms, gives an achiral quasi-nucleoside which could be formed, in principle, by the condensation of glycerol with formaldehyde followed by the addition of a purine or pyrimidine base (Fig. 13.6). The free hydroxyl groups of the glycerol moiety in the achiral analogue, corresponding to the 3'- and the 5'-hydroxy groups of ribose in a nucleoside, can be phosphorylated and activated with imidazole. The activated diphosphate of the glycerol analogue of guanosine condenses on a poly(C) template to form analogues of oligo-guanylic acid with a backbone of achiral repeat units of glycerol-1,3-pyrophosphate diester (Schwartz and Orgel 1985).

The poly(C) template itself is a complex chiral macromolecule, and the ancestral genetic template was probably achiral and less selective. Possibly replication by specific A–U and C–G base pairing developed on the surface of phosphate mineral microcrystals suspended in the primordial seas. These, or other suspended mineral particles, may have served as the structural support system for the purine and pyrimidine bases before the sugar–phosphate backbone of RNA was evolved (Orgel 1986*a*).

13.3 The RNA and other worlds

The central dogma of molecular biology, laid down by Francis Crick in 1958, specified a unidirectional and irreversible flow of chemical instruction from DNA to RNA and from RNA to protein: 'The transfer of information from nucleic acid to nucleic acid, or from nucleic acid to protein, may be possible, but transfer from protein to protein, or from protein to nucleic acid is impossible.' The new set of explicit assumptions replaced the old and often implicit preconception, dominant over the previous century, that proteins are the primary agents of metabolism, biosynthesis, and reproduction. It was now asserted that 'once information has passed into a protein it cannot get out again'. Darwin had envisaged a protein specifically as the primary prebiotic product of a 'warm little pond'. The biochemical reformation now implied an initial world of self-replicating nucleic acids, possibly sustained by polypeptides and carbohydrates but uninstructed by them.

The counter-reformation against the central dogma began with the discovery of the retroviruses, containing RNA as the genetic molecule and the enzyme, reverse transcriptase, which catalyses the production of a complementary secondary DNA strand from the RNA genetic template (Temin and Mizutani 1970). Clearly information could be transferred from RNA to DNA in the reverse direction to the observed general trend, which can be nothing more than an evolved relative propensity. Less clearly, the discovery of the *prions* (proteinaceous infective particles) questioned the forbidden transfer of instructions from proteins to the nucleic acids. The prions, responsible for the degenerative disorder of the central nervous system in sheep (scrapie) and in other animals, appear to be wholly protein, deactivated by proteolytic enzymes but not by either ribonuclease or de-oxyribonuclease enzymes (Prusiner 1982). Similarly the second stage in the charging of a tRNA with its specific amino acid involves a degree of in-struction of the tRNA by a protein, the tRNA synthetase enzyme (Fig. 12.10).

Despite the central dogma, the traditional prebiotic protein worlds retain their place, now as possible precursors of the more elaborate nucleic acid epoch. Virtually all protein models rest on the generally accepted inference, from studies of carbonaceous meteorite compositions and prebiotic re-action simulations, that amino acids and polypeptides were more readily available and abundant on the early Earth than the nucleotides and their polymers.

The thermal polymerization of the amino acids under geologically feasible conditions is facile and the proteinoids formed are not wholly random copolymers. The degree of order found in the thermal proteinoids expresses an informational content demonstrated by their catalytic activity and the selective recognition of nucleotides. The microspheres formed by the thermal proteinoids in aqueous solution have cellular attributes: they reproduce by budding and exhibit enzyme-like properties. The nucleotides first served the microspheres in a bioenergetic role, namely the photophosphorylation of AMP to ATP through photoactivation by a pigment found in all thermal proteinoids. Initially the informational recognition between the polypep-tides and the polynucleotides was mutual and reciprocal, but the nucleic acids took over the assembly and instruction became unidirectional with the development of the proto-ribosome (Fox and Dose 1977).

A more abstract model of a precursor protein world derived from an analysis of the possible survival and development by random genetic drift of a dynamic network of interdependent polymers and their monomers in an isolated assembly (a proto-cell), during an archaebiotic period before the proliferation of the proto-cells resulted in Darwinian natural selection between them (Dyson 1982). The analysis, based upon the formalism of population statistics applied to molecular systems, considered the conditions for a viable ordered turnover of monomers among the self-sustaining polymers in the assembly, where the macromolecules catalyse the replication of each other with a degree of random error. The ordered viability of such an assembly requires no great fidelity of replication (an error rate of one part in a hundred turned out to be acceptable), but at least nine distinct species of monomer are necessary for the preservation of dynamic order against the pressures of entropic disorder. The four nucleotide monomers alone are inadequate for an ordered turnover in the cell-like assembly, but the 20 species of amino acid, or even only 10 of these, readily serve the purpose and can be accommodated within the model. Thus the polymers of an isolated proto-cell were protein enzymes, possibly some 100 polypeptide catalysts of moderate specificity, in an assembly of some 2000 monomers. The nucleic acids came later, initially as energy carriers, such as ATP. Subsequently a protein enzyme evolved polymerase activity, producing RNA strands from ATP and its analogues. The role of RNA in the cell, initially parasitic, became symbiotic (with tRNA, rRNA, mRNA) and ultimately genetic. Cells came before the enzymes, and enzymes before the genes (Dyson 1982, 1985).

Critics of the protein world doubt the self-organizing capacity of a set of near-random polypeptides produced abiotically. While modern enzymes catalyse the formation and the cleavage of specific peptide bonds, there is as yet no evidence that abiotic polypeptides can do so with any significant degree of sequence specificity. A network of mutually interacting polypeptides in a proto-cell would require a distinctive coordination agency and, more particularly, a genetic mechanism of replication. In the period before the appearance of the proto-cell, the requirements for a separate coordination agency and a self-replicating repository of genetic information could be met by the layered clay minerals (Cairns-Smith 1982).

The process of crystal nucleation and growth provided an early model for the mechanism of gene replication (e.g. Haldane 1929). While the perfect crystal embodies little information, no finite crystal can be perfect and most real crystals contain structural features ('defects') which, serving as origins for crystal growth, are replicated. Crystal cleavage provides new nuclei containing the information-bearing defects, and these microcrystalline seeds serve as the starting point for another generation of crystal growth. Most clay minerals are made up of stacks of layers, the surfaces of which catalyse a variety of reactions, including the polymerization of activated amino acids to polypeptides. The catalysis has a degree of specificity, dependent upon the distribution of charges over the surfaces of the layers. The absorbed organic molecules in turn modify the clay crystal growth by facilitating the preferential incorporation of particular inorganic components in the crystal lattice, and by promoting the growth of some crystal faces at the expense of others.

The more efficient symbiotic organizations of a clay crystal genotype and its organic phenotype were picked out by natural selection, which promoted the particular elaboration of the organic component, with its vast structural and functional potential. Initially the clay mineral component served as a sole repository of the genetic information carried by the symbiotic combination but, with the incorporation of an RNA-like polymer into the organization, the organic moiety began first to read, then to copy, and ultimately to elaborate the genetic information of the partnership for its own independent replication. The organic section of the assembly had now become chemically sophisticated and ready for autonomous propagation. The clay mineral 'scaffolding' of the organic edifice could be disposed of, like the ladder of Wittgenstein which, once climbed, may be thrown away. Such a precellular origin of membrane-enclosed organisms was unique and could never be repeated. Like Darwin's warm-pond protein, the organic moiety of any subsequent clay–organic symbiosis would be 'instantly devoured' by the organisms generated initially (Cairns-Smith 1982, 1985). As to the clay mineral component of the original symbiosis, the lower strata of the vast deposits of China clay possibly represent the graveyards of our utterly remote ancestors.

Both the clay mineral and the protein worlds lead to a nucleic acid epoch where RNA-like polymers metabolize and replicate, possibly bound to the surface of phosphate mineral microcrystals in suspension, or possibly enclosed already in bounded proto-cellular assemblies. The specialized limitation of DNA to self-duplication and transcriptive instruction makes the 'master molecule' a poor candidate for the primordial self-replicating entity.

Further, the biosynthetic evidence suggests that DNA was a later development from the more versatile RNA molecule. In contemporary organisms the deoxyribose sugar of the DNA nucleotides and the thymine base are not directly synthesized. The deoxyribonucleotides are derived enzymatically by replacing the $2'$-hydroxy group in the ribose of the corresponding RNA nucleotide by a hydrogen atom. Similarly the thymine base unit of DNA is derived from the uracil base of RNA by 5-methylation. Each of these steps indicates, on the general principle that biosynthetic pathways retain traces of their development, that RNA was prior to DNA in biochemical evolution. Other indications are the requirement of an RNA primer for DNA replication, and the pervasive role of RNA in the transfer of information and the realization of specific coded instructions (Joyce 1989).

The biochemical versatility of RNA and the ribonucleotides makes plausible a stage in organic evolution where nucleic acids replicated directly on RNA templates from activated monomers, diversified into tRNA, mRNA, and rRNA functionality and 'invented' ribosomal protein synthesis and the genetic code. Uncoded polypeptides may have been involved in the earliest replication mechanism but, as the laboratory template-directed oligomerizations indicate such polypeptides were not necessarily essential (Orgel 1986b). The primary mediator of biochemical free-energy changes, ATP, belongs to the set of ribonucleotides, and so do most of the common coenzymes, e.g. nicotinamide adenine dinucleotide and its phosphate (NAD^+, $NADPH^+$), flavin adenine dinucleotide (FAD), coenzyme A, coenzyme B_{12}, etc.

The originator of the concept of the 'high-energy phosphate bond' perceived the origin and evolution of ATP as the salient development in proto-biotic chemistry (Lipmann 1971). A supply of free energy in the form of ATP and its analogues (GTP, UTP, CTP) was a prerequisite to the storage, transport, replication, and transcription of information embodied in their copolymers, the RNA molecules, since information is a representation of negative entropy or positive free energy. The discovery in 1982 of ribozymes (RNA molecules with enzyme-like catalytic activities) strongly supported the conjecture that organic evolution had passed through a stage based upon RNA alone, combining information and function in a single molecular species. Even before the discovery of ribozymes, the coenzymes often had been regarded as molecular fossils, dating back to a period of RNA catalysis prior to the appearance of the more versatile protein enzymes.

The initially produced ribosomal RNA of the protozoan *Tetrahymena* is found to process itself, following initiation with guanosine or a derivative, such as GTP, into mature rRNA form by the removal of a superfluous intervening sequence of 413 nucleotides through the sequence-specific hydrolysis and transesterification of phosphate diester bonds. The excised intervening sequence in turn has self-splicing properties, forming circular polynucleotides of initially 399 and then 395 residues by site-specific transesterification reactions that eliminate first a 15 nucleotide segment and then a tetramer. In addition the intervening sequence catalyses the splicing reactions of other RNA molecules. After ring-opening, the polynucleotide of 395 residues promotes the dismutation, through transesterification, of pentacytidylic acid into shorter and longer oligomers, conserving the total number of phosphodiester bonds (Cech 1986a).

Subsequently a number of other newly transcribed RNA strands were shown to self-splice into mature form by sequence-specific transesterification reactions, or to undergo maturation through reactions catalysed specifically by the RNA moiety of ribonucleoprotein enzymes. The ribosome is the largest of the ribonucleoprotein enzymes, and the discovery of ribozymes opens up the perspective that coded protein synthesis through specific ester-to-peptide bond conversion was a development from the sequence-specific transesterification reactions of catalytic RNA strands (Westheimer 1986).

Studies of the evolution of RNA molecules under conditions of laboratory selection became possible in 1965 when Sol Spiegelman, then at the University of Illinois, characterized the reproductive mechanism of the

RNA bacteriophage Qβ, which infects *E. coli* and other bacteria. The protein enzyme Qβ-replicase is virtually specific for the template replication of the genetic Qβ-RNA of the bacteriophage, a molecule of about 4500 ribonucleotides. The enzyme does not replicate the RNA molecules of the host bacterium, although it does produce a side product—a small satellite RNA molecule, a 'minivariant' some 220 nucleotides long, under natural conditions. In the laboratory the purified Qβ-replicase, supplied with the nucleotide triphosphates ATP, GTP, UTP, and CPT, replicates infectious daughter Qβ-RNA from a parent Qβ-RNA template, directing the production of the normal virus in the bacterial host.

The infectivity is progressively lost over a succession of generations of the replicated RNA, produced by the serial transfer of a sample of the RNA product from a previous batch to a new batch of purified Qβ-replicase and the four nucleotide triphosphates. The artificial selection of laboratory serial transfer promotes the evolution of variant RNA strands adapted to fast replication. These variants are smaller molecules than the infectious Qβ-RNA, typically some 500 nucleotides long ('Spiegelman's monsters'). The variant RNA molecules are adaptable to changes in the conditions of growth. The addition of substances inhibiting the replication of Qβ-RNA, such as the ATP analogue tuberdicin, or ethidium bromide, which is intimately bound by nucleic acids, initially depresses the replication rate of the variant RNA molecules, but they recover by evolving into drug-resistant template-replicating RNA species. Fast evolution is allowed by the relatively high error rate of RNA replication, about one part in 10^4 for Qβ-polymerase under physiological conditions (Orgel 1979).

Without a template, the purified Qβ-replicase in a medium containing the four ribonucleoside triphosphates spontaneously forms RNA molecules after an induction period of variable duration, dependent upon the particular experimental conditions (Biebricher *et al.* 1981). Although produced without a template, the RNA strands can serve as templates for replication. The extent of the induction period is highly sensitive to the concentration both of the enzyme and of the nucleoside triphosphates, being considerably longer for the more dilute solutions. The sensitivity suggests that monomeric nucleotides, bound to one enzyme molecule in a particular sequence, function as a quasi-template for the polymerase reaction catalysed by another enzyme molecule. The RNA strands produced spontaneously without a template by Qβ-replicase turned out to have a uniform composition closely similar to that of the minivariant RNA side product formed by the Qβ-virus in a bacterial host cell. The similarities extended to the repetitive features of the nucleotide sequences of the RNA molecules from the two sources: the tetramers UUCG and CCCC, together with their complements, AAGC and GGGG; and the trimer CCC with its complement GGG.

The trimer CCC and the tetramer UUCG are sequences required for recognition in all RNA molecules that interact specifically with the

Qβ-replicase enzyme. The selection of these sequences and their complements by Qβ-replicase in the spontaneous production of RNA molecules represents the transmission of a set of instructions from a protein enzyme to a nucleic acid during the course of the latter's elaboration. This mode of instruction is a clear counter-example to the tenet of the central dogma that information cannot flow from a protein to a nucleic acid and that, once inside a protein, information cannot escape (Eigen *et al.* 1981).

The character of the RNA molecules produced spontaneously by Qβ-replicase has a bearing on the possible nature of the nucleic acid species and their mode of replication in the early RNA world. Serial transfer of these RNA molecules through many generations of replication under constant conditions establishes an optimal polyribonucleotide product with chain lengths between 150 and 250 monomers. The composition, structure, and properties of the optimal RNA strand, uniform for a given set of experimental conditions, can be altered by an environmental change. Serial transfer experiments conducted at a high salt concentration lead to an optimal polyribonucleotide adapted to replication in a saline environment. The addition to the serial growth medium of ribonuclease, an enzyme that cleaves RNA, leads to the evolution of an RNA molecule with hairpin folds that protect the sensitive sites from enzymatic cleavage.

The nucleotide sequences of the optimal RNA molecules produced under constant environmental conditions are found to be variants, due to nucleotide base changes, of a single 'master sequence' which remains unchanged from one replication generation to another, provided the growth conditions remain unaltered. The population of the master sequence and its set of variants, termed a 'quasi-species' by Manfred Eigen and Peter Schuster (1978), can remain stable over successive generations only if the 'mutation' or 'error' rate is less than one per molecular replication. At a higher mutation rate the information embodied in the optimal master sequence is lost and the population becomes a collection of polynucleotides with random sequences. If the mutation rate goes to zero, the master sequence becomes the dominant or sole type of RNA molecule in the population, but the species cannot then adapt by natural selection to an altered environment.

The mutation rate for informational RNA duplication at an early stage, during the transition from enzyme-free to enzymatic template-directed replication, is expected to lie at the level of approximately one base error for every 100 to 500 ribonucleotides polymerized, from the data for the oligomerization of activated mononucleotides on a cytosine-rich template (Inoue and Orgel 1983). The error rate limits the early informational RNA molecules with replication stability, a quasi-species, to some 100 base pairs, which could code for a polypeptide of only limited catalytic capacity, consisting of 30 or so amino acid monomers. The transition to the next stage, represented by the modern RNA bacteriophage Qβ, with a genome

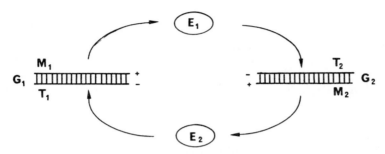

Fig. 13.7 The hypercycle organization of the early RNA world from a quasi-species with a genome G, consisting of an amino acid 'adapter' RNA $(-)$-strand, T, a precursor of tRNA, base paired to a complementary RNA messenger $(+)$-strand, M, a proto-mRNA, which codes for a polypeptide enzyme, E, catalysing the replication of the M and the T strands of G. Mutations produce two symbiotic subspecies, G_1 and G_2, such that M_1 codes for a modified enzyme E_1 that replicates more efficiently the two strands, T_2 and M_2, of the second genome G_2. In turn M_2 codes for a modified enzyme E_2 which efficiently catalyses the replication of the complementary T_1 and M_1 RNA strands of the first genome G_1. With no change in the length of the RNA strands, the information content of the simple hypercycle of G_1 and G_2 becomes enlarged to double that of the ancestral genome G with little change in the vulnerability to deleterious mutations. The information content increases up to n-fold with the addition of further symbiotic subspecies, up to G_n, where the complexity of the cooperative hypercycle organization requires compartmentalization into a membrane-bounded cell to preserve integrity. At the compartmentalization stage, the external competition between cells enhances, through natural selection, the internal cooperative biochemical efficiency of the cell.

of a few thousand base pairs and a replication error rate of one part in about 10^4, could have been accomplished, according to Eigen and Schuster (1978), by a cooperative assembly of interdependent quasi-species, termed a 'hypercycle'.

The simplest hypercycle consists of two double-stranded RNA quasi-species living symbiotically in a common medium, and so probably resembling one another as descendants of the same ancestral quasi-species adapted to the medium (Fig. 13.7). The polyribonucleotide genome of a quasi-species consists of a T-strand, a precursor of tRNA, which couples with specific amino acids, and a M-strand, a precursor of mRNA, which codes for a protein enzyme E. The T- and the M-strands are complementary in their nucleotide sequences, so that each one serves as a template for the replication of the other. Two quasi-species in a simple hypercycle become symbiotic when the protein enzyme E_1, coded and adapted by the RNA double-strand M_1T_1 of the first species, serves as the catalytic replicase for the duplication of the RNA genome M_2T_2 of the second species. In turn, the double-strand M_2T_2 of the second species codes and adapts the protein enzyme E_2 catalysing the mutual template replication of the complementary strands M_1T_1 of the first species.

Competition between the two quasi-species would lead to an evolutionary dead-end, with the informational content of the surviving RNA genome limited to the primitive level of 100 base pairs or so. The symbiosis of the hypercycle combines the instructional capacities of the two quasi-species, and the addition of further cooperative quasi-species to the assembly increases linearly the available information of the hypercycle, independent of the internal mutations leading to better (or to worse) catalytic replication. At some stage, probably an early one, the hypercycle assembly must become compartmentalized into a membrane-bound proto-cell to preserve the integrity of its increasingly complex chemical ecosystem. The division of the proto-cell, by budding or fission, at an optimum size determined by the growth rate of the enclosed hypercycle, produces a succession of individual organisms. Natural selection now acts positively on the individual organisms, favouring the mutations which increase the efficiency of the biochemical symbiosis within the enclosed cooperative hypercycle assembly of the bounded cell.

An RNA genome of some thousands of base pairs with an error rate in replication of one part in about 10^4 becomes feasible at this juncture. The next step, accomplished by the prokaryotes, brought natural selection to bear on the functional efficiency of the genotype, as well as that of the phenotype. The step consisted in the specialization of informational nucleic acid into DNA and the enzymatic recognition in replication of a distinction between the parent template strand and the newly forming daughter strand. The recognition allowed for proof-reading and the correction of mismatched bases in the new strand. The consequent reduction in the error rate of replication to one part of about 10^7 permitted genome sizes of several million base pairs, increased by two or more orders of magnitude by recombinative reproduction in the eukaryotes (Eigen 1981).

In the early days of the RNA world, when genome size was restricted to some 100 base pairs, the genetic code for protein synthesis could not have had its present-day complexity, with each of the 64 possible triple combinations of the four RNA nucleotides corresponding specifically to one of the 20 protein amino acids, or to a *start* or *stop* signal for protein synthesis. The early polyribonucleotides were probably rich in cytosine and guanine, since the error rate of the non-enzymatic polymerization of activated guanylic acid (ImpG) on a poly(C) template (some 1 per 100 bases) is lower by an order of magnitude than that of activated adenylic acid (ImpA) on a poly(U) template (1 in about 10). The greater fidelity of the guanylic acid polymerization derives from the stronger GC base pairing, the GC pair being maintained by three hydrogen bonds but the AU pair by only two (Fig. 12.4).

Modern tRNA molecules, with chain lengths of 73–93 nucleotides (the magnitude surmised for the early informational RNA strands), have compositions much richer in G and C than in A and U, together with a number of 'anomalous' bases which may have survived from a more diverse 'non-standard' primitive origin. The nucleotide sequences of tRNA molecules are highly conserved, and analyses of those sequences suggest that they are all variants of master sequences, like the individual RNA molecules in the population of a quasi-species. The derived master sequences show a repeating pattern of nucleotide base triplets of the form RNY, where R stands for a purine (A or G), Y for a pyrimidine (U or C), and N for any of the nucleotide bases (A, G, U, or C). The triplet RNY pattern and the higher frequency of (G + C) over (A + U) found in the tRNA molecules extend to the ribosomal 5S rRNA sequences (~120 nucleotides long). These features of non-coding RNA are more prominent among the prokaryote than the later-evolved eukaryote species (Winkler-Oswatitsch *et al.* 1986). A similar triplet repeat pattern RNY in coding DNA sequences emerged from analyses of the nucleotide frequencies as a function of sequential position in viral, bacterial, and eukaryote DNA (Shepherd 1986).

From the probable dominance of guanine and cytosine in the early polyribonucleotides, and the prominence of the RNY base triplet in the tRNA and the 5S rRNA molecules, it is conjectured that the earliest messenger RNA was limited to GNC codons, developing subsequently through the RNY set to the RNN group and, finally, to the modern 64 triplet NNN code. The four representations of the GNC group are glycine (GGC), alanine (GCC), aspartic acid (GAC), and valine (GUC). Remarkably, these four amino acids are the most abundant of the protein amino acids found in the organic material of the Murchison carbonaceous meteorite and also in the products generated from mixtures of reducing gases by electric discharges (Eigen *et al.* 1981).

14

Biomolecular handedness

14.1 Molecular dissymmetry

In 1860 Louis Pasteur reviewed in lectures to the Chemical Society of Paris his studies of the relationship between molecular shape and optical activity, the clockwise or counter-clockwise rotation of the plane of polarized light by fluid or dissolved substances, and the consequences of his conclusion that molecules of biological origin are uniquely handed in form. Before the development of testable theories of internal molecular structure, the main guiding principle to the overall architecture of molecules had been the dictum, postulated in 1809 from crystal cleavage observation by René Haüy (1743–1822), that a crystal and each constituent space-filling molecule must be morphologically 'images of each other'. Haüy's principle, and studies guided by it, enabled Pasteur to surmise in 1848 that a substance forming two enantiomorphous sets of crystals, related as an object to its non-superposable mirror image like the right and the left hand, must be composed of two similarly shape-related molecules (enantiomers), which separated out on crystallization into the one or the other crystal set.

The extension of Haüy's morphological analogy to the chemical and physical properties of compounds led Eilhard Mitscherlich (1794–1863) to his law of isomorphism (1819), which prescribed an equivalent composition to different substances with the same crystal shape, and to the connection made in 1822 by John Herschel (1792–1871) between the morphological handedness of quartz crystals and the sign of the crystal optical activity. Natural quartz crystals divide into two enantiomorphous sets, distinguished by the right- or the left-handed screw sequence of hemihedral facets which reduce the crystal symmetry from hexagonal to trigonal (Fig. 14.1). Herschel found that all crystals of the left-handed morphological set are laevorotatory, while those of the right-handed set are dextrorotatory, producing a respective anticlockwise or clockwise rotation of the plane of polarized light propagated along the trigonal crystal axis.

Herschel supposed that the particular morphological handedness of a quartz crystal and the sign of its optical activity had a common molecular basis, the supposition being developed and generalized by Augustin Fresnel (1788–1827) in 1824. Fresnel, developing the then new transverse wave theory of light, showed that optically active substances have a circular birefringence, i.e. unequal refractive indices for left-circular (n_L) and right-circular (n_R) polarized radiation, the difference $n_L - n_R$ being positive for dextrorotatory and negative for laevorotatory media. Arguing by analogy,

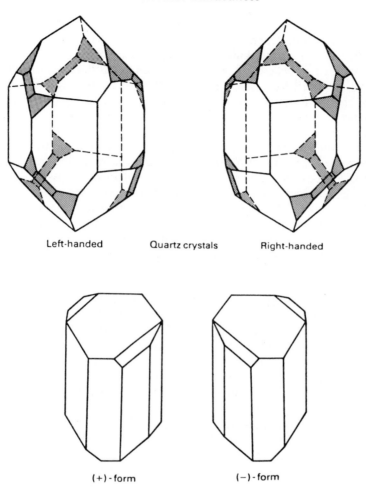

Fig. 14.1 The enantiomorphous crystal sets of quartz and of sodium ammonium tartrate. The minor crystal facets of a quartz crystal follow either a left- or a right-handed screw sequence viewed on end, along the direction of the three-fold crystal axis. A section of the quartz crystal cut perpendicular to the trigonal axis is laevorotatory for plane polarized light propagated along the crystal axis if the crystal is morphologically left-handed, or dextrorotatory for the right-handed morphological form. Racemic sodium ammonium tartrate crystallized below 27°C forms two sets of crystals, similarly distinguished by their non-superposable mirror-image morphology and their oppositely signed optical rotation.

Fresnel proposed that the molecules of an optically active medium have a left- or a right-handed helical form, like the envelope traced out by the oscillating vector of a circularly polarized light wave.

The studies of Herschel and Fresnel enabled Pasteur to resolve the paradox reported by Mitscherlich in 1844, that the sodium ammonium salts of the optically active (+)-tartaric acid and the inactive racemic or paratartaric acid appear to be isomorphous and identical in all chemical and physical respects save optical activity. Both acids are natural products, isolated from the tartars deposited by maturing wines; the active (+)-tartrate forms the major product while the racemic acid (Latin, *racemus*, a cluster of grapes) is a side product or an artefact of the work-up process. Repeating the crystallization of the inactive racemic salt, Pasteur obtained in 1848 two sets of hemihedral crystals, related as mirror-image enantiomorphs, like the two sets of quartz crystals (Fig. 14.1). One set proved to be wholly isomorphous with the crystals of the corresponding natural (+)-tartrate and, moreover, gave an identical specific optical rotation in solution, positive in sign. In addition to the enantiomorphous crystal facets, the other set exhibited a specific optical rotation of the same magnitude, but negative in sign. Pasteur inferred that the individual molecules of (+)- and (−)-tartaric acid are morphologically dissymmetric, related as non-superposable mirror-image forms like the macroscopic crystals of the corresponding sodium ammonium salts, and that the inactivity of the racemic acid arose from the mutual cancellation of the opposed optical activity of the two enantiomeric molecules.

Subsequently, in 1853, Pasteur discovered diastereomeric substances, which contain two or more inequivalent chiral subunits and lack the non-superposable mirror-image relation overall. The finding that 'the absolute identity of the physical and chemical properties of left and right non-superposable substances ceases to exist in the presence of another active substance' gave him a second method of resolving the individual enantiomers from a racemic mixture. An optically active alkaloid base, quinine or cinchonine, with (+)- and (−)-tartaric acid gave salts differing in solubility and other properties, allowing a ready separation of the two optical isomers from racemic tartaric acid.

His third method of optical resolution, demonstrating chiral discrimination between enantiomeric substrates by living organisms, had particular significance for Pasteur. In 1858 he discovered that a mould of the *Penicillium* family, grown on racemic ammonium tartrate with added phosphate, preferentially used the (+)-tartrate isomer as a carbon source, leaving the (−)-enantiomer, which Pasteur isolated as the ammonium salt. Pasteur emphasized in his 1860 lectures that all organic optically active substances then known originated from the biosynthetic activity of living organisms and that the laboratory syntheses of the time afforded only inactive compounds or mixtures. In consequence living systems must have access to some universal chiral agency, and 'dissymmetric forces exist at the moment of the elaboration of natural organic products; forces which are absent or ineffectual in the reactions of our laboratories'. These natural chiral forces might be magnetic or electrical in character, or due to irradiation with light in combination with rotatory motion.

More than 20 years later, in his autobiographical address to the Chemical Society of Paris in 1883, Pasteur outlined some investigations on the chiral forces of nature he had carried out in his youth at Strasbourg (1849–54) and Lille (1854–7) before returning to the Paris École Normale Supérieure, where he had studied in the mid-1840s, as director of scientific studies. The discovery of magnetically induced optical activity in glass and other isotropic media by Michael Faraday in 1846 suggested to Pasteur the experiment of growing normally symmetric (holohedral) crystals in a magnetic field with the expectation that dissymmetric (hemihedral) facets would develop. Plants were rotated during growth, or exposed to a clockwork heliostat presenting an appearance of the Sun rising in the west and setting in the east, in order to determine whether natural products of antimeric dissymmetry were formed thereby.

The negative outcome of these experiments had led Pasteur to his 1860 view that optical activity provides an empirical demarcation criterion between the chemistry of the laboratory and the chemistry of life. But the division could not be fundamental. 'Not only have I refrained from posing as absolute the existence of a barrier between the products of the laboratory and those of life,' Pasteur declared, 'but I was the first to prove that it was merely an artificial barrier, and I indicated the general procedure necessary to remove it, by recourse to those forces of dissymmetry never before employed in the laboratory.' The spontaneous separation of the optical isomers of racemic sodium ammonium tartrate on crystallization may well have been due, Pasteur supposed, to the dissymmetric forces at the surface of his crystallization dishes. The solar system as a whole is dissymmetric on account of the particular sense of the orbital and spin rotations of the planets, for it is not superposable on its mirror image. Chiral interactions mediate the transformations of the organic and the inorganic world alike, possibly even the generation of organisms. 'Life is dominated by these dissymmetric actions of whose enveloping and cosmic existence we have some indication.'

Pasteur's speculations on the ubiquity of dissymmetric structures and forces in the universe influenced Le Bel who, like van't Hoff independently, provided a structural interpretation for molecular dissymmetry in 1874, with the postulate that the four valencies of the carbon atom have a tetrahedral orientation. At a meeting of the Chemical Society of France in 1924, celebrating the half-century of organic stereochemistry since its foundation in 1874, Le Bel pointed out that the then new Bohr theory of atomic structure implied that the chemical atom itself is a chiral system.

According to Bohr's theory, the electrons of an atom circulate round the nucleus like the planets round the Sun. In general an assembly of atoms constitutes a racemic mixture, the electronic circulation being clockwise in some cases and counter-clockwise in others. Thus sodium chlorate crystallizes to give equal numbers of dextrorotatory and laevorotatory crystals, as F. S. Kipping and W. J. Pope had shown in 1898.

But, Le Bel conjectured, an assembly of carbon atoms must be enriched in a particular enantiomer, since many native petroleums, including those from the oil wells of the Le Bel family at Pechelbronn, Alsace, are optically active. Following the view of Mendeleev, Le Bel held, like many other chemists of the period, that the native petroleums have an inorganic origin from the action of water on metallic carbides. The atoms of the heavier group IV element, silicon, must be similarly non-racemic, Le Bel supposed, since potassium silicotungstate and the corresponding silicomolybdate appeared to form predominantly dextrorotatory crystals, according to the claim of G. Wyrouboff in 1896 and its apparent confirmation by H. Copaux in 1910.

The continuing influence of Haüy's morphological analogy between the primitive form of a crystal and the shape of the constituent molecular building units led Le Bel for a brief period (1890–4) to the aberrant view that the four valencies of the carbon atom have an orientation of lower symmetry than tetrahedral. Haüy's principle implied that, if the valencies of the carbon atoms are directed to the apices of a regular tetrahedron, a compound containing four identical groups bonded to a carbon atom should form isotropic crystals with cubic symmetry. Crystals of carbon tetrabromide obtained at ambient temperature turned out to be anisotropic, exhibiting birefringence, although they appeared to become cubic in form at a 'certain temperature'. (Carbon tetrabromide undergoes a transition at 47°C to a cubic plastic crystal in which the molecules become effectively spherical, rotating on their lattice sites.)

If the carbon valencies have a square pyramidal orientation the atoms of the ethylene molecule could not share a common plane, and appropriately substituted derivatives should be resolvable into optical isomers. Le Bel attempted an optical resolution of citraconic and mesaconic acid by Pasteur's third method, i.e. the growth of microbial cultures using preferentially one of the enantiomers in a racemic mixture as a carbon source. Optically active substances were isolated from the cultures, but the products were not the residual unsaturated acid substrates. From the culture containing citraconic acid, Le Bel isolated (−)-methylmalic acid, formed by the enantioselective addition of the elements of water to the carbon–carbon double bond of the substrate, catalysed by the microbial enzymes (Fig. 14.2). Despite the negative outcome of his experiments, Le Bel maintained that the four valencies of carbon are not necessarily directed towards the apices of an exactly regular tetrahedron.

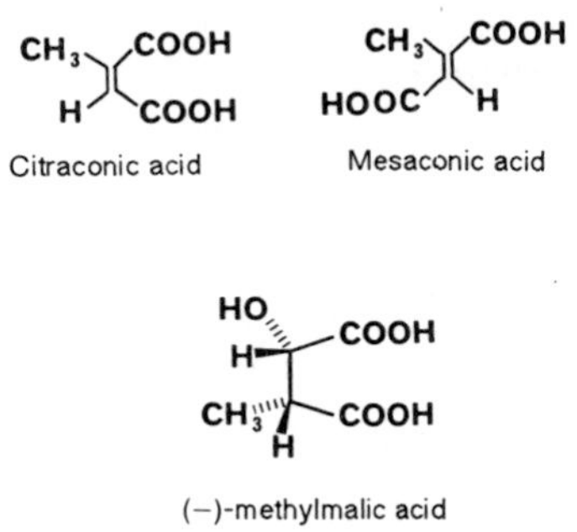

Citraconic acid   Mesaconic acid

(−)-methylmalic acid

Fig. 14.2 The structures of citraconic and mesaconic acid, which Le Bel (1892) attempted to resolve optically on the assumption that the atoms of the ethylene molecule are not coplanar. The microbial cultures employed for the resolution gave optically active products ((−)-methylmalic acid from the culture in which citraconic acid provided the carbon source) owing to the stereospecific addition of the elements of a water molecule across the carbon–carbon double bond of the substrate, catalysed by a microbial enzyme.

Le Bel's adoption of Haüy's morphological analogy was egregious, for the crystal was no longer generally regarded as an assembly of congruent space-filling polyhedral units, an individual molecule, or an ordered set of a small number of molecules. Through the geometrical analyses of Auguste Bravais (1811–63) at the École Polytechnique in Paris the crystal came to be considered as a reticular array of lattice points, symmetrically related to one another by translational repetition. A lattice point corresponds to the centre of gravity of a molecular unit, but the molecular shape has few constraints on the macroscopic crystal morphology: the crystal form became related primarily to the network symmetry of the reticulation. The 14 Bravais lattices, based upon translational repetition alone, account for 7 of the 32 crystal point groups classifying the external morphology of crystals, first identified by J. F. C. Hessel (1795–1872) at Marburg in 1831 and subsequently rediscovered in Bravais in 1851.

In 1879 L. Sohncke (1842–97) at Karlsruhe introduced the screw axis and the glide plane as lattice symmetry operations, extending to 65 the number of possible reticulation symmetries of lattice points. Finally, in the early 1890s, E. S. Fedorov (1853–1919) in St Petersburg, A. Schoenflies (1853–1928) at Göttingen, and W. Barlow (1845–1934), a British amateur crystallographer, each independently added the rotation–inversion lattice symmetry operation, completing the set of 230 space groups.

By the end of the century the main significant connection remaining between crystal morphology and molecular shape was the requirement, upon which the success of Pasteur's first method of optical resolution depends, that a single enantiomer necessarily crystallizes in an enantiomorphous space group, where the lattice points are related by pure rotations and translations alone. Dissymmetric hemihedral facets, distinguishing morphologically the crystals of one enantiomer from those of the other, are not always developed in cases where the rotational lattice symmetry is high, e.g. the enantiomers of 1,1'-binaphthyl which crystallizes in the tetragonal enantiomorphous space groups $P4_12_12$ and $P4_32_12$ (Mason 1982).

It was fortunate for Pasteur in 1848 that he was led to the problem of molecular dissymmetry by studying the crystallization of a racemic *salt* (rather than a non-polar racemate) and that he worked at a temperature (below 27°C) where the two sets of homochiral and enantiomorphous crystals of (+)- and (−)-[Na(NH$_4$)tartrate].4H$_2$O separate out. As van't Hoff showed in 1899, holohedral crystals of the less hydrated racemate [Na(NH$_4$) (±)tartrate].H$_2$O are formed above 27°C, following his general rule that the less solvated salt is the more stable at the higher temperature.

Statistical surveys indicate that, while some 15 per cent of racemic salts spontaneously resolve on crystallization into a conglomerate of individual (+)- and (−)-homochiral crystals, only 6 per cent of racemic non-polar substances display a similar behaviour (Jacques *et al.* 1981). It has been shown that, in the two-dimensional space of Langmuir–Blodgett monolayers, the interactions dominant between non-polar chiral molecules (the van der Waals' forces) favour the formation of heterochiral D–L molecular assemblies, whereas electrostatic interactions between the corresponding polar substances promote the homochiral D–D or L–L intermolecular assemblies (Andelman 1989). While an extrapolation to the three-dimensional crystal may not be secure, the identification of the general factors leading to preferred heterochiral or homochiral molecular assemblies in monolayers helps to explain the higher incidence of spontaneous optical resolution in the crystallization of racemic salts and dipolar racemic substances. Thus the α-amino acids are highly dipolar (zwitterionic) in solution or in the solid state, and racemates of most of the protein α-amino acids, or simple derivatives, spontaneously resolve on crystallization into the two homochiral sets of individual D- and L-crystals.

14.2 Homochirality from common ancestry

Emil Fischer's studies of the sugars (1884–1908) and peptides (1908–19) led not only to the identification of the D-sugars and the L-amino acids as the enantiomeric series dominating the biochemistry of living organisms, but also to the eclipse of Pasteur's conjecture that universal dissymmetric forces are in constant action throughout the inorganic and the organic world. The ascent of the sugar series, using achiral laboratory reagents, proved to be highly stereoselective, differing only in degree from the stereospecificity of enzyme reactions. Fischer's 'key and lock' reaction mechanism appeared to account just as well for the discrimination between the two diastereomeric products observed in the synthetic reactions of an enantiomer as for enzyme specificity. The similarities eliminated the distinction Pasteur had drawn between the chemistry of the laboratory and the chemistry of life, and removed the need for either a vital force or Pasteur's handed forces of nature to explain the chiral purity of biosynthesis and metabolism.

The 'key and lock' hypothesis offered 'a simple solution to the enigma of natural asymmetric synthesis'. Starting with a single enantiomer, or even with an enantiomeric excess in an otherwise-racemic mixture, synthetic reactions lead inevitably to a dominant diastereomeric product favoured by the steric congruence of the reaction intermediates: 'once a molecule is asymmetric, its extension proceeds also in an asymmetric sense'. The 'key and lock' mechanism, in distinguishing between alternative enantiomeric or diastereomeric biochemical intermediates in living organisms, serves as an agent of natural selection. Chiral homogeneity is a precondition for an efficient and economic biosynthesis and metabolic turnover, providing a Darwinian advantage that became refined from stereoselective reactivity to stereospecific enzyme-catalysed reactions during the course of biochemical evolution.

Early in his study of the sugars, in 1890, Fischer supposed that the first product of plant photosynthesis is a racemic mixture of L- and D-glucose, which the plant optically resolves into the two enantiomers. The D-glucose is used to build up starch and cellulose, while the L-isomer has other metabolic or biosynthetic roles. As he could find no trace of L-glucose in natural products, Fischer came round to the view by 1894 that the sole initial product of plant photosynthesis must be the D-sugar, owing to the assimilation of carbon dioxide by means of a chiral catalyst, chlorophyll. Subsequent materials produced by the plant include further molecules of the chiral chlorophyll catalyst, which are passed on to the daughter cells. The homochiral biochemistry of plants thus derives from the ancestral inheritance of a stereospecific biosynthetic apparatus, confined to the D-sugar series, extending back in time to an uncertain origin.

Fischer's early model of a plant producing both L- and D-glucose was taken up and extended by the Cambridge stereochemist W. H. Mills (1873–1959) in an address to the British Association for the Advancement of Science at York in 1932. Mills supposed that a plant producing a racemic mixture of glucose would be equipped with two stereospecific enzyme systems, the L- and the D-system, for the further metabolism of the sugar enantiomers initially formed. Such a hypothetical 'racemic' plant would be less efficient than the standard homochiral plant, producing D-glucose alone, in proportion to the kinetic order of each reaction step mediated by the enzyme systems: one-half if first order, one-quarter if second order, and so on. Moreover the 'racemic' plant would be highly metastable to small deviations from equimolecularity in the concentrations of the L- and the D-glucose initially formed. A slight excess of one enantiomer, D-glucose say, would provide an advantage to the D-enzyme system, which would propagate competitively in a Darwinian fashion at the expense of the L-enzyme system. In such circumstances the L-enzyme system would be eliminated after a brief period of competition by the natural selection of the more fecund D-system, and the latter would become dominant universally because of its intrinsic kinetic advantage over the racemic assembly of concurrent L- and D-systems.

Mills argued that, in the production of a racemate from achiral starting materials with no chiral agencies, the formation of a small excess of one of the enantiomers has a significant probability. In the synthesis of N molecules of a racemate, with an equal chance of forming a D- or an L-enantiomer in each elementary reaction, the product distribution follows the Gaussian normal error curve and an equimolecular outcome, $L = N/2$ and $D = N/2$, is only the most probable. In general the product has a 'degree of dissymmetry', expressed by the relative enantiomeric excess, $k = (L - D)/N$. The normal error curve has an area dividing into equal portions inside and outside of the limits set by the particular values of the degree of dissymmetry, $k_b = \pm 0.674(N)^{-1/2}$. That is, a greater or a smaller degree of dissymmetry than these boundary values is equally likely in a run of repeated syntheses of the racemate. For the production of a relatively small number of racemic molecules the expected boundary degree of dissymmetry is quite large (± 0.21 per cent for the synthesis of 10^5 molecules) although vanishingly small for laboratory-scale production. Mills regarded the small-number value of approximately 10^5 molecules as appropriate for a chiral catalytically active compound with a molecular weight of about 35 000, i.e. an enzyme, at a concentration of 0.1 per cent in a microscopic organism, such as a blue–green alga (cyanobacterium) only 3 μm in diameter.

The evolutionary mechanism for a homochiral enantiospecific reaction sequence by natural selection, developed by Mills from Fischer's model, was placed on a formal reaction kinetics basis by F. C. Frank at Bristol in 1953. Frank envisaged an open flow reactor system, in which each enantiomer serves as a 'catalyst for its own production and an anticatalyst for the production of its optical antimer'. Darwin's 'warm little pond', equipped with an input stream for substrates and an output stream for products, models such as a flow reactor.

The system is fed by an input of achiral substrate molecules A, which react reversibly with each enantiomer, L or D, with common rate constants k_1 and k_{-1} to duplicate the enantiomer molecule:

$$A + L(D) \underset{k_{-1}}{\overset{k_1}{\rightleftharpoons}} 2L\ (2D) \qquad (14.1)$$

In addition the two enantiomers react together irreversibly, eliminating one another as the inactive side product, *P*, which constitutes, together with excess enantiomers, the output of the flow reactor system:

$$L + D \xrightarrow{\ k_2\ } P \tag{14.2}$$

The dynamic steady state of the two competing autocatalytic sequences is stable, giving a racemic output with some inactive side product as long as the input solution of achiral substrate molecules remains dilute. Under these conditions, where all of the molecular populations are small, both of the molecular reproduction processes (equation (14.1)) have more significance than the mutual annihilation of the enantiomers (equation (14.2)). At a larger critical substrate input concentration, the racemic production becomes metastable, owing to the general increase in the molecular populations and the resulting greater competition between the enantiomers for the achiral substrate needed for self-propagation. Now the mutual elimination of the enantiomers (equation (14.2)) assumes a greater importance relative to their autocatalytic self-duplication (equation (14.1)).

Under the conditions where the overall racemic production becomes metastable, in a domain far from thermodynamic chemical equilibrium, a minor perturbation suffices to tip the precarious balance between the two competing enantiomeric reaction sequences and to switch the entire system over to single-enantiomer production (Fig. 14.3). Small departures from the equimolecular production of the enantiomers at the metastability stage, notably the statistical fluctuations envisaged by Mills, provide such a perturbation. Once the open flow reactor system has switched to homochiral production there can be no return to racemic production, nor any reversal to the now extinct antimeric production channel, for the surviving enantiomeric reaction sequence dominates both substrate utilization (equation (14.1) and competitor elimination (equation (14.2)).

There are no Darwinian grounds of 'fitness advantage' for the choice between the L and the D autocatalytic reaction sequences. The two enantiomers have an equal probability of gaining dominance under conditions of metastability, and the selection of the surviving reaction channel from the 'bifurcation catastrophe' must be wholly a matter of chance. On this view the particular outcome, the D-sugars and the L-amino acids universally adopted in the organic world, appears to be a 'frozen accident', devoid of any specific causation or connection with other chiral inequalities in the universe. The origin of life itself, and specific biomolecular handedness with it, were expressions of a miraculous chance event, according to Jacques Monod (1972), and only the ensuing Darwinian natural selection reimposed necessity.

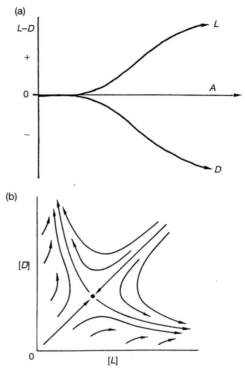

Fig. 14.3 The kinetic mechanism of Frank (equations (14.1) and (14.2)) for the evolution of biomolecular homochirality in an open flow reactor system, fed by an input of achiral substrate molecules (A) and drained by an output of the enantiomeric molecules (L and D) and the inactive side product (P). (a) The output of the reactor remains racemic, $[L] = [D]$, at low substrate inputs ($[A]$ small) where enantiomer autocatalytic production with competition for the achiral substrate is dominant (equation (14.1)). At higher substrate inputs, the concentration of all species in the reactor increases, and the mutual annihilation of the two enantiomers to give the inactive side product (P) irreversibly becomes relatively more important (equation (14.2)). The system now becomes metastable and hypersensitive to minor perturbations that trigger a switch to homochiral production ('bifurcation catastrophe'), either all-L or all-D with an equal probability if parity (mirror-image equivalence) is conserved. (b) The time evolution of enantiomer production in the open flow reactor system (from the integration over time of the equations connecting the time derivatives of the species concentrations in equations (14.1) and (14.2)). For an increasing achiral substrate input from low values, the system moves along the diagonal of racemic output, $[L] = [D]$, to the point of singularity where the system becomes metastable. In the metastable domain, a small perturbation suffices to switch production into one of the two homochiral reaction sequences to give either an all-L or an all-D output. The relations beyond the point of metastability, around the diagonal at high values of $[L] = [D]$, represent only formal solutions of the integrated kinetic equations (after Frank 1953).

14.3 Chiral fields

The 'key and lock' hypothesis for the propagation of biomolecular homochirality left unexplained the origin of the initial prebiotic enantiomer, or enantiomeric excess, from which diastereoselective synthesis began, as Emil Fischer himself appreciated in 1894. A solution to the problem proposed by the Aberdeen organic chemist Francis Japp (1848–1925) at the Bristol meeting of the British Association for the Advancement of Science in 1898, the traditional vital force, provoked a heated controversy in the correspondence columns of the journal *Nature* for several months after the meeting. Japp's numerous critics divided into two categories: the advocates of a chance provenance for optically active natural products, like the London statistician Karl Pearson (1857–1936); and those supporting a physical cause, such as the Birmingham chemist Percy Frankland (1856–1946), who proposed that terrestrial optical activity originated from the dissymmetry of the solar radiation.

Developments earlier during the 1890s had given substance to the suggestion made by Le Bel, in his 1874 stereochemical interpretation of optical isomerism, that circularly polarized light provides a chiral force 'which favours the formation of one of the dissymmetric isomers'. Le Bel's surmise was supported by a symmetry analysis of the natural forces Pasteur had taken to be dissymmetric by Pierre Curie (1859–1906) at the Sorbonne in Paris. From the symmetry relations observed between the physical properties and the morphology of a crystal, Curie arrived in 1894 at the general principle that a physical cause cannot have a higher symmetry than any of its effects, although the cause may have the same or a lower symmetry than the effect. The forces considered possibly dissymmetric by Pasteur, an electric or magnetic field, orbital or spin rotation, all have a centre or plane of symmetry, and thus none of these forces could give rise to an optically active molecule, which is necessarily devoid of a centre or plane of symmetry, or a rotation–inversion symmetry element in general.

Curie divided forces into two types: one 'axial' with a rotational form, like a magnetic field; and the other 'polar' with a vectorial translatory form, such as an electric field. In a collinear combination of an axial with a polar field, the general rotation–inversion symmetry elements of the individual fields vanish, and the two possible combinations, one parallel the other antiparallel, are truly dissymmetric fields, i.e. non-superposable mirror images of each other. The collinear combination of a rotation with a translation gives a helical motion in one of two dissymmetric forms, left-handed or right-handed. Similarly, right-handed or left-handed circularly polarized electromagnetic radiation is represented by the respective parallel or antiparallel combination of an electric and a magnetic field oscillating at a common frequency (Fig. 14.4). Curie conjectured that such a chiral field may be expected to generate an optically active substance from an achiral substrate.

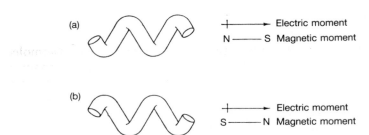

Fig. 14.4 Chiral fields and optical activity. The stereochemistry of a chiral molecule requires its constituent charges (negative electrons and positive nuclei) to move through helical paths on interaction with the electromagnetic radiation field. The helical motion of a charge (positive or negative) gives rise to a collinear electric moment (polar vector) and magnetic moment (axial vector), resulting in a chiral electromagnetic field. (a) If the helical path is right-handed the electric and the magnetic dipole moments have a parallel collinear orientation, and the molecule interacts more strongly with left-circularly polarized (LCP) than with right-circularly polarized (RCP) light of the same frequency. At a resonance frequency the molecule gives a positive circular dichroism, and at lower frequencies exhibits a positive circular birefringence (dextrorotatory optical activity). (b) In the enantiomeric molecule the corresponding helical path is left-handed and the electric and the magnetic dipole moments have an antiparallel mutual orientation. The enantiomer interacts more strongly with RCP than LCP radiation, giving a negative circular dichroism at the resonance frequency and a negative circular birefringence at a lower frequency (laevorotatory optical activity).

A realization of the conjecture became feasible with the discovery of 1895, by Aimé Cotton (1869–1951) at the Paris École Normale Supérieure, of circular dichroism, the differential absorption of left- and right-circularly polarized light by an optically active substance in homogeneous solution. The circular dichorism ($\triangle\epsilon = \epsilon_L - \epsilon_R$, where ϵ_L and ϵ_R are the respective molar extinction coefficients for left- and right-circularly polarized light) represents the absorption counterpart of the circular birefringence, or optical rotation, observed at wavelengths of transparency. Like the optical rotation, $\triangle\epsilon$ is equal in magnitude but opposite in sign for two enantiomers.

Thus a solution of the racemic mixture of two enantiomers, irradiated with one of the two forms of circularly polarized light at a wavelength producing photochemical change, is expected to become optically enriched in the enantiomer with the smaller extinction coefficient for the particular circular radiation component employed. Cotton himself searched in vain for the optical photoresolution of a racemate implied by his discovery of circular dichroism. The first authentic enantioselective photolysis of a racemic mixture with circularly polarized light was achieved only in 1929, by Werner Kuhn (1899–1963) at Heidelberg.

Kuhn and subsequent investigators showed that an enantioselective photolysis or photosynthesis requires the irradiation of a racemic mixture with monochromatic circularly polarized light at a wavelength corresponding to one particular circular dichroism absorption band, so that one of the two optical isomers absorbs more radiation and undergoes more photochemical change than the other. In the circular dichroism (CD) spectrum of an optical isomer the CD bands or band systems alternate in sign along the wavelength ordinate, and the sum of the frequency-weighted CD band areas vanishes over the spectrum as a whole (the Kuhn–Condon sum rule). Thus the broad-band irradiation of a racemic substance with left- or right-circularly polarized 'white' light, a continuous range of wavelengths as in sunlight, necessarily produces racemic photoproducts with no enantioselection.

A proof of the rule requiring the sum of the optical activity of an enantiomer over the electromagnetic spectrum as a whole to go to zero was established classically by Kuhn in 1929 and quantum mechanically by E. U. Condon at Princeton in 1937. Before these proofs of the sum rule, and for many years afterwards, it was commonly supposed that the net circular polarization of broad-band sunlight or of 'white' radiation from other celestial sources served as the chiral agency priming the original prebiotic enantiomeric excess from which biochemical homochirality evolved by the 'key and lock' mechanism of diastereoselective biosynthesis.

Measurements during the 1980s established that the solar radiation has indeed a net circular polarization over the visible, near-infrared, and near-ultraviolet regions of up to 0.5 per cent at twilight when the contribution from multiple aerosol scattering is the most significant. The time average is very small since the right-circular component in excess at sunrise is opposite in sign and almost equal in magnitude to the left-circular component in excess at sunset. The spatial average is small too, as the excess of right-circular light from the north pole of the Sun is almost balanced out by the excess of left-circular radiation from the south pole. The overall net excess of right-circular polarization in the radiation from the Sun lies at the level of one part per million, and it is broad band covering the whole photochemically active wavelength range.

On an overall time and space average, the electromagnetic and the other chiral fields deriving from Curie's analysis are even-handed. An origin for the prebiotic enantiomeric excess required to prime the evolution of biomolecular chiral homogeneity through the agency of a classical chiral field is feasible only under special conditions, a particular place or time, or circumstances giving rise to narrow-band circularly polarized radiation which is photochemically active with chiral discrimination.

Thus right-circularly polarized radiation over the 200–230 nm ultraviolet wavelength region produces an enantioselective photolysis of racemic amino acids in solution, leaving an excess of the L-isomer, and longer ultraviolet wavelengths and visible radiation are not absorbed. Before the development of an appreciable amount of oxygen in the atmosphere and an ozone layer, the absorption of solar radiation by the atmospheric gases set a short-wavelength limit of about 200 nm to the ultraviolet radiation reaching the Earth's surface, transmitting a significant flux in the 200–230 nm wavelength range. A pool of racemic amino acids located on an east-facing slope, and so exposed preferentially to the right-circular excess of solar radiation at sunrise, would be expected to develop an enantiomeric excess of the L-amino acids, owing to the more extensive photolysis of the D-enantiomers. While the set of circumstances postulated leads to the L-amino acids, other sets producing the D-enantiomers or leaving the race-mate unchanged appear equally probable (Mason 1988).

14.4 Party non-conservation and the electroweak interaction

Throughout the first half of the twentieth century the universe was generally taken to be isotropic and homogeneous, with a chiral symmetry expressed by the apparent even-handedness of the established forces of nature. In the new quantum mechanics the symmetry of the atomic world became dominated by the principle of the conservation of parity, proposed in 1927 by Eugene Wigner, then in Berlin, shortly before his emigration to Princeton. All physical causes and the laws linking them to the observable effects produced, Wigner held, are unchanged by space inversion (the parity operation) or, what is equivalent, by mirror-plane reflection. Non-observable theoretical entities, notably the wavefunctions of quantum mechanics, are either symmetric, retaining their sign ('gerade'), or antisymmetric, changing sign ('ungerade'), under the operation of spatial inversion through the centre of the system. Observable physical quantities emerge from the square modulus of the wavefunction and so they are necessarily symmetric and conserve parity.

One of the pioneers of the new quantum mechanics, Friedrich Hund at Göttingen, regarded the very existence of stable chiral molecules as para-doxical, for the operation of spatial inversion changes an optical isomer into a different molecule, namely its enantiomer. In 1927 Hund proposed that all chiral molecules are inherently unstable, since their wavefunctions must be of mixed parity, a superposition of symmetric and antisymmetric forms, and such molecules should undergo spatial inversion from one enantiomer to the other continuously, albeit on a very long time-scale in the apparently more stable cases. The alternative interpretation, that quan-tum mechanics based upon the parity-conserving electromagnetic inter-actions alone is incomplete, awaited developments over the following half-century (Quack 1989).

Among atomic and nuclear physicists, the principle of the conservation of parity soon became something of a 'sacred cow', sometimes with bizarre exemplifications. Pascual Jordan with Ralph Kronig at Bohr's Institute in

Copenhagen reported to the readers of *Nature* in 1927 the observation that, among Danish cattle, there are approximately equal numbers of left-circular and right-circular cud-chewers, providing a pastoral manifestation of universal parity conservation.

For the following 30 years it was generally assumed that the principle of parity conservation applied not only to the classical gravitational and electromagnetic interactions but also to the new strong and weak nuclear interactions, mediating α- and β-radioactivity, respectively. The use of β-emitting radionuclides as beam sources for electron polarization and scattering studies (1928–30) gave results which, at least in retrospect, provided evidence for parity violation in the weak interaction (Franklin 1979). But these results were set aside and forgotten until the accumulation of other anomalies led to the conclusion in 1956 that parity is not conserved in the weak nuclear interaction. From their conclusion, T. D. Lee at Columbia University and C. N. Yang at the Brookhaven National Laboratory in New York predicted parity-violating effects in β-radioactivity. These effects were soon verified experimentally (Fig. 14.5).

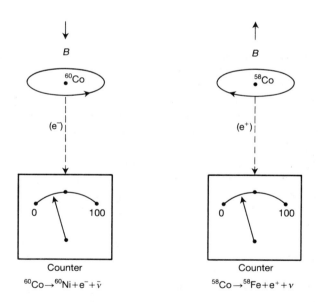

Fig. 14.5 Parity non-conservation in the weak interaction governing β-radioactivity. The β-decay emission of cobalt radionuclei aligned at liquid-helium temperatures by a magnetic field *B* lacks the equivalence along the two directions of the field axis, expected from parity conservation. A β-electron from ^{60}Co is projected preferentially in the direction antiparallel to its spin-axis vector, whereas a β-positron emitted from ^{58}Co has a preferred parallel orientation between its translation and spin-axis vector.

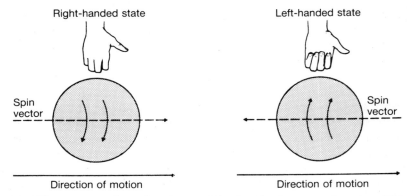

Fig. 14.6 The conservation of parity combined with charge conjugation (CP conservation) for particle–antiparticle enantiomers. The equivalence of a left-handed electron (preferred antiparallel linear momentum and angular momentum vectors) and the right-handed positron (preferred parallel linear momentum and angular momentum vectors) requires not only the parity operation (inversion through the centre) but also charge conjugation (the change from (−)-charged particle to (+)-charged antiparticle).

The β-decay experiments with radionuclides aligned in a magnetic field at a low temperature showed that the fundamental particles have an intrinsic handedness or helicity. The electrons emitted in β-decay are inherently left-handed, with a spin axis preferentially orientated antiparallel to the linear momentum direction, whereas the corresponding β-positrons are right-handed, with a parallel alignment of the spin axis and the momentum direction (Fig. 14.6). Although parity itself is not conserved in the weak nuclear interaction, the combination of parity (P) with charge conjugation (C), the conversion of a particle into the corresponding antiparticles, such as an electron into a positron, is conserved in good approximation (strictly, time reversal T must be included to give the more complete principle of CPT conservation).

According to the principle of CP conservation the negatively charged electron and the positively charged positron are CP mirror-image forms, and the hydrogen atom, composed of an electron and a proton, has a CP enantiomer made up of a positron and an antiproton. Similarly a chiral molecule composed of electrons and other particles, such as L-alanine, has a true CP mirror-image D-enantiomer, with equivalent properties, made up of positrons and other antiparticles in a counter-world of antimatter. But the natural terrestrial enantiomer related by the parity operation alone, D-alanine, with the standard particle composition, has properties dependent upon the weak interaction which are inequivalent.

The initial inequivalence investigated followed from the expectation that the β-decay of radionuclides, through which parity violation had been first observed, would differentially decompose the two enantiomers in a racemic mixture. Two effects are expected: the direct competitive radiolysis of the optical isomers by the β-electrons, and the indirect differential photolysis due to the left-circularly polarized electromagnetic braking radiation (Bremsstrahlung) produced as the β-electrons slow down from their initial relativistic velocities in the β-decay emission. Despite extensive studies over several decades, no reproducible enantioselective effects from the β-irradiation of racemates have been found as yet, nor in analogous studies employing spin-polarized electrons, positrons, or protons produced by particle accelerators.

The expected effect of the differential β-radiolysis of the enantiomers in a racemic mixture is small for a number of reasons. First, the initial energy of a β-electron is very large ($>10^6$ electron volts, eV) and lacks much discrimination in relation to the energies (2–5 eV) of the molecular valency electron changes involved in stereoselective chemical reactions. Second, the handedness (helicity) of a β-electron and the circular polarization of the associated braking radiation decrease as the β-electron is slowed down, and both become very small at the energy level of the molecular valency electrons. Third, the retardation of a β-electron by the reaction medium gives an energy spread to both the polarized photons of the braking radiation and the radiolytic electrons, but broad-band polarized photolysis or radiolysis lacks significant chiral discrimination. Monochromatic resonant radiation is essential for the photochemical enantiodifferentiation of a racemic mixture, as indicated by the Kuhn–Condon sum rule for the vanishing of the optical activity of an enantiomer over the spectrum as a whole.

The β-decay of radionuclides is governed by the component of the weak interaction involving charge changes, mediated by the charged massive bosons $W^\pm$. The parity-violating effects of the charged component of the weak interaction, such as the helicity of a β-electron, vanish in the non-relativistic limit of low energies. During the 1960s the neutral component of the weak interaction, mediated by the massive neutral boson Z^0, was unified with electromagnetism into the new electroweak interaction. The neutral boson Z^0 was detected, together with its charged counterparts $W^\pm$, at CERN in 1983. The parity-violating effects of the neutral electroweak interaction, in contrast to those of the charged weak interaction, remain finite at the low-energy limit and have significance for the ground electronic state and the other stationary states of atoms and molecules, as well as the transitional electronic states involved in thermal reactions and photoexcitations.

An atom in its electronic ground state, traditionally regarded as a centrosymmetric sphere, becomes a chiral system exhibiting optical activity according to the theory of the electroweak interaction. The expected optical activity, although small, increases in approximate proportion to

Z^6, where Z is the atomic number. Measurements of the optical rotation of heavy-metal atoms of bismuth, lead, and thallium in the gas phase, and related measurements for caesium, give an optical activity in agreement with the calculated sign and magnitude. The electroweak optical activity of a heavy atom is smaller by a factor of about 10^{-4} than the typical electromagnetic optical activity of a chiral molecule.

Since the natural enantiomers of a chiral molecule differ only in the spatial arrangement of the atoms they possess in common, the electroweak optical activity of the constituent atoms produces an optical enantiomeric inequivalence, adding to the molecular optical activity of one isomer and subtracting from that of the other. Thus a minor enantioselection is expected in the photolysis of a racemic mixture with monochromatic radiation which is tuned to a wavelength of optimum circular dichroism absorption but lacks a net polarization, containing equal numbers of left- and right-circularly polarized photons.

Similarly the neutral electroweak interaction differentiates between the ground-state electronic binding energy of two enantiomeric molecules, and between the energies of other corresponding states, stationary or transitional. A small electroweak energy increment adds to the binding energy of a given state in one enantiomer and subtracts from that of the corresponding state in the other, giving an electroweak binding energy difference ΔE_{ew} between the two optical isomers. The inclusion of the electroweak interaction in *ab initio* quantum-mechanical calculations of the ground-state electronic binding energy of salient biomolecules shows that the particular handed series selected during the course of biochemical evolution, the D-sugars and the L-amino acids, are in fact preferentially stabilized relative to their mirror-image forms by the neutral electroweak interaction.

The peptide unit of an L-polypeptide in either of the two regular conformations, the α-helix and the pleated β-sheet (fig. 12.1), and the parent L-amino acids (fig. 12.8) with the conformation preferred in aqueous solution are slightly more stable than their D-enantiomers in the ground electronic state. Similarly it turns out that the parent D-aldotriose, in the form of hydrated D-(+)-glyceraldehyde with its preferred solution conformation (Fig. 14.7), is energetically stabilized relative to its L-enantiomer, and so too is the biologically central furanose, D-ribose, in either of its two preferred envelope conformations, C_2-endo or C_3-endo (Fig. 14.8).

In all of these cases, the electroweak enantiomeric energy difference ΔE_{ew} amounts to no more than about 10^{-14} J mol^{-1}, corresponding to an excess of about 10^6 molecules of the stabilized enantiomer per mole of racemate in thermodynamic equilibrium at Earth-surface temperature. But racemic mixtures in thermodynamic equilibrium can no longer be regarded as truly equimolecular in the two enantiomers, or statistically spread in a Gaussian normal error curve around the equimolecular centre at L = $N/2$ and D = $N/2$, as Mills supposed in 1932. The centre of the distribution is

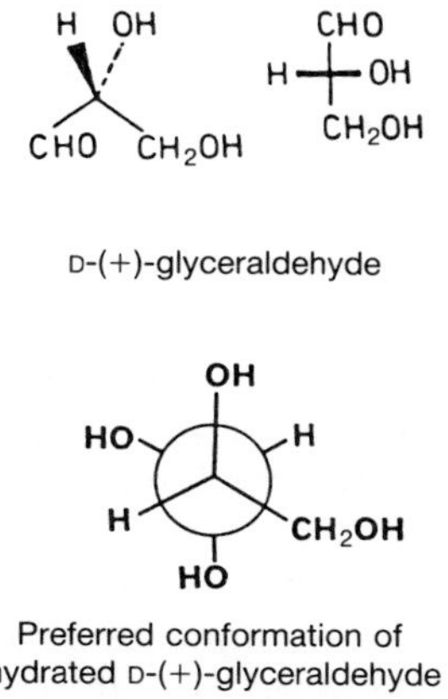

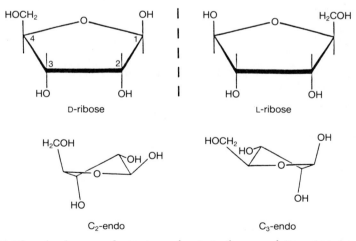

D-(+)-glyceraldehyde

Preferred conformation of
hydrated D-(+)-glyceraldehyde

Fig. 14.7 The absolute configuration of D-(+)-glyceraldehyde, with the corresponding Fischer projection structure, and the preferred conformation of hydrated D-(+)-glyceraldehyde in aqueous solution, stabilized relative to its L-(−)-enantiomer ($\sim 10^{-14}$ J mol^{-1}) by the electroweak interaction.

now displaced to a position registering the minor excess of the electroweak-stabilized enantiomer, and the fluctuations of a racemic mixture from exact equimolecularity are not strictly even-handed, being superimposed upon the electroweak bias.

Although small, the electroweak interaction has provided a universal

D-ribose

L-ribose

C_2-endo

C_3-endo

Fig. 14.8 The absolute configuration of D-(−)-ribose and its L-(+)-enantiomer, with the preferred solution conformations of D-ribose (C_2-endo and C_3-endo) in which it is stabilized relative to its enantiomer ($\sim 10^{-14}$ J mol^{-1}) by the electroweak interaction.

and uniform background chiral bias from the 'first 3 minutes' after the Big Bang and beyond, throughout the history of the Earth, constant in time and equivalent in the two hemispheres, unlike the classical chiral fields deriving from Curie's 1894 analysis, which have a handedness dependent upon particular times and places, as well as other restrictive conditions. As J. B. S. Haldane observed in 1960, the discovery of parity violation in the weak interaction has vindicated Pasteur's conjecture of universal dissymmetry to a degree, although the inherent chiral force of nature has neither the magnitude nor the biosynthetic role that Pasteur supposed.

The Darwinian flow reactor mechanism of Frank for the evolution of biochemical homochirality, formulated in 1953 before the discovery of parity non-conservation, envisaged the competitive reaction of autocatalysing L- and D-enantiomers for an input of achiral substrates, resulting in a metastable racemic output that led to a bifurcation catastrophe from which one of the enantiomeric reaction sequences survived and became predominant. In the original mechanism the two enantiomeric reaction sequences were taken to be equivalent, with equal values for the concentrations of corresponding intermediates and for the corresponding activation parameters of the unit reactions, and the outcome, the particular enantiomer perpetuated, appeared to be wholly a matter of chance (Section 14.2).

At the stage where the racemic output becomes metastable, as a result of increasing substrate input, the flow reactor system becomes hypersensitive to small perturbations which, if of a classical origin, are all even-handed. Such perturbations include fluctuations in the temperature or in the local concentrations of the reaction intermediates, and irradiation with white circularly polarized light, or monochromatic LCP and RCP radiation periodically alternating over time, as between dawn and dusk. But the parity-violating electroweak enantiomeric energy difference suffices to determine which of the two enantiomeric reaction sequences survives even if each competing classical perturbation individually is energetically 10^5-fold larger (Fig. 14.9). A computer simulation of the flow reactor system, with typical rate constants for the unit reactions of the Frank mechanism, indicates that the enantiomeric series preferentially stabilized by the electroweak interaction is selected with 98 per cent probability if the passage through the critical metastable stage occupies some 10 000 years, during which the input concentration of the substrate increases by up to 10^{-2} molar in a reservoir with a volume corresponding to a lake 1 kilometre in diameter and 4 metres deep.

The particular values of the flow reactor volume, time duration, and substrate input change indicated by the simulation are conservative limits, based upon the minimal electroweak advantage factor for the favoured enantiomeric reaction sequence. The ratio of the enantiomeric energy difference to the thermal energy at approximately 300 K, $\triangle E_{ew}/kT \sim 10^{-17}$, employed in the computer simulation, measures the electroweak advantage of a chiral system over its enantiomer in terms of the equilibrium enantiomeric excess or the relative activation parameters for reaction at Earth-surface temperature, at the level of the individual chiral molecule, or the monomer residue of a helical polymer, or the unit cell of an enantiomorphous crystal.

For a biopolymer, the electroweak advantage factor is amplified in proportion to the degree of polymerization, the number of monomer residues in the macromolecule. For a macroscopic enantiomorphous crystal, the amplification becomes proportional to the number of unit cells the crystal contains. At the level of the biomonomer or other small chiral molecule, a minor enantiomeric excess may be amplified to near chiral purity by a range of chemical reactions or physical processes, such as interrupted polymerization or partial crystallization (de Min *et al.* 1988).

Preceded by such an amplification of the electroweak advantage factor or of the ensuing minor enantiomeric excess, the bifurcation catastrophe of the Frank mechanism, from racemic production to one of the two enantiomeric reaction branches, is expected to require less restrictive conditions. That is, a smaller flow reactor volume, a shorter time interval for the critical transition through the metastable stage, and more latitude in other factors, such as the rate constants and the magnitude and the time-scale of the change in the substrate input concentration.

The basic significance of the discovery that the neutral electroweak interaction promotes the particular enantiomeric series selected during the course of biochemical evolution, the D-sugars and the L-amino acids, lies in the connection established between biomolecular handedness and other chiral inequivalencies in the universe. The central position of the D-sugars and the L-amino acids in the biochemical economy of living organisms no longer appears to be a mere matter of chance, decoupled and isolated from the interactions of the inorganic world. The principle of CP (or CPT) conservation provides a direct connection between terrestrial biomolecular handedness and the overwhelming predominance of particles over antiparticles in the universe. In a hypothetical counter-world of antimatter, it is expected that the biochemistry of the counter-organisms would be dominated by the L-anti-sugars and the D-anti-amino acids.

The universe as a whole appears to have less chiral symmetry and uniformity than was supposed during the first half of the twentieth century. Isotropy and homogeneity are confined to radiation: the black-body background from the Big Bang, now in the microwave wavelength region corresponding to 2.7 K, and the weak and more mysterious X-ray background radiation. The matter of the universe perceived by electromagnetic radiation appears to be far from homogeneous. The chiral galaxies aggregate into non-racemic clusters and superclusters, and the fine-tuning on their general red-shifts indicates that they all stream towards a location defining the Great Attractor.

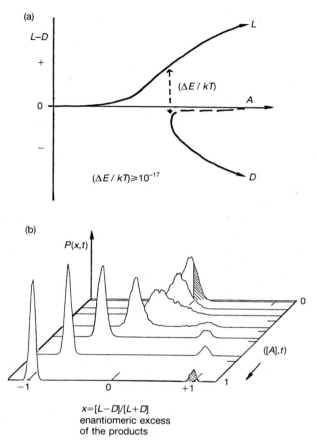

Fig. 14.9 The kinetic mechanism of Frank for the evolution of biomolecular homochirality in an open flow reactor system (equations (14.1) and (14.2)) with the inclusion of enantiomeric inequivalence due to the electroweak interaction. (a) As in the case of enantiomeric equivalence (Fig. 14.3(a)), the chiral output is racemic, but now with the minor enantiomeric excess of about 10^6 molecules per mole of the electroweak-stabilized enantiomer, for a low achiral substrate input. At the point of kinetic metastability, arising from the increase in the substrate input concentration $[A]$, the direction of the bifurcation catastrophe into a particular homochiral production sequence becomes determined with good probability by the electroweak energy difference $(\triangle E_{ew})$ between enantiomeric intermediates if the ratio $\triangle E_{ew}/kT$ equals or exceeds approximately 10^{-17} at the ambient temperature $(T \sim 300\,\mathrm{K})$. (b) A computer simulation of the passage through the domain of kinetic metastability as a function of time t and substrate input concentration $[A]$, based upon the time integration of the equations of the Frank mechanism (as in Fig. 14.3(b)) with the inclusion of the electroweak energy difference between enantiomeric intermediates at the minimum level $(\triangle E_{ew}/kT = 10^{-17})$ appropriate for the biomonomers, an L-amino acid, or a D-monosaccharide. The probability

$P(x,t)$ of an enantiomeric excess (x) in the chiral output of the flow reactor over time (t) is presented as a normalized function of the relative substrate input concentration and time $([A],t)$. With typical values for the kinetic rate constants (equations (14.1) and (14.2)), the probability $P(x,t)$ becomes 98 per cent for an increase in the enantiomeric excess, $x = [L - D]/[L + D]$, from zero to unity over about 10^4 years, during which the substrate input concentration increases by some 10^{-2} molar in a flow reactor with a volume corresponding to a lake approximately 1 kilometre in diameter and 4 metres deep (after Kondepudi and Nelson 1985).

The cosmic dissymmetry perceived by Pasteur in his later years was confined to the solar system: a century later the dissymmetry extends over the assemblies of galactic superclusters. Surveys of the winding directions of the spiral galaxies reveal a significant enantiomorphic excess of S-type counter-clockwise rotation over Z-type clockwise rotation. The rotational directions, in conjunction with the general translational recession of the galaxies from their red-shifts, specify the S-type as left-handed and the Z-type as right-handed. A general survey of 7563 galaxies finds a 4.6 per cent fractional excess of S-rotational forms, and a study of our local super-cluster reports a 9.8 per cent enantiomorphic excess of the S-rotational type. According to the standard Big Bang model, the particles outnumbered the antiparticles from an early stage, but galaxy formation began long after the primary separation of electromagnetism from the weak interaction. Possibly the phase transition of the separation supercooled, or possibly the residual electroweak interaction sufficed to confer a left-handed preference upon the motions of the proto-galaxies.

Certainly the electroweak preference had attenuated to a low level by the time that the solar system began to evolve some 4.6 billion years ago, judging by the near even-handed terrestrial distribution of chiral minerals. Over a total of 17 738 quartz crystals collected from a variety of locations, the fractional excess of left-handed $(-)$-quartz forms, $[L-D]/[L + D]$, averages only to 1.4 per cent, between the extremes of 2.4 per cent and zero for individual collections of a thousand or more crystals from a particular location. The electroweak stabilization of the unit cell of L-$(-)$-quartz relative to that of its D-$(+)$-enantiomorph amounts to an advantage ratio $\Delta E_{ew}/kT$ of about 10^{-17} at Earth-surface temperatures. This value corresponds to the global mean fractional excess (1.4 per cent) of L-$(-)$-quartz crystals when extended, on an additive basis, to an aggregate of some 10^{15} unit cells, forming a macroscopic single crystal with dimensions of approximately 0.1 mm along each main edge.

Throughout geohistory, the electroweak interaction has had only a minor effect on the relative distribution of the chiral products from un-selective processes, or even autocatalytic selective processes, such as enantiomorphous crystal growth, which are only passively competitive (for substrates) but not subject to active enantiomeric inhibition (mutual L/D annihilation). Darwin's mechanism of natural selection for the evolution of the organic species, when extrapolated to the prebiotic period gave the molecular scene envisaged by Mills (1932) and formalized by Frank (1953). In the primordial open flow reactor system autocatalytically generated enantiomers actively compete not only for substrates but also by mutual annihilation, producing the metastable steady state that is hypersensitive to small environmental perturbations, including the minor, but universal, chiral bias of the electroweak interaction. While the electroweak bias had little effect on the distribution of the enantiomorphous products of 'normal' processes not greatly distant from thermodynamic equilibrium, it was decisive in the irreversible 'bifurcation catastrophe' of the Frank mechanism, out in the symmetry-breaking and self-organizing domain of reaction kinetics, at a far remove from classical chemical equilibrium. The struggle for molecular existence in the prebiotic flow reactor resulted in the survival of the electroweak fittest, the ancestral molecules of the D-sugars and the L-amino acids (Mason 1988).

Bibliography

Achenbach-Richter, L., Stetter, K. O., and Woese, C. R. (1987). A possible bio-chemical missing link among the archaebacteria. *Nature, 327*, 348–9.

Ahrens, L. H. (ed.) (1979). *Origin and distribution of the elements*. Pergamon, Oxford.

Aller, L. H. (1961). *The abundance of the elements*. Interscience, New York.

Aller, L. H. (1986). Solar abundance of the elements. In *Spectroscopy and astrophysical plasmas*, (ed. A. Dalgarno and D. Layzer), Cambridge University Press.

Alpher, R. A. and Herman, R. C. (1950). Theory of the origin and relative abundance distribution of the elements. *Rev. Mod. Phys., 22*, 153–212.

Alpher, R. A., Bethe, H., and Gamow, G. (1948). The origin of the chemical elements. *Phys. Rev., 73*, 803.

Alvarez, L. W. (1987). Mass extinctions caused by large bolide impacts. *Phys. Today, 40*, No. 7, 24–33.

Andelman, D. (1989). Chiral discrimination and phase transitions in Langmuir monolayers. *J. Am. Chem. Soc., 111*, 6536–44.

Anders, E. (1968). Chemical processes in the early solar system, as inferred from meteorites. *Acc. Chem. Res., 1*, 289–98.

Anders, E. and Hayatsu, R. (1981). Organic compounds in meteorites and their origins. *Topics in current chemistry 99*. Springer, Berlin.

Anders, E., Hayatsu, R., and Studier, M. H. (1973). Organic compounds in meteorites. *Science, 182*, 781–90.

Anderson, D. L. and Dziewonski, A. M. (1984). Seismic tomography. *Sci. Am., 251*, No. 4, 58–66.

Anderson, R. B. (1984). *The Fischer–Tropsch synthesis*. Academic Press, New York.

Arnett, W. D. and Clayton, D. D. (1970). Explosive nucleosynthesis in stars. *Nature, 227*, 780–4.

Aston, F. W. (1927). A new mass-spectrograph and the whole number rule. *Proc. R. Soc.,* A 115, 487–514.

Bak, F. and Cypionka, H. (1987). A novel type of energy metabolism involving fermentation of inorganic sulphur compounds. *Nature, 326*, 891–2.

Baldwin, J. E. and Krebs, H. (1981). The evolution of metabolic cycles. *Nature, 291*, 381–2.

Bally, J. (1986). Interstellar molecular clouds. *Science, 232*, 185–93.

Baltscheffsky, H., Jörnvall, H., and Rigler, R. (eds) (1986). *Molecular evolution of life*. Cambridge University Press.

Baly, E. C. C., Davis, J. B., Johnson, M. R., and Shanassy, H. (1927). The photosynthesis of naturally occurring compounds. *Proc. R. Soc.,* A 116, 197–226.

Bania, T. M., Stark, A. A., and Heiligman, G. M. (1986). Clump 1: An unusual molecular cloud complex near the galactic centre. *Astrophys. J., 307*, 350–66.

Banks, B. E. C. and Vernon, C. A. (1970). Reassessment of the role of ATP *in vivo*. *J. Theor. Biol., 29*, 301–26.

Barber, J. (1987). Photosynthetic reaction centres: a common link. *Trends Biochem. Sci., 12*, 321–26.

Barnes, B. (1983). *T. S. Kuhn and social science*. Columbia University Press, New York.

Barron, L. D. (1986). Symmetry and molecular chirality. *Chem. Soc. Rev.*, **15**, 189–223.

Beadle, G. W. (1946). Genes and the chemistry of the organism. *Am. Sci.*, **34**, 31–53.

Beadle, G. W. and Tatum, E. L. (1941). Genetic control of biochemical reactions in *Neurospora*. *Proc. Natl. Acad. Sci. USA*, **27**, 499–506.

Beichman, C. and Harris, S. (1981). The formation of a T Tauri star: Observations of the infrared source in L1551. *Astrophys. J.*, **245**, 589–92.

Beichman, C. A., Myers, P. C., Emerson, J. P., Harris, S., Mathieu, R., Benson, P. J., and Jennings, R. E. (1986). Candidate solar-type protostars in nearby molecular cloud cores. *Astrophys. J.*, **307**, 337–49.

Bentley, R. (1969, 1970). *Molecular asymmetry in biology*, (2 vols). Academic Press, London.

Bernal, J. D. (1949). The Goldschmidt Memorial Lecture. *J. Chem. Soc.*, 2108–14.

Bernal, J. D. (1951). *The physical basis of life*. Routledge and Kegan Paul, London.

Bernal, J. D. (1967). *The origin of life*. Weidenfeld and Nicholson, London.

Bethe, H. (1939). Energy production in stars. *Phys. Rev.*, **55**, 434–56.

Biebricher, C. K. (1986). Darwinian evolution of self-replicating RNA. *Chem. Scr.*, **26B**, 51–7.

Biebricher, C. K., Eigen, M., and Luce, R. (1981). Product analysis of RNA generated *de novo* by Qβ replicase. *J. Mol. Biol.*, **148**, 369–90.

Bishop, M. J. and Friday, A. E. (1985). Evolutionary trees from nucleic acid and protein sequences. *Proc. R. Soc.*, B **226**, 271–302.

Blackett, P. M. S., Clegg, J. A., and Stubbs, P. H. S. (1960). An analysis of rock magnetic data. *Proc. R. Soc.*, A **256**, 291–322.

Blitz, L., Fich, M., and Kulkarni, S. (1983). The new Milky Way. *Science*, **220**, 123–40.

Bohr, N. and Wheeler, J. A. (1939). The mechanism of nuclear fission. *Phys. Rev.*, **56**, 426.

Bouchiat, M.-A. and Pottier, L. (1986). Optical experiments and weak interactions. *Science*, **234**, 1203–10.

Bretscher, M. S. (1985). The molecules of the cell membrane. *Sci. Am.*, **253**, No. 4, 86–90.

Brock, T. D. (1978). *Thermophilic microorganisms and life at high temperatures*. Springer, Berlin.

Brock, W. H. (1969). Lockyer and the chemists: the first dissociation hypothesis. *Ambix*, **16**, 81–99.

Brock, W. H. (1985). *From protyle to proton: William Prout and the nature of matter, 1785–1985*. Adam Hilger, Bristol.

Broda, E. (1978). *The evolution of the bioenergetic processes*. Pergamon, Oxford.

Brooke, J. H. (1971). Organic synthesis and the unification of chemistry—A reappraisal. *Br. J. Hist. Sci.*, **5**, 363–92.

Brooks, J. and Shaw, G. (1973). *Origin and development of living systems*. Academic Press, London.

Brown, H. (1986). *The wisdom of science: Its relevance to Culture and Religion*. Cambridge University Press.

Brush, S. G. (1980). Discovery of the Earth's core. *Am. J. Phys.*, **48**, 705–24.

Brush, S. G. (1988). *The history of modern science: A guide to the second scientific revolution 1800–1950.* Iowa State University Press, Ames, IA.

Budyko, M. I. (1986). *The evolution of the biosphere*, (trans. M. I. Budyko, S. F Lemesko, and V. G. Yanuta). Reidel, Dordrecht.

Bulloch, W. (1938). *The history of bacteriology.* Oxford University Press.

Burbidge, E. M., Burbidge, G. R., Fowler, W. A., and Hoyle, F. (1957). Synthesis of the elements in the stars. *Rev. Mod. Phys.*, **29**, 547–650.

Burbidge, G. (1971). Was there really a Big Bang? *Nature*, **233**, 36–40.

Burchfield, J. D. (1975). *Lord Kelvin and the age of the Earth.* Macmillan, London.

Burke, J. G. (1986). *Cosmic debris: Meteorites in history.* University of California Press, Berkeley, CA.

Cairns-Smith, A. G. (1982). *Genetic takeover and the mineral origins of life.* Cambridge University Press.

Cairns-Smith, A. G. (1985). The first organisms. *Sci. Am.*, **252**, No. 6, 74–82.

Cairns-Smith, A. G. (1986). Chirality and the common ancestor effect. *Chem. Br.*, **22**, 559–61.

Cairns-Smith, A. G. and Hartman, H. (1986). *Clay minerals and the origin of life.* Cambridge University Press.

Calvin, M. (1956). The photosynthetic carbon cycle (1955 Centenary Lecture). *J. Chem. Soc.*, 1895–1915.

Calvin, M. (1969). *Chemical evolution: Molecular evolution towards the origin of living systems on the Earth and elsewhere.* Clarendon Press, Oxford.

Cameron, A. G. W. (1982). Elemental and nuclide abundances in the solar system. In *Essays in nuclear astrophysics*, (ed. C. A. Barnes, D. D. Clayton, and D. N. Schramm), pp. 23–43. Cambridge University Press.

Carbo, R. and Ginebreda, A. (1985). Interstellar chemistry. *J. Chem. Educ.*, **62**, 832–6.

Carlile, M. J., Collins, J. F., and Moseley, B. E. B. (eds) (1981). *Molecular and cellular aspects of microbial evolution*, Symp. 32, the Society for General Microbiology. Cambridge University Press.

Carrington, A. and Ramsey, D. A. (eds) (1982). *Molecules in interstellar space.* The Royal Society, London. ((1981). *Phil. Trans. R. Soc.*, A 303 463–631.)

Cavalier-Smith, T. (1987*a*). The origin of eukaryote and archebacterial cells. In *Endocytobiology III, Ann. NY Acad. Sci.*, **503**, 17–54.

Cavalier-Smith, T. (1987*b*). Eukaryotes with no mitochondria. *Nature*, **326**, 332–3.

Cavalier-Smith, T. (1989). Archebacteria and archezoa. *Nature*, **339**, 100–1.

Cech, T. R. (1986*a*). RNA as an enzyme. *Sci. Am.*, **255**, No. 5, 76–84.

Cech, T. R. (1986*b*). The generality of self-splicing RNA: Relationship to nuclear mRNA splicing. *Cell*, **44**, 207–10.

Chandrasekhar, S. (1937). *An introduction to the study of stellar structure.* University of Chicago Press, Dover reprint, New York, 1957.

Chandrasekar, S. (1984). On stars, their evolution and their stability. *Rev. Mod. Phys.*, **56**, 137–47.

Chargaff, E. (1979). How genetics got a chemical education. *Ann. NY Acad. Sci.*, **325**, 345–60.

Cherdyntsev, V. V. (1961). *Abundance of chemical elements*, (trans. W. Nichiporuk). University of Chicago Press.

Clarke, B. F. C. (1984). *The genetic code and protein biosynthesis*. Edward Arnold, London.

Clayton, D. D. (1983). *Principles of stellar evolution and nucleosynthesis*. University of Chicago Press.

Clayton, D. D. and Woosley, S. E. (1974). Thermonuclear astrophysics. *Rev. Mod. Phys.*, **46**, 755–71.

Clayton, R. N. (1961). Oxygen isotope fractionation between calcium carbonate and water. *J. Chem. Phys.*, **34**, 724–6.

Clayton, R. N., Grossman, L., and Mayeda, T. K. (1973). A component of primitive nuclear composition in carbonaceous meteorites. *Science*, **182**, 485–8.

Clayton, R. N., Hinton, R. W., and Davis, A. M. (1988). Isotope variations in rock-forming elements in meteorites. *Philos. Trans. R. Soc.*, A **325**, 483–501.

Collard, P. (1976). *The development of microbiology*. Cambridge University Press.

Compston, W. and Pidgeon, R. T. (1986). Jack Hills, evidence of more very old detrital zircons in Western Australia. *Nature*, **321**, 766–9.

Cowan, G. A. (1976). A natural fission reactor. *Sci. Am.*, **253**, No. 1, 36–47.

Cramer, F. and Freist, W. (1987). Molecular recognition by energy dissipation, a new enzymatic principle: The example isoleucine–valine. *Acc. Chem. Res.*, **20**, 79–84.

Creighton, T. E. (1984). *Proteins: Structures and molecular principles*. Freeman, New York.

Crick, F. H. C. (1958). On protein synthesis. *Symp. Soc. Exp. Biol.*, **12**, 138–63.

Crick, F. C. H. (1988). *What mad pursuit: A personal view of scientific discovery*. Weidenfeld and Nicholson, London.

Crick, F. H. C., Brenner, S., Klug, A., and Pieczenik, G. (1976). A speculation on the origin of protein synthesis. *Origin of Life*, **7**, 389–97.

Cronin, J. R. and Pizzarello, S. (1986). Amino acids of the Murchison meteorite. *Geochim. Cosmochim. Acta*, **50**, 2419–27.

Crookes, W. (1886). On the nature and origin of the so-called elements. *Reports of the British Association for the Advancement of Science for 1886*, pp. 558–76, London.

Crookes, W. (1887). Genesis of the elements; Friday evening discourse held at the Royal Institution. Reprinted in (1970) *The Royal Institution Library of Science: Physical Sciences*, (ed. W. L. Bragg and G. Porter), Vol. 3, pp. 403–26. Elsevier, Amsterdam.

Crookes, W. (1888). Annual General Meeting: Presidential Address, elements and meta-elements. *J. Chem. Soc.*, **53**, 487–504.

Crookes, W. (1898). On the position of helium, argon, and krypton in the scheme of the elements. *Proc. R. Soc.*, **63**, 408–11.

Crowe, M. J. (1986). *The extraterrestrial life debate 1750–1900: The idea of a plurality of worlds from Kant to Lowell*. Cambridge University Press.

Crowfoot Hodgkin, D. (1979). Crystallographic measurements and the structure of protein molecules as they are. In *The origins of modern biochemistry: A retrospect on proteins*, (ed. P. R. Srinivasan, J. S. Fruton, and J. T. Edsall), *Ann. NY Acad. Sci.*, **325**, 121–48.

Crowfoot Hodgkin, D. and Riley, D. P. (1968). Some ancient history of protein X-ray analysis. In *Structural chemistry and molecular biology: A volume dedicated to Linus Pauling*, (ed. A. Rich and N. Davidson), pp. 15–28. Freeman, San Francisco.

Cushny, A. R. (1926). *Biological relations of optically isomeric substances.* Williams and Wilkins, Baltimore.

Dauvillier, A. (1965). *The photochemical origin of life.* Academic Press, New York.

Dawes, E. A. (1986). *Microbial energetics.* Blackie, Glasgow.

Dayhoff, M. O. (ed.) (1972). *Atlas of protein sequence and structure,* Vol. 5. National Biomedical Research Foundation, Washington, DC (Supplements 1–3, 1973–8; Vol. 4, 1969; Vols 1–3, with R. V. Eck, 1968).

Dearnley, R. (1965). Orogenic fold-belts and continental drift. *Nature,* **206,** 1083–7.

de Duve, C. (1988). The second genetic code. *Nature,* **333,** 117–18.

Deisenhofer, J., Epp, O., Miki, K., Huber, R., and Michel, H. (1985). Structure of the protein subunits in the photosynthetic reaction centre of *Rhodopseudomonas viridis* at 3 Å resolution. *Nature,* **318,** 618–24.

de Jong, T. and van Soldt, W. H. (1989). The earliest known solar eclipse record redated. *Nature,* **338,** 238–40.

DeKosky, R. K. (1973). Spectroscopy and the elements in the late nineteenth century: The work of Sir William Crookes. *Br. J. Hist. Sci.,* **6,** 400–23.

Delsemme, A. H. (1988). The chemistry of comets. *Philos. Trans. R. Soc.,* A **325,** 509–23.

de Min, M., Levy, G., and Micheau, J. C. (1988). Chiral resolutions, asymmetric synthesis and amplification of enantiomeric excess. *J. Chim. Phys.,* **85,** 603–19.

Dick, S. J. (1982). *Plurality of worlds: The origins of the extraterrestrial life debate from Democritus to Kant.* Cambridge University Press.

Dicke, R. H., Peebles, P. J. E., Roll, P. G., and Wilkinson, D. T. (1965). Cosmic black-body radiation. *Astrophys. J.,* **142,** 414–19.

Dill, R. F., Shinn, E. A., Jones, A. T., Kelly, K., and Steinen, R. P. (1986). Giant subtidal stromatolites forming in normal salinity waters. *Nature,* **324,** 55–8.

Dobell, C. (1932). *Antony van Leeuwenhoek and his 'little animals'.* Constable, London. Dover Reprint, 1960.

Döbereiner, J. (1818). *Gilberts Ann. Chem.,* **58,** 210. From Bauer, H. (1980) Die ersten organisch-chemischen Synthesen. *Naturwissenschaften,* **67,** 1–6.

Dobzhansky, T., Ayala, F. J., Stebbins, G. L., and Valentine, J. W. (1977). *Evolution.* Freeman, San Francisco.

Dodd, R. T. (1981). *Meteorites: A petrologic-chemical synthesis.* Cambridge University Press.

Duley, W. W. and Williams, D. A. (1984). *Interstellar chemistry.* Academic Press, London.

Du Toit, A. L. (1937). *Our wandering continents.* Oliver and Boyd, Edinburgh.

Dyson, F. J. (1979). Time without end: Physics and biology in an open universe. *Rev. Mod. Phys.,* **51,** 447–60.

Dyson, F. J. (1982). A model for the origin of life. *J. Mol. Evol.,* **18,** 344–50.

Dyson, F. J. (1985). *Origins of life.* Cambridge University Press.

Eaton, N. (1986). Recent developments in asteroid science. *Phys. Rep.,* **132,** 261–76.

Eigen, M. (1971). Self-organisation of matter and the evolution of biological macromolecules. *Naturwissenschaften,* **58,** 465–526.

Eigen, M. (1981). Darwin and molecular biology. *Angew. Chem. Int. Ed. Eng.,* **20,** 233–41.

Eigen, M. (1986). The physics of molecular evolution. *Chim. Scr.*, **26B**, 13–26.

Eigen, M. and Schuster, P. (1977, 1978). The hypercycle: A principle of natural self-organisation. *Naturwissenschaften*, **64**, 541–565, **65**, 7–41, 341–369.

Eigen, M., Gardiner, W., Schuster, P. and Winkler-Oswatitsch, R. (1981). The origin of genetic information. *Sci. Am.*, **244**, No. 4, 78–94.

Elsasser, W. M. (1934). Sur le principe de Pauli dans les noyaux. *J. Phys. Radium*, **4**, 549, **5**, 389, 635.

Emmons, T. P., Reeves, J. M., and Fortson, E. N. (1983). Parity-nonconserving optical rotation in atomic lead. *Phys. Rev. Lett.*, **51**, 2089–92 and (1984) **52**, 86.

Epstein, S., Krishnamurthy, R. V., Cronin, J. R., Pizzarello, S., and Yuen, G. U. (1987). Unusual stable isotope ratios in amino acid and carboxylic acid extracts from the Murchison meteorite. *Nature*, **326**, 477–9.

Evans, M. C. W., Buchanan, B. B., and Arnon, D. I. (1966). A new ferredoxin-dependent carbon reduction cycle in a photosynthetic bacterium. *Proc. Natl Acad. Sci. USA*, **55**, 928–34.

Farber, E. (1964). Theory of the elements and nucleosynthesis in the nineteenth century. *Chymia*, **9**, 181–200.

Farley, J. (1974). *The spontaneous generation controversy from Descartes to Oparin*. Johns Hopkins University Press, Baltimore.

Farrer, W. V. (1965). Nineteenth-century speculations on the complexity of the chemical elements. *Br. J. Hist. Sci.*, **2**, 297–323.

Feher, G., Allen, J. P., Okamura, M. Y., and Rees, D. C. (1989). Structure and function of photosynthetic reaction centres. *Nature*, **339**, 111–16.

Fersht, A. (1985). *Enzyme structure and mechanism*. Freeman, New York.

Florkin, M. and Stoz, E. H. (eds) (1975–86). *Comprehensive biochemistry; section VI: A history of biochemistry*, Vols 30–6. Elsevier, Amsterdam.

Fodor, S. P. A., Ames, J. B., Gebhard, R., van den Berg, E. M. M., Stoeckenius, W., Lugtenburg, J., and Mathies, R. A. (1988). Chromophore structure in bacteriorhodopsin's N intermediate: Implications for the proton-pumping mechanism. *Biochemistry*, **27**, 7097–101.

Fogg, G. E. (1986). Picoplankton. *Proc. R. Soc.*, B **228**, 1–30.

Fowler, W. A. (1978). Nuclear cosmochronology. In *Proceedings of the Robert A. Welch Foundation Conferences on Chemical Research: XXI Cosmochemistry*, (ed. W. O. Mulligan), pp. 61–93, Welch Foundation, Houston, TX.

Fowler, W. A. (1984). Experimental and theoretical nuclear astrophysics: The quest for the origin of the elements. *Rev. Mod. Phys.*, **56**, 149–79.

Fox, S. W. and Dose, K. (1977). *Molecular evolution and the origin of life*. Marcel Dekker, New York.

Frank, F. C. (1953). On spontaneous asymmetric synthesis. *Biochim. Biophys. Acta*, **11**, 459–63.

Franklin, A. (1979). The discovery and nondiscovery of parity nonconservation. *Stud. Hist. Philos. Sci.*, **10**, 201–67.

Frederick, J. F. (ed.) (1981). Origins and evolution of eukaryote intracellular organelles. *Ann. NY Acad. Sci.*, **361**, 1–510.

Freudenberg, K. (1966). Emil Fischer and his contribution to carbohydrate chemistry. *Adv. Carbohydrate Chem.*, **21**, 1–38.

Fruton, J. S. (1972). *Molecules and life: Historical essays on the interplay of chemistry and biology*, Wiley-Interscience, New York.

Galison, P. (1987). *How experiments end*. University of Chicago Press.

Gamow, G. (1946). Expanding universe and the origin of elements. *Phys. Rev.*, **70**, 572.

Gamow, G. (1948). The origin of elements and the separation of galaxies. *Phys. Rev.*, **74**, 505.

Garland, P. B. (1981). The evolution of membrane-bound bioenergetic systems: The development of vectorial oxidoreductions. In *Molecular and cellular aspects of microbial evolution*, (ed. M. J. Carlile, J. F. Collins and B. E. B. Moseley), Symp. 32, the Society for General Microbiology. Cambridge University Press.

Gause, G. F. (1941). *Optical activity and living matter*. Biodynamica, Normandy, MO.

Gest, H. (1980). The evolution of biological energy-transducing systems. *FEMS Microbiol. Lett.*, **7**, 73–7.

Gest, H. (1987). Evolutionary roots of the citric acid cycle in prokaryotes. In *Krebs' citric acid cycle—Half a century and still turning*, (ed. J. Jay and P. D. J. Weitman), *Biochem. Soc. Symp.*, **54**, 3–16.

Gingerich, O. (ed.) (1984). *Astrophysics and twentieth-century astronomy to 1950*, Vol. 4A, *The general history of astronomy*. Cambridge University Press.

Goldanskii, V. I. (1986). Quantum chemical reactions in the deep cold. *Sci. Am.*, **254**, No. 2, 38–44.

Goldanskii, V. I. and Kuzmin, V. V. (1988). Spontaneous mirror symmetry breaking in nature and the origin of life. *Z. Phys. Chem. Leipzig*, **269**, 216–74.

Goldschmidt, V. M. (1937). The principles of distribution of chemical elements in minerals and rocks. *J. Chem. Soc.*, 655–73.

Goldschmidt, V. M. (1958). *Geochemistry*, (ed. A. Muir). Clarendon Press, Oxford.

Goodwin, A. M. (1981). Precambrian perspectives. *Science*, **213**, 55–61.

Graham, A. L., Bevan, A. W. R., and Hutchison, R. (1985). *Catalogue of meteorites*, (4th edn). British Museum (Natural History), London.

Greenberg, J. M. (1984). The structure and evolution of interstellar grains. *Sci. Am.*, **250**, No. 6, 96–107.

Greenberg, J. M., van de Bult, C. E. M. P., and Allamandola, L. J. (1983). Ices in space. *J. Phys. Chem.*, **87**, 4243–60.

Gutfreund, H. (ed.) (1981). *Biochemical evolution*. Cambridge University Press.

Habing, H. J. and Neugebauer, G. (1984). The infrared sky. *Sci. Am.*, **251**, No. 5, 42–51.

Haldane, J. B. S. (1929). The origin of life. *Rationalist Annual*, **148**, 3–10. Reprinted in (1985) *On being the right size, and other essays*, (ed. J. Maynard Smith), pp. 101–12. Oxford University Press.

Haldane, J. B. S. (1930). *Enzymes*. Longmans, Green, London.

Hall, D. O. and Rao, K. K. (1987). *Photosynthesis*, (4th edn). Edward Arnold, London.

Hallam, A. (1973). *A revolution in the earth sciences: From continental drift to plate tectonics*. Clarendon Press, Oxford.

Hallam, A. (1983). *Great geological controversies*. Oxford University Press.

Hamilton, J. H. and Maruhn, J. A. (1986). Exotic atomic nuclei. *Sci. Am.*, **255**, No. 1, 74–83.

Harkins, W. D. (1917). The evolution of the elements and the stability of complex atoms. *J. Am. Chem. Soc.*, **39**, 858–79.

Harkins, W. D. (1931). The periodic system of atomic nuclei and the principle of regularity and continuity of series. *Phys. Rev.*, **38**, 1270–88.

Harold, F. M. (1986). *The vital force: A study of bioenergetics*, Freeman, New York.

Hartman, H., Lawless, J. G., and Morrison, P. (1985). *Search for universal ancestors.* NASA SP-477, Washington, DC.

Hartung, J. B. (1976). Was the formation of a 20-km-diameter impact crater on the Moon observed on June 18, 1178? *Meteoritics*, **11**, 187–94.

Hearnshaw, J. B. (1986). *The analysis of starlight: One hundred and fifty years of astronomical spectroscopy.* Cambridge University Press.

Hegstrom, R. A. and Kondepudi, D. K. (1990). The handedness of the universe. *Sci. Am.*, **262**, No. 1, 98–105.

Herrmann, D. B. (1984). *The history of astronomy from Herschel to Hertzsprung*, (trans. K. Krisciunas). Cambridge University Press.

Hey, J. S. (1973). *The evolution of radio astronomy*. Elek Science, London.

Hill, R. and Bendall, F. (1960). Function of the two cytochrome components in chloroplasts: A working hypothesis. *Nature*, **186**, 136–7.

Hinkle, P. C. and McCarty, R. E. (1978). How cells make ATP. *Sci. Am.*, **238**, No. 3, 104–23.

Holland, H. D. (1984). *The chemical evolution of the atmosphere and oceans.* Princeton University Press.

Holmes, F. L. (1985). *Lavoisier and the chemistry of life.* University of Wisconsin Press, Madison, WI.

Holton, G. (1975). The mainsprings of scientific discovery. In *The nature of scientific discovery*, (ed. O. Gingerich). Smithsonian Institution Press, Washington, DC.

Horowitz, N. H. (1945). On the evolution of biochemical synthesis. *Proc. Natl Acad. Sci. USA*, **31**, 153–7.

Hou, Y. M. and Schimmel, P. (1988). A simple structural feature is a major determinant of the identity of a transfer RNA. *Nature,* **333**, 140–5.

Howard, E. (1802). Experiments and observations on certain stony and metalline substances, which at different times are said to have fallen on the Earth; also on various kinds of native iron. *Philos. Trans. R. Soc.*, **92**, 168–212.

Hoyle, F. and Wickramsinghe, C. (1981). *Evolution from space*. Dent, London. (Paladin reprint, Granada, 1983.)

Hubble, E. (1936), *The Realm of the nebulae*. Yale University Press, New Haven, CT. (Reprinted 1982.)

Hudson, C. S. (1941). Emil Fischer's discovery of the configuration of glucose. *J. Chem. Educ.*, **18**, 353–7.

Hudson, C. S. (1948). Historical aspects of Emil Fischer's fundamental conventions for writing stereo-formulas in a plane. *Adv. Carbohydrate Chem.*, **3**, 1–22.

Huebner, W. F. (1987). First polymer in space identified in Comet Halley. *Science*, **237**, 628–30.

Huggins, W. and Miller, W. A. (1864). On the spectra of some of the fixed stars, and supplement by Huggins, On the spectra of some of the nebulae. *Philos. Trans. R. Soc.*, **154**, 413–435, 437–44.

Hughes, D. W. (1981). The influx of comets and asteroids to the Earth. *Philos. Trans. R. Soc.*, A **303**, 353–68.

Hund, F. (1974). *The history of quantum theory*. Harrap, London.

Hunt, J. M. (1979). *Petroleum geochemistry and geology.* Freeman, San Francisco.

Hutchison, R. (1983). *The search for our beginning: An enquiry, based on meteorite research, into the origin of our planet and life.* Oxford University Press.

Ihde, A. J. (1964). *The development of modern chemistry.* Harper and Row, New York.

Inoue, T. and Orgel, L. E. (1983). A nonenzymatic RNA polymerase model. *Science,* **219,** 859–62.

Jacques, J., Collet, A., and Wilen, S. H. (1981). *Enantiomers, racemates and resolutions.* Wiley-Interscience, New York.

Japp, F. R. (1898). Stereochemistry and vitalism. *Nature,* **58,** 452–60.

Jeffreys, H. (1924). *The Earth: Its origin, history and physical constitution.* Cambridge University Press.

Joyce, G. F. (1989). RNA evolution and the origins of life. *Nature,* **338,** 217–24.

Joyce, G. F., Visser, G. M., van Boeckel, C. A. A., van Boom, J. H., Orgel, L. E., and van Westrenen, J. (1984). Chiral selection in poly(C)-directed synthesis of oligo(G). *Nature,* **310,** 602–4.

Kamminga, H. (1980). *Studies in the history of ideas on the origin of life from 1860.* Ph.D. Thesis, University of London.

Kamminga, H. (1982). Life from space—A history of panspermia. *Vistas Astron.,* **26,** 67–86.

Kavanagh, R. W. (1982). Solar power. In *Essays in nuclear astrophysics,* (eds C. A. Barnes, D. D. Clayton, and D. N. Schramm), pp. 159–70. Cambridge University Press.

Keilin, D. (1966). *The history of cell respiration and cytochrome.* Cambridge University Press.

Kenyon, D. H. and Steinman, G. (1969). *Biochemical predestination.* McGraw-Hill, New York.

Kimura, M. (1969). The rate of molecular evolution considered from the viewpoint of population genetics. *Proc. Natl Acad. Sci. USA,* **63,** 1181–8.

Kimura, M. (1983). *The neutral theory of molecular evolution.* Cambridge University Press.

Kissel, J. and Krueger, F. R. (1987). The organic component in dust from Comet Halley as measured by the PUMA mass spectrometer on board Vega 1. *Nature,* **326,** 755–60.

Knapp, D. B. (1988). Reaction centres of photosynthetic bacteria. *Trends Biochem. Sci.,* **13,** 157–8.

Knight, D. M. (1970). *Atoms and elements: A study of theories of matter in England in the nineteenth century.* Hutchinson, London.

Kondepudi, D. K. and Nelson, G. W. (1985). Weak neutral currents and the origin of biomolecular chirality. *Nature,* **314,** 438–41.

Krauskopf, K. B. (1982). *Introduction to geochemistry.* McGraw-Hill, New York.

Krebs, H. (1981a). *Otto Warburg: Cell physiologist, biochemist and eccentric.* Clarendon Press, Oxford.

Krebs, H. (1981b). *Reminiscences and reflections.* Clarendon Press, Oxford.

Kueppers, B. O. (1983). *Molecular theory of evolution: Outline of a physico-chemical theory of the origin of life* (trans. P. Woolley). Springer, Berlin.

Kuhn, T. (1957). *The Copernican revolution.* Harvard University Press, Cambridge, MA.

Kuhn, T. (1962). *The structure of scientific revolutions*, (2nd edn, 1970). University of Chicago Press.

Kuhn, T. (1978). *Black-body theory and the quantum discontinuity 1894–1912.* Clarendon Press, Oxford.

Lakatos, I. (1976). *Proofs and refutations*, (ed. J. Worrall and E. Zahar). Cambridge University Press.

Lakatos, I. (1978). *The methodology of scientific research programmes*, (ed. J. Worrall and G. Currie). Cambridge University Press.

Laszlo, P. (1986). *Molecular correlates of biological concepts*, Vol. 34A, *Comprehensive biochemistry*. Elsevier, Amsterdam.

Lawless, J. C., Folsome, C. E., and Kvenvolden, K. A. (1972). Organic matter in meteorites. *Sci. Am.*, **226**, No. 6, 38–46.

Lechevalier, H. A. and Solotorovsky, M. (1974). *Three centuries of microbiology.* Dover, New York.

Lee, J. J. and Frederick, J. F. (eds) (1987). *Endocytobiology III. Ann. NY Acad. Sci.*, **503**, 1–589.

Le Grand, H. E. (1988). *Drifting continents and shifting theories.* Cambridge University Press.

Leicester, H. M. (1974). *Development of biochemical concepts from ancient to modern times.* Harvard University Press, Cambridge, MA.

Lemmon, R. M. (1970). Chemical evolution. *Chem. Rev.*, **70**, 95–109.

Lewis, G. N. and Randall, M. (1923). *Thermodynamics and the free energy of chemical substances.* McGraw-Hill, New York.

Lewis, J. S. and Prinn, R. G. (1984). *Planets and their atmospheres: Origin and evolution.* Academic Press, London.

Lewis, R. S. and Anders, E. (1983). Interstellar matter in meteorites. *Sci. Am.*, **249**, No. 2, 54–66.

Lipmann, F. (1971). *Wanderings of a biochemist.* Wiley-Interscience, New York.

Lockyer, N. (1900). *Inorganic evolution as studied by spectrum analysis.* Macmillan, London.

Lockyer, N. (1914). Notes on stellar classification. *Nature*, **94**, 282–4, 618–19, 644–5.

Lohrmann, R. and Orgel, L. E. (1973). Prebiotic activation processes. *Nature*, **244**, 418–20.

Lovejoy, A. O. (1950). *The great chain of being.* Harvard University Press, Cambridge, MA.

Lynden-Bell, D. (ed.) (1982). The Big Bang and element creation. *Philos. Trans. R. Soc.*, A **307**, 1–148.

McCarty, M. (1985). *The transforming principle: Discovering that genes are made of DNA.* Norton, New York.

McDonnell, J. A. M. (ed.) (1978). *Cosmic dust.* Wiley, New York.

McGucken, W. (1969). *Nineteenth-century spectroscopy: Development of the understanding of spectra, 1802–1897.* Johns Hopkins University Press, Baltimore.

McMenamin, M. A. A. (1987). The emergence of animals. *Sci. Am.*, **256**, No. 4, 94–102.

Marglin, A. and Merrifield, R. B. (1966). The synthesis of bovin insulin by the solid phase method. *J. Am. Chem. Soc.*, **88**, 5051–52.

Margulis, L. (1981). *Symbiosis in cell evolution: Life and its environment on the early Earth.* Freeman, San Francisco.

Mason, B. (1975). The Allende meteorite—Cosmochemistry's Rosetta Stone? *Acc. Chem. Res., 8,* 217–24.

Mason, B. and Moore, C. B. (1982). *Principles of geochemistry,* (4th edn). Wiley, New York.

Mason, S. F. (1953). *A history of the sciences: Main currents of scientific thought.* Routledge and Kegan Paul, London.

Mason, S. F. (1963). Optical rotatory power. *Q. Rev. Chem. Soc., 17,* 20–66.

Mason, S. F. (1976). The foundations of classical stereochemistry. *Top. Stereochem., 9,* 1–34.

Mason, S. F. (1982). *Molecular optical activity and the chiral discriminations.* Cambridge University Press.

Mason, S. F. (1987). From molecular morphology to universal dissymmetry. In *Essays on the history of organic chemistry,* (ed. J. G. Traynham). Louisiana State University Press, Baton Rouge, LA.

Mason, S. F. (1988). Biomolecular homochirality. *Chem. Soc. Rev., 17,* 347–59.

Mason, S. F. (1989). The development of concepts of chiral discrimination. *Chirality,* 1, 183–91.

Mason, S. F. and Tranter, G. E. (1985). The electroweak origin of biomolecular handedness. *Proc. R. Soc.,* A 397, 45–65.

Mattauch, J. (1934). Zur Systematik der Isotopen. *Z. Phys., 91,* 361–71.

Matthews, B. W. and Fenna, R. E. (1980). Structure of a green bacteriochlorophyll protein. *Acc. Chem. Res., 13,* 309–17.

Mayer, M. G. (1948). On closed shells in nuclei. *Phys. Rev., 74,* 235–9.

Mayer, M. G. (1964). The shell model. *Science, 145,* 999–1006.

Mayer, M. G. and Jensen, H. (1955). *Elementary theory of nuclear shell structure.* Wiley, New York.

Mayer, M. G. and Teller, E. (1949). On the origin of elements. *Phys. Rev.* 76, 1226–31.

Maynard Smith, J. (1979). Hypercycles and the origin of life. *Nature, 280,* 445–6.

Mendeleev, D. (1905). *The principles of chemistry,* (3rd English edn., trans. G. Kamensky, ed. T. H. Pope), 2 Vols. Longmans, Green, London. Appendix I, Friday Evening Discourse at the Royal Institution (1889), An attempt to apply to chemistry one of the principles of Newton's natural philosophy. Appendix II, Faraday Lecture to the Chemical Society (1889), The periodic law of the chemical elements. Appendix III, An attempt towards a chemical conception of the ether (1902).

Miller, S. L. (1955). Production of some organic compounds under possible primitive earth conditions. *J. Am. Chem. Soc., 77,* 2351–61.

Miller, S. L. (1984). The prebiotic synthesis of organic molecules and polymers. *Adv. Chem. Phys., 55,* 85–107.

Miller, S. L. (1986). Current status of the prebiotic synthesis of small molecules. *Chem. Scr., 26B,* 5–11.

Miller, S. L. and Orgel, L. E. (1974). *The origins of life on the Earth.* Prentice-Hall, Englewood Cliffs, NJ.

Mitchell, P. (1961). Coupling of phosphorylation to electron and hydrogen transfer by a chemi-osmotic type of mechanism. *Nature, 191,* 144–8.

Mitchell, P. (1979). Keilin's respiratory chain concept and its chemiosmotic consequences. *Science*, **206**, 1148–59.

Mitchell, P. (1981). Bioenergetic aspects of unity in biochemistry: Evolution of the concept of ligand conduction in chemical, osmotic, and chemiosmotic reaction mechanisms. In *Of oxygen, fuels, and living systems*, (ed. G. Semenza), pp. 1–160. Wiley, New York.

Mizuno, T. and Weiss, A. H. (1974). Synthesis and utilization of formose sugars. *Adv. Carbohydrate Chem. Biochem.*, **29**, 173–227.

Monod, J. (1972). *Chance and necessity.* Collins, London.

Moorbath, S. (1983). The dating of the earliest sediments on Earth. In *Cosmochemistry and the origin of life*, (ed. C. Ponnamperuma), pp. 213–33. Reidel, Dordrecht.

Moorbath, S. (1985). Crustal evolution in the early Precambrian. *Origins of Life*, **15**, 251–61.

Moorbath, S., Taylor, P. N., Orpen, J. L., Treloar, P., and Wilson, J. F. (1987). First direct radiometric dating of Archaean stromatolitic limestone. *Nature*, **326**, 865–7.

Morton, D. C. (1974). Interstellar abundances toward zeta-Ophiuchi. *Astrophys. J.*, **193**, L35–9.

Mullie, C. and Reisse, J. (1987). Organic matter in carbonaceous chondrites. *Top. Curr. Chem. 139*, pp. 83–117. Springer, Berlin.

Myers, P. C., Ho, P. T. P., Schneps, M. H., Chin, G., Pankonin, V., and Winnberg, A. (1978). Atomic and molecular observations of the rho-Ophiuchi dark cloud. *Astrophys. J.*, **220**, 864–82.

Nagy, B. (1975). *Carbonaceous meteorites.* Elsevier, Amsterdam.

Ninio, J. (1983). *Molecular approaches to evolution.* Princeton University Press.

Oesterhelt, D. and Tittor, J. (1989). Two pumps, one principle: Light-driven ion transport in halobacteria. *Trends Biochem. Sci.*, **14**, 57–61.

Olby, R. (1974). *The path to the double helix.* Macmillan, London.

O'Nions, R. K. and Parsons, B. (eds) (1989). Seismic tomography and mantle circulation. *Philos. Trans. R. Soc.*, A **328**, 289–442.

Oparin, A. I. (1924). *The origin of life*, (trans. A. Synge). Appendix in J. D. Bernal (1967) *The origin of life*, pp. 199–234, Weidenfeld and Nicholson, London.

Oparin, A. I. (1938). *The origin of life*, (trans. S. Morgulis). Dover reprint (1965), New York.

Open University (1974). *Science and belief: From Copernicus to Darwin.* Interfaculty Second Level Course AMST 283.

Open University (1976). *The nature of chemistry.* Third Level Course S304.

Open University (1985). *Matter in the universe.* Second Level Course S256.

Open University (1986). *Biochemistry and cell biology.* Third Level Course S325.

Orgel, L. E. (1973). *The origins of life: Molecules and natural selection.* Wiley, New York.

Orgel, L. E. (1979). Selection *in vitro*. *Proc. R. Soc.*, B **205**, 435–42.

Orgel, L. E. (1986a). Did template-directed nucleation precede molecular replication? *Origins of Life*, **17**, 27–34.

Orgel, L. E. (1986b). RNA catalysis and the origins of life. *J. Theor. Biol.*, **123**, 127–149.

Orgel, L. E. and Lohrmann, R. (1974). Prebiotic chemistry and nucleic acid replication. *Acc. Chem. Res.,* **7**, 368–77.

Owen, T. (1985). Life as planetary phenomenon. *Origins of Life,* **15**, 221–34.

Ozima, M. (1981). *The Earth: Its birth and growth.* Cambridge University Press.

Ozima, M. (1987). *Geohistory: Global evolution of the Earth.* Springer, Berlin.

Pagel, B. E. J. (1982). Abundances of elements of cosmological interest. In *The Big Bang and element creation,* (ed. D. Lynden-Bell). *Philos. Trans. R. Soc.,* A **307**, 19–35.

Pagel, B. E. J. (1987). Observing the Big Bang. *Nature,* **326**, 744–5.

Pais, A. (1982). *'Subtle is the Lord. . .': The science and life of Albert Einstein.* Oxford University Press.

Pais, A. (1986). *Inward bound: Of matter and forces in the physical world.* Clarendon Press, Oxford.

Partington, J. R. (1961–70). *A history of chemistry,* 4 Vols. Macmillan, London.

Pauling, L. (1948*a*). Molecular structure and biological specificity. *Chem. Ind.,* Supplement, 1–4.

Pauling, L. (1948*b*). *Molecular architecture and the processes of life.* Jesse Boot Foundation, Nottingham.

Pauling, L. and Delbruck, M. (1940). The nature of intermolecular forces operative in biological systems. *Science,* **92**, 77–9.

Penzias, A. A. (1979). The origin of the elements. *Rev. Mod. Phys.,* **51**, 425–31.

Penzias, A. A. and Wilson, R. W. (1965). A measurement of the excess antenna temperature at 4080 Mc/s. *Astrophys. J.,* **142**, 419–21.

Perutz, M. F. (1964). The hemoglobin molecule. *Sci. Am.,* **211**, No. 5, 64–76.

Perutz, M. F., Fermi, G., Luisi, B., Shaanan, B., and Liddington, R. C. (1987). Stereochemistry of cooperative mechanisms in hemoglobin. *Acc. Chem. Res.,* **20**, 309–21.

Pflug, H. D. (1987). Chemical fossils in early minerals. *Top. Curr. Chem. 139,* pp. 1–55. Springer, Berlin.

Phillips, D. C. (1966). The three-dimensional structure of an enzyme molecule. *Sci. Am.,* **215**, No. 5, 78–90.

Pinto, J. P., Gladstone, G. R., and Yung, Y. L. (1980). Photochemical production of formaldehyde in Earth's primitive atmosphere. *Science, 210,* 183–5.

Planck, M. (1933). *Where is science going?,* (trans. and ed. J. Murphy). Allen and Unwin, London.

Planck, M. (1948). *Wissenschaftliche Selbstbiographie,* Barth, Leipzig: (1950) *Scientific autobiography,* (trans. F. Gaynor). Williams and Norgate, London.

Ponnamperuma, C. (ed.) (1983). *Cosmochemistry and the origin of life.* Reidel, Dordrecht.

Popper, K. R. (1972). *Objective knowledge: An evolutionary approach.* Clarendon Press, Oxford.

Portugal, F. H. and Cohen, J. S. (1977). *A century of DNA: A history of the discovery of the structure and function of the genetic substance.* MIT Press, Cambridge, MA.

Postgate, J. R. (1984). *The sulphate-reducing bacteria,* (2nd edn). Cambridge University Press.

Postgate, J. R. (1986). *Microbes and man.* Penguin Books, Harmondsworth.

Prigogine, I. and Stengers, I. (1984). *Order out of chaos: Man's new dialogue with Nature*. Heinemann, London.

Prince, R. C. (1986). Manganese at the active site of the chloroplast oxygen-evolving complex. *Trends Biochem. Sci.*, **11**, 491–2.

Prout, W. (1834). *Chemistry, meteorology and the function of digestion considered with reference to natural theology*, Bridgewater Treatise VIII. William Pickering, London.

Prout's hypothesis (1932). *Alembic club reprint no. 20*, Papers by William Prout (1815–16), J. S. Stas (1860), and C. Marignac (1860). Oliver and Boyd, Edinburgh.

Prusiner, S. B. (1982). Novel proteinaceous infectious particles cause scrapie. *Science*, **216**, 136–44.

Quack, M. (1989). Structure and dynamics of chiral molecules. *Angew. Chem. Int. Ed. Engl.*, **28**, 571–86.

Quayle, J. R. and Ferenci, T. (1978). Evolutionary aspects of autotrophy. *Microbiol. Rev.*, **42**, 251–73.

Racker, E. (1979). Transport of ions. *Acc. Chem. Res.*, **12**, 338–44.

Racker, E. (1980). From Pasteur to Mitchell: A hundred years of bioenergetics. *Fed. Proc.*, **39**, 210–15.

Rao, K. K., Cammack, R., and Hall, D. O. (1985). Evolution of light energy conversion. In *Evolution of the prokaryotes*, (ed. K. H. Schleifer and E. Stackebrandt). Academic Press, London.

Retey, R. and Robinson, J. A. (1982). *Stereospecificity in organic chemistry and enzymology*. Verlag Chemie, Weinheim.

Ringwood, A. E. (1984). The Earth's core: Its composition, formation and bearing upon the origin of the Earth. *Proc. R. Soc.*, A **395**, 1–46.

Rosenberg, G. D. and Runcorn, S. K. (eds) (1975). *Growth rhythms and the history of the Earth's rotation*. Wiley, New York.

Ross, J. E. and Aller, L. H. (1976). The chemical composition of the Sun. *Science*, **191**, 1223–9.

Rowe, M. W. (1985). Radioactive dating: A method for geochronology. *J. Chem. Educ.*, **62**, 580–4.

Rowe, M. W. (1986). The age of the elements. *J. Chem. Educ.*, **63**, 300–3.

Runcorn, S. K., Turner, G., and Woolfson, M. M. (eds) (1988). The solar system: Chemistry as a key to its origin. *Philos. Trans. R. Soc.*, A **325**, 389–641.

Russell, H. N. (1914). Relations between spectra and other characteristics of the stars. *Nature*, **93**, 227–30, 252–8, 281–6.

Russell, H. N. (1921). A superior limit to the age of the Earth's crust. *Proc. R. Soc.*, A **99**, 84–6.

Russell, H. N. (1929). On the composition of the Sun's atmosphere. *Astrophys. J.*, **70**, 11–82.

Russell, H. N. and Saunders, F. A. (1925). New regularities in the spectra of the alkaline earths. *Astrophys. J.*, **61**, 38–69.

Rutherford, E. (1904). *Radio-activity*, (2nd edn, 1905). Cambridge University Press.

Rutherford, E. (1962–5). *The collected papers of Lord Rutherford of Nelson*, 3 Vols, (ed. J. Chadwick). Allen and Unwin, London.

Saenger, W. (1984). *Principles of nucleic acid structure*. Springer, Berlin.

Salpeter, E. E. (1974). Dying stars and reborn dust. *Rev. Mod. Phys.*, **46**, 433–6.

Sauer, K. (1978). Photosynthetic membranes. *Acc. Chem. Res.*, **11**, 257–64.

Schidlowski, M. (1988). A 3,800-million-year isotopic record of life from carbon in sedimentary rocks. *Nature*, **333**, 313–18.

Schleifer, K. H. and Stackebrandt, E. (eds) (1985). *Evolution of prokaryotes.* Academic Press, London.

Schonland, B. (1968). *The atomists (1805–1933).* Clarendon Press, Oxford.

Schopf, J. W. (ed.) (1983). *Earth's earliest biosphere: Its origin and evolution.* Princeton University Press.

Schopf, J. W. and Packer, B. M. (1987). Early Archean (3.3-billion to 3.5-billion-year-old) microfossils from Warrawoona group, Australia. *Science*, **237**, 70–3.

Schulze, G. E. (1981). Protein differentiation: Emergence of novel proteins during evolution. *Angew. Chem. Int. Ed. Engl.*, **20**, 143–51.

Schulze, G. E. and Schirmer, R. H. (1979). *Principles of protein structure.* Springer, Berlin.

Schuster, P. (1986). The physical basis of molecular evolution. *Chem. Scr.*, **26B**, 27–41.

Schwartz, A. W. and Orgel, L. E. (1985). Template-directed syntheses of novel, nucleic acid-like structures. *Science*, **228**, 585–7.

Schwartz, R. M. and Dayhoff, M. O. (1978). Origins of prokaryotes, eukaryotes, mitochondria, and chloroplasts. *Science*, **199**, 395–403.

Sears, D. W. (1975). Sketches in the history of meteoritics 1: The birth of the science. *Meteoritics*, **10**, 215–25.

Selbin, J. (1973). The origin of the chemical elements. *J. Chem. Educ.*, **50**, 306–10, 380–7.

Shepherd, J. C. W. (1986). Origins of life and molecular evolution of present-day genes. *Chem. Scr.*, **26B**, 75–83.

Smith, D. (1987). Interstellar molecules. *Philos. Trans. R. Soc.*, A **323**, 269–86.

Smith, D. (1988). Formation and destruction of molecular ions in interstellar clouds. *Philos. Trans. R. Soc.*, A **324**, 257–73.

Soddy, F. (1904). *Radio-activity: An elementary treatise from the standpoint of the disintegration theory.* Electrician Publishing, London.

Soddy, F. (1911, 1914). *Chemistry of the radio-elements*, 2 Vols. Longmans, Green, London.

Stanier, R. Y., Ingraham, J. L., Wheelis, M. L., and Painter, P. R. (1987). *General microbiology.* Macmillan, London.

Stephenson, F. R. and Clark, D. H. (1978). *Applications of early astronomical records.* Adam Hilger, Bristol.

Stoeckenius, W. (1980). Purple membrane of halobacteria: A new light-energy converter. *Acc. Chem. Res.*, **13**, 337–44.

Stryer, L. (1988). *Biochemistry*, (3rd edn). Freeman, New York.

Suess, H. E. (1987). *Chemistry of the solar system: An elementary introduction to cosmochemistry.* Wiley, New York.

Suess, H. E. and Urey, H. C. (1956). Abundances of the elements. *Rev. Mod. Phys.*, **28**, 53–74.

Sun, S.-S. (1984). Geochemical characteristics of Archaean ultramafic and mafic volcanic rocks: Implications for mantle composition and evolution. In *Archaean*

Geochemistry, (ed. A. Kröner, G. N. Hanson and A. M. Goodwin), pp. 25–46. Springer, Berlin.

Tayler, R. J. (1988). Nucleosynthesis and the origin of the elements. *Philos. Trans. R. Soc.*, A **325**, 391–403.

Temin, H. and Mizutani, S. (1970). RNA-dependent DNA polymerase in virons of Rous sarcoma virus. *Nature*, **226**, 1211–13.

Thielemann, F. K., Metzinger, J., and Klapdor, H. V. (1983). Beta-delayed fission and neutron emission: Consequences for the astrophysical *r*-process and the age of the galaxy. *Z. Phys.*, A **309**, 301–17.

Todd, A. (1983). *A time to remember: The autobiography of a chemist.* Cambridge University Press.

Towe, K. M. (1985). Habitability of the early Earth: Clues from the physiology of nitrogen fixation and photosynthesis. *Origins of Life*, **15**, 235–50.

Tranter, G. E. (1985). Parity-violating energy differences of chiral minerals and the origin of biomolecular homochirality. *Nature*, **318**, 172–73.

Tranter, G. E. (1986). Preferential stabilisation of the D-sugar series by the parity-violating weak interactions. *J. Chem. Soc. Chem. Comm.*, 60–1.

Tranter, G. E. (1987). The enantio-preferential stabilisation of D-ribose from parity violation. *Chem. Phys. Lett.*, **135**, 279–82.

Trenn, T. J. (1977). *The self-splitting atom: The history of the Rutherford–Soddy collaboration.* Taylor and Francis, London.

Tribe, M. A., Morgan, A. J., and Whittaker, P. A. (1981). *The evolution of eukaryotic cells.* Edward Arnold, London.

Trimble, V. (1975). The origin and abundances of the chemical elements. *Rev. Mod. Phys.*, **47**, 877–976.

Trimble, V. (1982, 1983). Supernovae. Part I: The events; Part II: The aftermath. *Rev. Mod. Phys.*, **54**, 1183–1224; **55**, 511–63.

Turner, G. and Pillinger, C. T. (eds) (1987). Diffuse matter in the solar system: Comet Halley and other studies. *Philos. Trans. R. Soc.*, A **323**, 247–447.

Unsöld, A. and Baschek, B. (1983). *The new cosmos*, (trans. R. C. Smith), (3rd edn). Springer, New York.

Urey, H. C. (1952*a*). Chemical fractionation in the meteorites and the abundance of the elements. *Geochim. Cosmochim. Acta*, **2**, 269–82.

Urey, H. C. (1952*b*). On the early chemical history of the Earth and the origin of life. *Proc. Natl Acad. Sci. USA*, **38**, 351–63.

Urey, H. C. (1952*c*). *The planets: Their origin and development*, Yale University Press, New Haven, CT.

Urey, H. C. and Bradley, C. A. (1931). On the relative abundance of isotopes. *Phys. Rev.*, **38**, 318.

Urey, H. C. and Craig, H. (1953). The composition of the stone meteorites and the origin of meteorites. *Geochim. Cosmochim. Acta*, **4**, 36–82.

Vaucouleurs, G. de. (1957). *Discovery of the Universe: An outline of the history of astronomy from the origins to 1956.* Faber, London.

Walker, J. C. G., Klein, C., Schidlowski, M., Schopf, J. W., Stevenson, D. J., and Walter, M. R. (1983). Environmental evolution of the Archean–early Proterozoic Earth. In *Earth's earliest biosphere*, (ed. J. W. Schopf), pp. 260–90. Princeton University Press.

Walker, J. E. and Cozens, A. L. (1986). Evolution of ATP synthase. *Chem. Scr.,* **26B**, 263–72.

Wanke, H. and Dreibus, G. (1988). Chemical composition and accretion history of terrestrial planets. *Philos. Trans. R. Soc.,* A **325**, 545–7.

Wanke, H., Dreibus, G., and Jagoutz, E. (1984). Mantle chemistry and accretion history of the Earth. In *Archaean geochemistry,* (ed. A. Kröner, G. N. Hanson and A. M. Goodwin), pp. 1–24. Springer, Berlin.

Wasson, J. T. (1985). *Meteorites: Their record of early solar-system history.* Freeman, New York.

Watanabe, T., Kobayashi, M., Hongu, A., Nakazat, M., Hiyama, T., and Murata, N. (1985). Evidence that a chlorophyll *a'* dimer constitutes the photochemical reaction centre 1 (P_{700}) in photosynthetic apparatus. *FEBS Lett.,* **191**, 252–6.

Watson, J. D., Hopkins, N. H., Roberts, J. W., Steitz, J. A., and Weiner, A. M. (1987). *Molecular biology of the gene,* (4th edn), 2 Vols, Benjamin/Cummings, Menlo Park, etc.

Wayne, R. P. (1985). *Chemistry of atmospheres: An introduction to the chemistry of the atmospheres of the Earth, the planets, and their satellites.* Clarendon Press, Oxford.

Weeks, M. E. (1960). *Discovery of the elements,* (6th edn.). Journal of Chemical Education, Easton, PA.

Wegener, A. (1966). *The Origin of continents and oceans,* (trans. J. Biram from 4th German edn of 1929). Methuen, London.

Weinberg, S. (1983). *The first three minutes: A modern view of the origin of the universe.* Fontana, London.

Westheimer, F. H. (1986). Polyribonucleic acids as enzymes. *Nature,* **319**, 534–6.

Wetherill, G. W. (1979). Apollo objects. *Sci. Am.,* **240**, No. 3, 38–49.

Wetherill, G. W. (1980). Formation of the terrestrial planets. *Ann. Rev. Astron. Astrophys.,* **18**, 77–113.

Whipple, F. L. (1981). *Orbiting the Sun: Planets and satellites of the solar system.* Harvard University Press, Cambridge, MA.

Wigner, E. (1927). Einige Folgerungen aus der Schrödingerschen Theorie für die Termstrukturen. *Z. Phys.,* **43**, 624–52.

Wigner, E. (1979). *Symmetries and reflections.* Ox Bow Press, Woodbridge, CT.

Williams, R. J. P. (1978). The multifarious couplings of energy transduction. *Biochim. Biophys. Acta,* **505**, 1–44.

Williams, R. J. P. (1981). Natural selection of the chemical elements. *Proc. R. Soc.,* B **213**, 361–97.

Williams, R. J. P. (1985). The symbiosis of metal and protein functions. *Eur. J. Biochem.,* **150**, 231–48.

Wilson, A. C. (1985). The molecular basis of evolution. *Sci. Am.,* **253**, No. 4, 148–57.

Wilson, R. W. (1979). The cosmic microwave background radiation. *Rev. Mod. Phys.,* **51**, 433–45.

Winkler-Oswatitsch, R., Dress, A., and Eigen, M. (1986). Comparative sequence analysis exemplified with tRNA and 5S RNA. *Chem. Scr.,* **26B**, 59–66.

Winnewisser, G. and Herbst, E. (1987). Organic molecules in space. *Top. Curr. Chem. 139,* pp. 119–72. Springer, Berlin.

Woese, C. R. (1981). Archaebacteria. *Sci. Am.,* **244,** No. 6, 94–106.

Woese, C. R. (1987). Bacterial evolution. *Microbiol. Rev.,* **51,** 135–77, 221–71.

Wolfenstein, L. and Beier, E. W. (1989). Neutrino oscillations and solar neutrinos. *Phys. Today,* **42,** No. 7, 28–36.

Wolstencroft, R. D. (1985). Astronomical sources of circularly polarized light and their role in determining molecular chirality on Earth. In *The search for extraterrestrial life: Recent developments,* (ed. M. D. Papagiannis), IAU, Symp. 112, pp. 171–5. Reidel, Dordrecht.

Wolstencroft, R. D. and Walker, H. J. (1988). Dust discs around low-mass main-sequence stars. *Philos. Trans. R. Soc.,* A **325,** 423–37.

Wood, J. A. (1978). Ancient chemistry and the formation of the planets. In *Proceedings of the Robert A. Welch Foundation Conferences on Chemical Research: XXI Cosmochemistry,* (ed. W. O. Milligan), pp. 323–62. Houston, TX.

Wood, J. A. (1979). *The solar system.* Prentice-Hall, Englewood Cliffs, NJ.

Wood, J. A. and Chang, S. (1985). *The cosmic history of the biogenic elements and compounds.* NASA SP-476, Washington, DC.

Woolfson, M. M. (1984). The evolution of rotation in the early history of the solar system. *Philos. Trans. R. Soc.,* A **313,** 5–18.

Wurtz, A. (1862). On oxide of ethylene, considered as a link between organic and mineral chemistry. *J. Chem. Soc.,* **15,** 387–406.

Zhao, M. and Bada, J. L. (1989). Extraterrestrial amino acids in Cretaceous/Teriary boundary sediments at Stevns Kint, Denmark. *Nature,* **339,** 463–65.

Zuckerkandl, E. and Pauling, L. (1962). Molecular disease, evolution, and genic heterogeneity. In *Horizons in biochemistry: Albert Szent-Györgyi dedicatory volume,* (ed. M. Kasha and B. Pullman), pp. 189–225. Academic Press, New York.

Zuckerkandl, E. and Pauling, L. (1965a). Molecules as documents of evolutionary history. *J. Theor. Biol.,* **8,** 357–66.

Zuckerkandl, E. and Pauling, L. (1965b). Evolutionary divergence and convergence in proteins. In *Evolving genes and proteins,* pp. 97–166. (ed. V. Bryson and H. J. Vogel). Academic Press, New York.

Author Index

Subject Index